Internet of Things and Sensors Networks in 5G Wireless Communications

Internet of Things and Sensors Networks in 5G Wireless Communications

Special Issue Editors

Lei Zhang
Guodong Zhao
Muhammad Ali Imran

MDPI • Basel • Beijing • Wuhan • Barcelona • Belgrade

Special Issue Editors

Lei Zhang
University of Glasgow
UK

Guodong Zhao
University of Glasgow
UK

Muhammad Ali Imran
University of Glasgow
UK

Editorial Office
MDPI
St. Alban-Anlage 66
4052 Basel, Switzerland

This is a reprint of articles from the Special Issue published online in the open access journal *Sensors* (ISSN 1424-8220) from 2019 to 2020 (available at: https://www.mdpi.com/journal/sensors/special_issues/IOT_SN5G).

For citation purposes, cite each article independently as indicated on the article page online and as indicated below:

LastName, A.A.; LastName, B.B.; LastName, C.C. Article Title. *Journal Name* **Year**, *Article Number*, *Page Range*.

ISBN 978-3-03928-148-0 (Pbk)
ISBN 978-3-03928-149-7 (PDF)

© 2020 by the authors. Articles in this book are Open Access and distributed under the Creative Commons Attribution (CC BY) license, which allows users to download, copy and build upon published articles, as long as the author and publisher are properly credited, which ensures maximum dissemination and a wider impact of our publications.

The book as a whole is distributed by MDPI under the terms and conditions of the Creative Commons license CC BY-NC-ND.

Contents

About the Special Issue Editors vii

Rabeea Basir, Saad Qaisar, Mudassar Ali, Monther Aldwairi, Muhammad Ikram Ashraf, Aamir Mahmood and Mikael Gidlund
Fog Computing Enabling Industrial Internet of Things: State-of-the-Art and Research Challenges
Reprinted from: *Sensors* **2019**, *19*, 4807, doi:10.3390/s19214807 1

Collins Burton Mwakwat, Hassan Malik, Muhammad Mahtab Alam, Yannick Le Moullec, Sven Parand and Shahid Mumtaz
Narrowband Internet of Things (NB-IoT): From Physical (PHY) and Media Access Control (MAC) Layers Perspectives
Reprinted from: *Sensors* **2019**, *19*, 2613, doi:10.3390/s19112613 39

Ahmed Adel Aly, Hussein M. ELAttar, Hesham ElBadawy and Wael Abbas
Aggregated Throughput Prediction for Collated Massive Machine-Type Communications in 5G Wireless Networks
Reprinted from: *Sensors* **2019**, *19*, 3651, doi:10.3390/s19173651 73

Wenjun Hou, Song Li, Yanjing Sun, Jiasi Zhou and Nannan Lu
Interference-Aware Subcarrier Allocation for Massive Machine-Type Communication in 5G-Enabled Internet of Things
Reprinted from: *Sensors* **2019**, *19*, 4530, doi:10.3390/s19204530 93

Muhammad Asad Ullah, Junnaid Iqbal, Arliones Hoeller, Richard Demo Souza and Hirley Alves
K-Means Spreading Factor Allocation for Large-Scale LoRa Networks
Reprinted from: *Sensors* **2019**, *19*, 4723, doi:10.3390/s19214723 106

Shuang Zhang and Guixia Kang
User Association and Power Control for Energy Efficiency Maximization in M2M-Enabled Uplink Heterogeneous Networks with NOMA
Reprinted from: *Sensors* **2019**, *19*, 5307, doi:10.3390/s19235307 125

Jingyun Sun, Rongke Liu and Enrico Paolini
A Dynamic Access Probability Adjustment Strategy for Coded Random Access Schemes
Reprinted from: *Sensors* **2019**, *19*, 4206, doi:10.3390/s19194206 143

M. Carmen Lucas-Estañ, Javier Gozalvez and Miguel Sepulcre
On the Capacity of 5G NR Grant-Free Scheduling with Shared Radio Resources to Support Ultra-Reliable and Low-Latency Communications
Reprinted from: *Sensors* **2019**, *19*, 3575, doi:10.3390/s19163575 161

Jiewen Deng, Wanrong Sun, Lei Guan, Nan Zhao, Muhammad Bilal Khan, Aifeng Ren, Jianxun Zhao, Xiaodong Yang and Qammer H. Abbasi
Noninvasive Suspicious Liquid Detection Using Wireless Signals
Reprinted from: *Sensors* **2019**, *19*, 4086, doi:10.3390/s19194086 179

Mohammad Kazem Chamran, Kok-Lim Alvin Yau, Rafidah M. D. Noor and Richard Wong
A Distributed Testbed for 5G Scenarios: An Experimental Study
Reprinted from: *Sensors* **2020**, *20*, 18, doi:10.3390/s20010018 190

About the Special Issue Editors

Lei Zhang (Dr.) is a Lecturer at the University of Glasgow, UK. He received his Ph.D. from the University of Sheffield, UK. He worked as a research engineer in the Huawei Communication Technology Laboratory (CT Lab), and a research fellow in the 5G Innovation Centre (5GIC), Institute of Communications (ICS), University of Surrey, UK. His research interests broadly lie in communications and networks, including wireless blockchain networks, radio access network slicing (RAN slicing), new air interface designs, Internet of Things (IoT), multi-antenna signal processing, and massive MIMO systems. He has 19 US/UK/EU/China granted/filed patents on wireless communications and has published over 100 peer-reviewed papers. Dr. Lei Zhang also holds a visiting position in 5GIC at the University of Surrey. He is an associate editor of IEEE ACCESS and a senior member of IEEE.

Guodong Zhao (Dr.) received his B.E. degree from Xidian University, Xi'an, China, in 2005, and his Ph.D. degree from Beihang University, Beijing, China, in 2011, both in Electrical Engineering. From 2011 to 2018, he was an associate professor at the University of Electronic Science and Technology of China (UESTC), Chengdu, China. In 2018, he joined the University of Glasgow in the UK as a lecturer (assistant professor). He has 10 years of experience working on wireless communications with international partners and he is a senior member of IEEE. He has published one book with Springer press and more than 50 peer-reviewed research papers (including more than 10 IEEE transaction papers), had over 1700 citations in Google Scholar, and won a best paper award in IEEE Globecom, 2012, and a best poster award in IEEE WCNC, 2018. His current research interests are within the areas of wireless communication, control, and robotics.

Muhammad Ali Imran (Prof.) is a Fellow of IET, Senior Member of IEEE and Senior Fellow of the Higher Education Academy UK. He is a Professor of Wireless Communication Systems, with research interests in self-organised networks, wireless networked control systems and wireless sensor systems. He heads the Communications, Sensing and Imaging (CSI) research group at the University of Glasgow and is Dean at the University of Glasgow, UESTC. He is an Affiliate Professor at the University of Oklahoma, USA and a visiting Professor at the 5G Innovation Centre, University of Surrey, UK. He has over 20 years of combined academic and industry experience, with several leading roles in multi-million GBP-funded projects. He has filed 15 patents; has authored/co-authored over 400 journal and conference publications; was editor of five books and author of more than 20 book chapters; and successfully supervised over 40 postgraduate students at the Doctoral level. He has been a consultant for international projects and local companies in the area of self-organised networks. He has been interviewed by the BBC, Scottish television and many radio channels on the topic of 5G technology.

Review

Fog Computing Enabling Industrial Internet of Things: State-of-the-Art and Research Challenges

Rabeea Basir [1], Saad Qaisar [1], Mudassar Ali [1,2,*], Monther Aldwairi [3], Muhammad Ikram Ashraf [4], Aamir Mahmood [5] and Mikael Gidlund [5]

[1] School of Electrical Engineering and Computer Science, National University of Science and Technology, Islamabad 44000, Pakistan; rbasir.dphd17@seecs.edu.pk or rabeeabasir@gmail.com (R.B.); saad.qaisar@seecs.edu.pk (S.Q.)
[2] Department of Telecommunication Engineering, University of Engineering and Technology, Taxila 47050, Pakistan
[3] College of Technological Innovation, Zayed University, Abu Dhabi 144534, UAE; monther.aldwairi@zu.ac.ae
[4] Centre for Wireless Communication, University of Oulu, 90014 Oulu, Finland; ikram.ashraf@oulu.fi or ikramashraf@gmail.com
[5] Department of Information Systems and Technology, Mid Sweden University, 85170 Sundsvall, Sweden; aamir.mahmood@miun.se (A.M.); mikael.gidlund@miun.se (M.G.)
* Correspondence: mudassar.ali@hotmail.com or mudassar.ali@seecs.edu.pk

Received: 14 August 2019; Accepted: 23 October 2019; Published: 5 November 2019

Abstract: Industry is going through a transformation phase, enabling automation and data exchange in manufacturing technologies and processes, and this transformation is called Industry 4.0. Industrial Internet-of-Things (IIoT) applications require real-time processing, near-by storage, ultra-low latency, reliability and high data rate, all of which can be satisfied by fog computing architecture. With smart devices expected to grow exponentially, the need for an optimized fog computing architecture and protocols is crucial. Therein, efficient, intelligent and decentralized solutions are required to ensure real-time connectivity, reliability and green communication. In this paper, we provide a comprehensive review of methods and techniques in fog computing. Our focus is on fog infrastructure and protocols in the context of IIoT applications. This article has two main research areas: In the first half, we discuss the history of industrial revolution, application areas of IIoT followed by key enabling technologies that act as building blocks for industrial transformation. In the second half, we focus on fog computing, providing solutions to critical challenges and as an enabler for IIoT application domains. Finally, open research challenges are discussed to enlighten fog computing aspects in different fields and technologies.

Keywords: Industry 4.0; Internet of Things; Industrial Internet of Things; Cyber Physical System; cloud computing; fog computing; edge computing; smart devices; smart factory; industrial automation

1. Introduction

Revolution in any realm is required with the passage of time. Every field changes to go forward with better solutions dealing with the challenges of the era. Industrial Internet of Things (IIoT) is revolutionizing the classical communication methodologies. With the emergence of smart devices (mobile, machines, sensors) coupled with a diverse range of applications requirements, IIoT is the way forward. It is expected that 26 billion IoT devices of heterogeneous capabilities will be installed to perform functions with different Quality-of-Service (QoS) requirements by 2020 [1]. IIoT gives rise to 4*th* industrial revolution based on Cyber-Physical Systems (CPS) with the need arising back in 2015 originated basically in Germany [2]. Industry 4.0 defines diverse use cases ranging from interconnected digital technologies, CPS, Mobile Cloud Computing (MCC) and Internet of Things (IoT) for promoting the whole industry in terms of efficiency, effectiveness, supporting heterogeneous data, higher production, automation, and integrating knowledge [3]. These key enabling technologies have been deployed to some extent in industrial domains such as healthcare, transportation, smart cities, micro-grids, and smart factory. This trend gives rise to intelligent, distributed and self-organizing solutions to support these application domains.

Deploying industry 4.0 involves three-layer implementation; physical layer, network layer, and intelligent-application layer [4]. The physical layer comprises identification and location awareness entities i.e. actuators, sensors, and terminal devices; the network layer comprises of the development of a network that can support industrial automation, network can be cellular, indoor, cloud or private. Factory automation and coordination are processed on the application layer. Infrared (IR), Radio-Frequency Identification (RFID), Bluetooth, 6LoWPAN, IEEE 802.11 af, IEEE 802.11 a/b/n/ac for short range connectivity; Ultra-Wideband (UWB), cellular (2G, 3G, 4G, LTE-MTC, 5G), Sigfox, Long range (LoRa) for long range connectivity, are a few of the majorly used communication standards for IIoT [5,6].

The future of automation is based on decentralized intelligence in which all machines can communicate with one another to arrive at independent or consensus inference, called Machine-to-Machine (M2M) communication. These decentralized intelligent solutions play a vital role in industry 4.0 digital transformation. The decentralized solutions provide flexibility and quick decision assistance over centralized solutions. For M2M communication, 802.11ah technology has evolved in the recent past. Exchanging machine data demands real-time communication ensuring latency, security, reliability, bandwidth and privacy measures in all IIoT domains. To satisfy these critical requirements, there is a need to explore new enabling solutions that support these applications. In the future, 5G cellular technology will support such heterogeneous networks with massive number of IIoT devices. It is anticipated that future 5G networks not only provide flexibility but can optimize the usage of available resources of bandwidth, power, energy, connectivity to different applications at the same time [7].

In the last decade, computation and processing requirements of end users have increased exponentially. It has become increasingly challenging for designers to scale the processing and data storage capabilities for users within the given device size and battery constraints. To meet these growing requirements, researchers have come up with the solution to offload services to a centralized location known as the cloud. Cloud computing is an alternative for data computation, storage and management. It supports intensive computation and manages heterogeneous devices of next generation networks [8–10]. Additionally, cloud computing architecture involves the direct connection between devices and the cloud server. Practically, we are beginning to understand the connection between and the enormous number of IIoT devices and a single cloud server. However, cloud-based systems are unable to meet the requirement such as heavy data computation, real-time device control, security and management results in insufficient support of IIoT application requirements [11]. Considering a wide variety of IoT scenarios, some of the challenges [10–16] in cloud computing are listed below:

- Large distance between the cloud and edge devices causes propagation and transmission delays.
- Large computational load on a single cloud server causes processing and queuing delays.
- Increased number of smart devices has hindered meeting the bandwidth requirements.
- Enormous number of smart devices will bring scalability, speed, and computational issues.
- Wireless medium between cloud and smart devices brings resource management issues.
- Heterogeneity property of smart devices in terms of accessing technology will bring difficulty in handling at the cloud.
- Mobility of IoT devices bring service availability issues, cloud server may not be able to provide services due to network congestion and failure.

- Security is a very critical thread, as the cloud is exposed to the whole world over the public internet.
- Computing offloading every-time at cloud causes a loss in energy and battery lifetime.
- Although data storage at cloud brings benefits to application developers, they should be careful of integrity and authentication demands of IIoT applications.
- Cloud computing is a centralized and complex architecture for real-time applications of IIoT.

All these limitations require a change, how and where we process data. These challenges motivate us to explore new decentralized approaches/solutions in IIoT domains. A new concept of fog computing is introduced by Bonomi et al. [15] for handling data locally at the network edge in order to overcome the limitations of cloud architecture. Fog computing complements existing cloud architecture and has addressed the issue of latency and bandwidth efficiency [17]. Because of its distributed architecture, it calls for a strong check on QoS requirements to make it useful. Fog is mainly based on distributed networking with ubiquitous pervasive computing. It comprises small scale data centers or a group of computers known as cloudlets (fog clouds) that provide services to devices located in close proximity [17,18]. The initial installation cost, latency, and energy consumption is far less as compared to that of the cloud, but the operational cost varies. Fog architecture can leverage computations either from dedicated edge servers or adhoc infrastructure. For promoting IIoT architecture with fog computing as a key enabler technology, a group of fog clouds can also be used.

In fog computing, data processing in single server (fog cloud) helps in achieving real-time and reliable communication. It puts the safety and security of personal data back into our premises. Furthermore, a cost effective approach can be used in fog computing such that data transmission and storage fees can be reduced based on service premises. Therefore, fog computing has the potential to provide affordable solutions for large IIoT projects. Instead of being restricted to only one expensive cloud connection, fog computing gives the freedom to choose any hardware from Information Technology (IT) solutions. It supports all existing legacy devices and non-IIoT devices that never intended to be the part of IIoT application. This is not only economical but also more flexible. When it comes to speed, fog computing allows real-time processing and supports to process data as fast as our local system. Fog can be managed securely from remote places. It can be scaled and updated dynamically. It gives more security, better performance, and lower costs. Fog incorporates positive attributes of cloud and provides benefits that may support future IIoT applications [18–23].

Fog computing and edge computing being extended form of cloud computing gives solutions to the challenges faced by cloud computing that is attractive for IIoT real-time applications. The terms fog computing and edge computing are often used by industry interchangeably. Both these computing technologies bring computing and processing capabilities near the vicinity where data originates. Edge computing complements fog computing by bringing computation to one of the devices of a network. This device is named as E-node and is close to the data. E-node has more power, computation capabilities and intelligent controllers, such as programmable automation controllers (PAC). Presence of E-node in edge computing improves latency, reliability, security and privacy issues [24,25]. E-node acts as an interface/bridge between the data sources and the cloud. The basic architecture for fog network is given in Figure 1 depicting fog cloud serving as a middle layer between the cloud server and smart end-devices. Figure 1 demonstrates a basic idea of cloud, fog and edge computing promoting different IIoT application domains.

Fog is a relatively new paradigm that brings new challenges in terms of efficient and scalable network architecture. It is expected that it will gradually develop over the next few years for realizing the Industry 4.0. Challenges, such as energy conservation, real-time communication, efficient spectrum use, cache memory on edge devices and optimized allocation of resources are open issues that need to be addressed for future automation. Without such considerations, guaranteed QoS requirements of IoT devices may not be fulfilled. In the future, solutions to these challenges must be provided by researchers for the development of the industrial revolution. This paper is written with an aim to give a summarized version of existing solutions using fog computing acting as an enabler for IIoT applications.

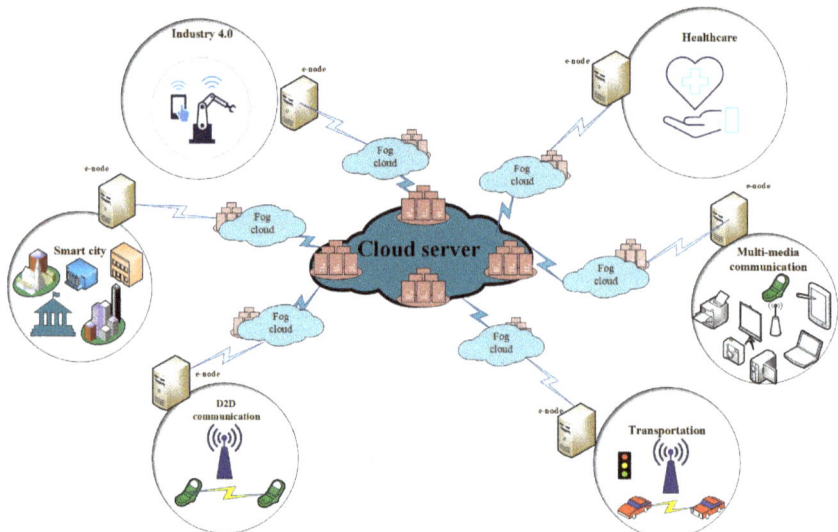

Figure 1. Generalized view: IIoT application domains with cloud, fog and edge computing.

The paper is organized as follows; Section 2 briefly introduces IIoT. Benefits of IIoT applications in daily life and their critical requirements are briefly explained in Section 3. Section 4 presents protocol/solution proposed by various researchers promoting fog computing as an enabling technology for IIoT development. Section 5 describes challenges and solutions in communication and networking proposed in the literature to use fog computing in IIoT. Section 6 lists down several open research issues in fog computing. Finally, the paper is concluded in Section 7. The flow of this survey paper is shown in Figure 2.

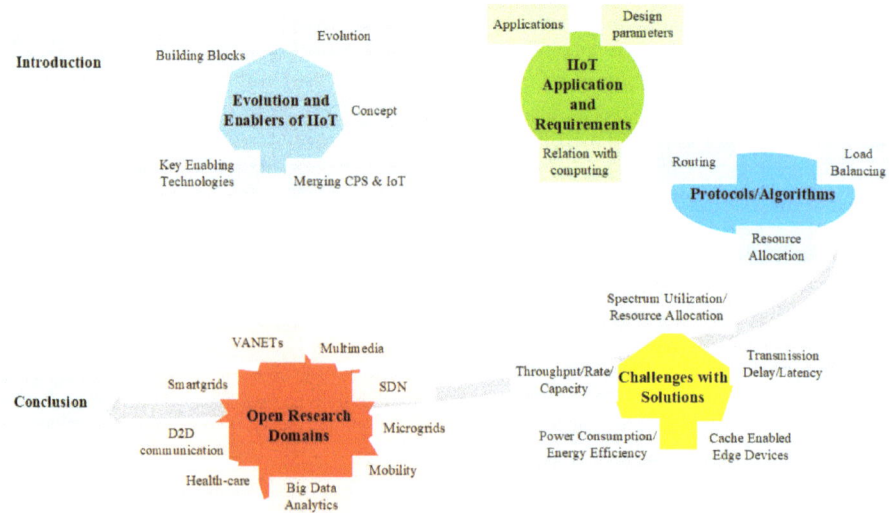

Figure 2. Flow of the paper.

2. Evolution and Enablers of Industrial Internet of Things

As discussed earlier, IIoT or Industry 4.0 is a new emerging term for future industry, which involves many key enabling technologies and applications of IoT. In this section, Industry 4.0 evolutional phases, IoT connectivity technologies, benefits from IIoT and key enabling technologies that endorse industrial revolution are briefly explained.

2.1. Industry 4.0-Evolution

The industrial revolution with the passage of time has many phases according to the requirement and challenges of the respective era. Figure 3 gives an idea about evolution towards Industry 4.0 with its elements.

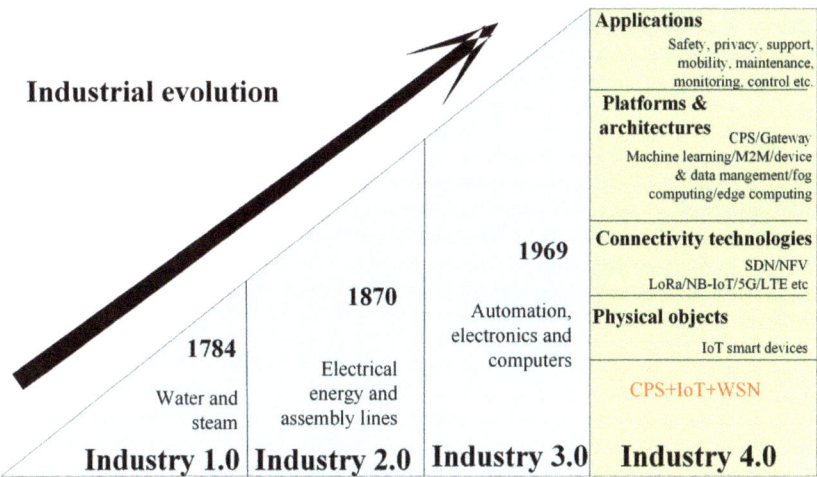

Figure 3. Evolution towards Industry 4.0/IIoT.

Industry 1.0: At the end of the 18th century, the 1st industrial revolution started with the help of water and steam power, which systematizes the factory floor. First, the mechanical weaving loom was established in 1784, and the first mechanical system was built thorough mechanical production facilities.

Industry 2.0: In the beginning of the 20th century, the 2nd industrial revolution started using electrical energy. The first assembly line using electrical energy was established in 1870. The introduction of mass production in industry 2.0 enhanced the industry.

Industry 3.0: Beginning of the 1970s i.e., in 1969, the first control system using programming language was established. Industry was slowly shifted to automation using information technology and micro-electronics's applications. This is the 3rd industrial revolution [26].

Industry 4.0: This previous industrial revolutions give rise to the development of industry 4.0. Industry 4.0 contributes a revolution to all domains comprising economic, academic, research, industrial and manufacturing sectors. There is a huge impact of the industrial revolution on the manufacturing processes of many fields. Implementing industry 4.0 demands change in many technologies namely automation, identification, computer, network communication, digital manufacturing, production process, production control management, decision making, judgment, sensing and analysis [27]. In the future, the manufacturing industry is expected to change on a large scale because of all new generation networks and interfaces offered by the environment of industry 4.0. This transformation is already in process in many industrial sectors. Up till now, for the fourth industrial revolution, exponentially growing technologies are sensor technology, artificial intelligence, machine learning, robotics, nanotechnology, and 3D printing [28]. These technologies were invented decades ago, but their minimum cost and exponential growth will shape industry 4.0. To change the industrial process, researchers are focusing on providing the evolved form of these technologies in terms of flexibility and fast computational process. All this automation in industry is very important for the economic growth of a country.

2.2. Industry 4.0-Concept

Increasing progress is witnessed in automation of industry using advancement in digitization, networking and new communication technologies, satisfying the market and consumer requirements [29]. Lenze SE, Corporate Communications Public Relations use idea of *machine modularization*. According to the demand of market/consumers, different modules are added or removed during the manufacturing process; machines are retooled smartly using smart communication technologies A cloud solution was given by Lenze, which is secure as no customer wants to share their demands and production details. Details about the customer's machine are saved in a cloud and he can investigate all the system details especially faults. This cloud solution is vulnerable to hackers, Lenze, along with another company, have provided a secure solution that is acceptable for today's industry [30].

Charlotta Johnsson explained the idea of industry 4.0 using four terms, these are smart devices and smart production processes with horizontally and vertically integrated manufacturing systems. Smart devices result in the production of intelligent products, these products do self-monitoring, self-controlling and self-manufacturing, have a uniquely identifiable ID, know how to solve and achieve goals [31]. The intelligent production process comprises smart starting and ending of manufacturing processes. Vertical and horizontal integration means all the steps during the smart/intelligent production process are integrated throughout the life cycle i.e., from starting phase to ending phase [30].

Industry 4.0 results in a faster manufacturing process, product development and improves the handling of complex environments inside an industry. The term first originated in Germany named as *Industrie 4.0*; in United States term used for this fourth generation is *Smart Manufacturing*, Chinese researchers have used term *China 2020*. *Industrial Digitisation* is the term used in Sweden for transformation of industry to automation [32]. It is believed that this industrial revolution will increase global competitiveness, preserve the domestic manufacturing industry and will have a huge impact on the business market as well. Now many countries around the globe have taken initiatives for automation in industries.

2.3. Industry 4.0-Merging CPS and IoT

The Industry 4.0 environment is comprised of the Internet of data, Internet of things, Internet of people and Internet of services. Interface of Industry 4.0 with existing smart infrastructure such as smart buildings, smart homes, smart grids, smart logistics, social web, and business web build a CPS system. This revolution will merge the real and virtual world on the basis of CPS. A CPS system has a computer-based algorithm that integrates the Internet and its users. It is simply digitization, in which these systems make connection of information technology with electronic/mechanical device components that exchange information among each other using a network. Using computer-based algorithms, CPS brings software and hardware components working in an automated and controlled manner to perform a certain task without human's assistance. The basic visualization of a CPS is given in Figure 4. After collection and analysis of big data, CPS can increase performance in terms of high-quality, low-cost goods production. With the advancement in sensors and computing technologies, various CPSs are emerging. CPS has evolved to Cyber Physical Production Systems (CPPS) to encourage the development and production process of Industry 4.0 [33]. CPPS combines physical smart IoT devices, networking technologies to compute in the production process. Robotics, remote machinery control and diagnosis, smart devices, heavy industry, transportation, health and condition monitoring, energy production, smart cities, and food manufacturing are IIoT services enable by CPS/CPPS architecture. Figure 5 represents the comparative analysis between CPS and IoT in the form of a Venn diagram. The similarities between the two give support for the development of Industry 4.0/IIoT.

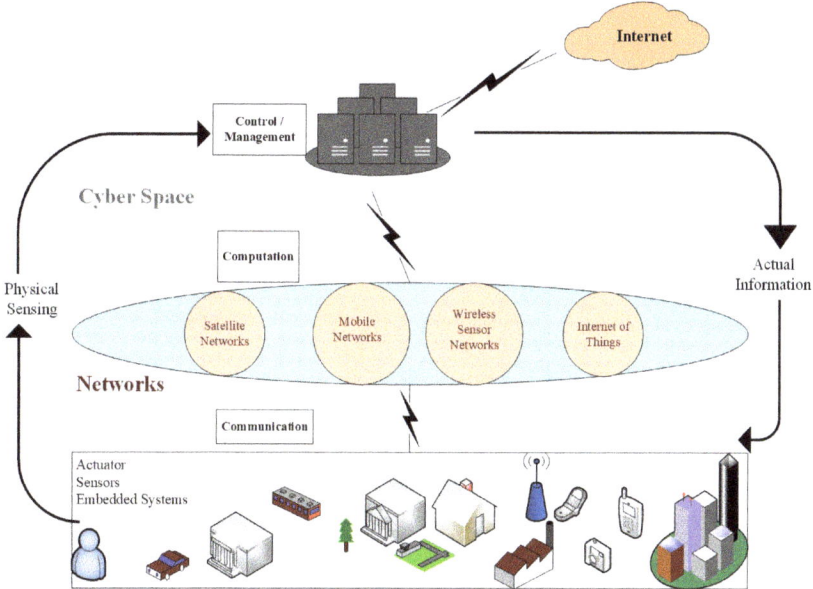

Figure 4. A cyber physical system architecture.

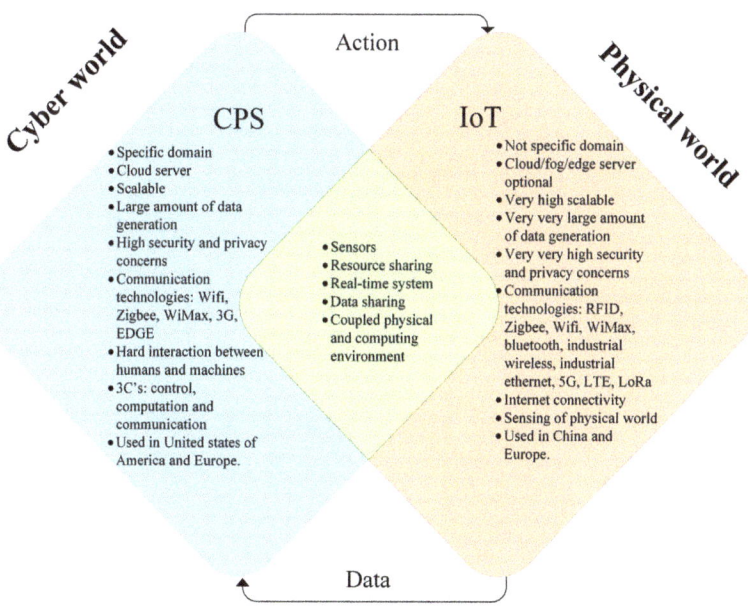

Figure 5. Comparison of CPS and IoT; supporting Industry 4.0 development.

The layered architecture of Industry 4.0 is given in Figure 3, which involves the common attributes of CPS and IoT. The physical or sensing layer should be designed to the extent that IIoT applications can sense/control information from physical environment and integrates with hardware sensors and actuators accordingly. The second layer should be optimized to provide a reliable connection to support data transfer over a communication medium (wired/wireless). So far, IIoT applications used wired medium to provide solutions. In the near future, the wireless medium is required because of shifting from centralized to decentralized solutions. Connectivity technologies, such as NB-IoT/5G and beyond 5G over different architecture such as SDN, NFV, cloud computing or fog computing will give solution to different applications. The intelligent-application layer, providing services to users has to be optimized in terms of service production, satisfaction, interaction, and management.

2.4. Industry 4.0-Key Enabling Technologies

With the use of advance technologies of wireless communication, diverse new emerging protocols and architectures are supporting automated industry 4.0 development. Resources can be efficiently used after integration of communication technology and big data processing in real time, this will result in better performance. Industry 4.0 development involves many communication technologies; however, big data, IoT, 5G, mobile computing and cloud/fog/edge computing are the key enabling technologies [2,34,35]. An extensive range of IIoT projects have been deployed in domains of building automation, manufacturing systems, health care systems, transportation systems, processing food and agricultural systems in the past few years. Reaching a common task in an IIoT application; sensing, integration, and communication are main steps. RFID tags are used for sensing, network topologies and protocols are used for communication. All these smart devices are associated with each other using internet. Many connectivity technologies are available for supporting IIoT applications. Critical requirements of IIoT applications have many open challenges in all domains (smart grid, smart cities, smart devices, D2D, healthcare) such as capacity, real-time connectivity, remote maintenance and topology of communication networks. Figure 6 gives a general overview of technologies to connect things to the Internet, representing short-range and long-range wireless technologies. All technologies work differently with aim of low-latency, low-power consumption, low-bandwidth requirement, and reliable communication.

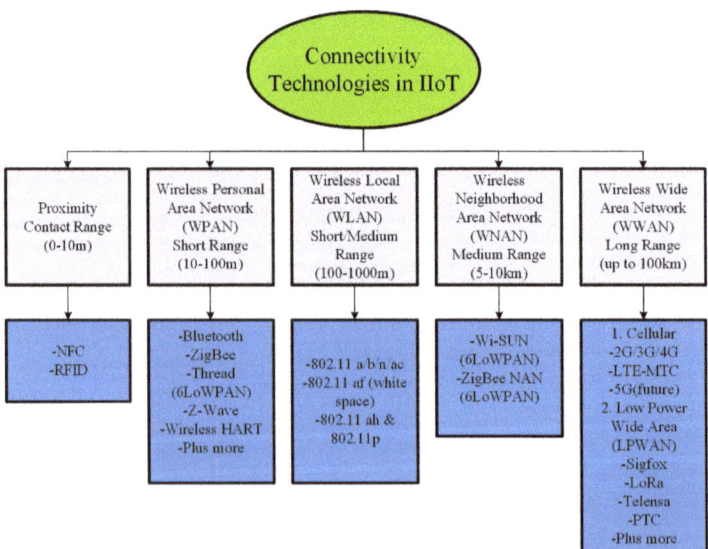

Figure 6. Different connectivity technologies in IIoT.

For connecting IIoT devices, all technologies have to work with the objective of maximum throughput, minimum power consumption, minimum transmission delay, and maximum transmission distance range. 2G, 3G, 4G, LTE are cellular technologies that were used for long range connectivity in wireless wide area networks (WWAN). IIoT application's critical requirements and the increasing number of smart devices need additional resources for connectivity. Increase in smart devices results in more data processing for which connectivity technology is moving towards 5G. The 3rd Generation Partnership Project (3GPP) proposed Extended Coverage-Global System for Mobile Communications for the Internet of Things (EC-GSM-IoT) and Narrowband-Internet of Things (NB-IoT) for supporting M2M, enhanced MTC (eMTC), massive MTC (mMTC) and critical MTC (cMTC) communication networks for IIoT applications [5]. 5G cellular technology gives super low latency, ultra-reliable and high availability to cMTC applications (industrial application and control, remote surgery, remote training, remote manufacturing, and traffic safety and control). Low cost, low energy, and the massive number of intelligent devices in smart agriculture, smart meter, tracking, fleet management, and logistics domain are supported by 5G as well. 5G is beneficial for IIoT applications comprising from mMTC, cMTC to enhanced mobile broadband. The distributed model of IIoT applications require a massive amount of data rate with minimum latency, 5G technology gives 10 Gbps with 1 ms latency. 5G is use case driven communication technology for upcoming IIoT applications.

For distributed ultra-low-latency and reliable connectivity in IIoT applications, 5G-IoT is an emerging solution. 5G-IoT scenario extends capabilities of IoT smart devices used in all domains. Recent research is focusing on low-latency, end-to-end reliability, and low energy consumption for both uplink and downlink communication. There is a lot of potential in research on IIoT with 5G communication technologies, to overcome challenges. This research will help in the industrial revolution. With the evolution of Industry 4.0, 5G is rapidly evolving in order to meet the requirement of IIoT applications mainly real time functioning, energy efficiency, less power consumption, shared spectrum regulation, reliable communication, and handing massive amount of data. Almost 90% of needs met using fixed line 3G and 4G cellular technologies, but need for deployment of industrial revolution can be fulfilled using 5G. 5G as an enabler of industry 4.0 gives multi-channel, capability, multi-network management, operating both local and global networks, supporting heterogeneous networks [27]. Mobile computing and cloud computing brings accurate data for IIoT application and provide efficiency to industry 4.0 infrastructure. Details of cloud computing in comparison with fog and edge computing is explained in the next section.

2.5. Industry 4.0-Building Blocks

Fourth manufacturing revolution, i.e., digital industrial technology provides services in industries that involves data exchange among machines making more efficient and fast processes. In words, the IIoT can be defined as: *Devices with centralized controllers, sensors, battery and memory attributes will interact with each other using Artificial Intelligence (AI) and Machine Learning (ML) algorithms.* Real time connection is possible using decentralized analytics and decision making of these devices. This section will give building blocks that are used in transforming industry 4.0 development.

2.5.1. Simulation, Autonomous Robots

A virtual model of a physical world which comprises machines, humans, and products can be interpreted in real time technology named as simulation. Every new product or updating process in available products for any machine can be verified, tested and optimized via simulation-based applications. It will result in increasing the quality of machinery and save the resources in the physical world. An autonomous robot collects information from its environment and learns from it and does work in the future without the involvement of humans using its self-learning algorithms (machine learning). These robots will transform the industries into automated industry. This technology will have a large impact on the industrial revolution. These robots are cheap and more capable of doing tasks efficiently.

2.5.2. Big Data and Analytics, Horizontal and Vertical System Integration

Different systems, ranging from customer to enterprise-level systems, have to collect, manage and evaluate the big amount of data. Three main goals of big data analytics result in the reduction of cost, efficient decision making and emerging new services and products. Industry 4.0 involves digital transformation in vertical and horizontal value chain networks. These networks result in the integration of customer and enterprise systems of companies, departments and business market-exchanging data. These value chain processes should be transparent and flexible with real-time functioning constraint.

2.5.3. Additive Manufacturing

Additive manufacturing is the process in which a 3D model is manufactured by joining the raw materials, usually layer by layer. It is the opposite of subtractive manufacturing in which raw material is carved to create a 3D model.

2.5.4. Augmented Reality

The idea of taking decisions remotely in real-time results in improving work procedures and will be implemented in the future as Augmented Reality (AR). In this augmented reality-based systems send repairing requirements or selection of new components.

2.5.5. Cyber-security

Exponential increase in connections among devices in industry 4.0 will increase threats to systems, networks, and processes. Cyber-security is a process that prevents unwanted intruders from accessing, destroying, interrupting or changing sensitive information about company/organization as well as business market networks; gives them secure and reliable communication systems.

2.5.6. Cloud Computing

Cloud computing is centralized and complex technology that supports high speed, high performance, flexible resource use and dynamic allocation in a network. As IIoT application requirements are low latency, high speed and reliable communication, privacy and security, efficient allocation of resources and energy-efficient communication technology. Some limitations regarding use of cloud computing for IIoT applications are:

- Confidential data and personal information of an industry should not be shared with outsiders.
- Security and privacy are in high demand by an industry from the cloud service provider.
- Data location on the basis of geographic follows rules and regulations. It also helps in securing the information.
- High load demands high-speed internet connectivity. This processing causes delays in communication.
- Memory and storage capacity may get exhausted because of many applications simultaneously accessing a single cloud server.
- Context awareness is required for speedy processes.
- Different standards cause problems in exchanging data, information, services, and applications among different clouds at different locations.
- Recovery and back-up update are required for industrial processing and decision making, cloud computing will cause delay.

2.5.7. Fog Computing

Fog computing or "fogging" is an extended form of cloud computing, in respect of industrial revolution giving applications and services (low latency and high processing) to autonomous heterogeneous devices inside an industry [36]. The idea is to bring processing, storage, maintenance and intelligence control to the proximity of data devices. Inside industry 4.0, there is critical requirement of real-time services with high data processing, maximum capacity and scalability. Fog computing gives the best solutions for such an environment because of its significant benefits over cloud computing. Extension of cloud computing, aims to minimize the burden on the cloud by introducing network edge computing concept.

For industrial automation, real-time services and decision making processes require low latency and enhanced cache memory. Required performance parameters are mobility, real time applications, low-latency, location-awareness, number of nodes and cache-enabled edge devices on this basis of geographical distribution. Virtualized nodes frequently known as cloudlets or fog nodes are placed between clouds of internet and end user devices. Fog computing provides services and applications as a cloud does with better QoS parameters performance covering critical requirements of IIoT. Important advantages of fog computing that influence its use for IIoT are:

- Data storage on network edge nodes eliminates the transmission delay by removing the need for accessing data from far-away clouds.
- Fog computing supports to process and analyze the data on faster speed for IIoT applications.
- Data storage on edge nodes will reduce the processing and computing delay.
- Cache enabled nodes will prevent transmission of irrelevant information over the network.
- Can give support to all IoT applications e.g., smart grids, smart cities, D2D, Vehicular Ad-hoc networks (VANETS) using edge networking concept.
- Provides filtered and required interaction between end devices and cloud service providers.

Fog computing is the building stone to provide solutions for more efficient, effective and manageable communication way for the massive number of smart IoT devices in the near future. Fog computing with extra features as compared to cloud computing in terms of latency, security, location awareness, location, and number of server nodes, real-time connectivity and mobility is a promising enabler for industrial automation.

2.5.8. Edge Computing

Introduction of enormous smart devices making an industrial revolution in all domains, causes extensive data processing, computation and burden of traffic on a single server either a cloud server or cloudlet. This motivates researchers to develop a new computing technology named as edge computing. The idea is to develop embedded automation controllers on devices named as edge-node (e-node) in the literature. This device is intelligent, with low processing power, better hardware security. Edge computing is an extending form of previous fog and cloud computing technologies. It comprises peer-to-peer networking, self-organizing network, and remotely manageable server. It gives following advantages:

- Encourages real-time connectivity.
- Overall network traffic reduces, as some computation is done on the edge of the network.
- Enhances security by encryption of data near to the network core.
- Optimize the resource usage.

IIoT applications have critical communication requirements. Cloud computing, fog computing, and edge computing platforms need to be optimized for better, efficient results. Cloud computing can be used where there is no high requirement of real-time connections, privacy, and security. On a local area network, fog computing uses a centralized system which interacts between the network and cloud server, whereas edge computing does computation on embedded systems of the network. Edge computing has direct interaction with sensors and actuators. The need for cloud, fog and edge computing architectures is increased with the growth of the IIoT application. To increase the use of IIoT smart devices, researchers are focusing on fog or edge computing paradigms which results in industrial development.

3. Industrial Internet of Things Applications and Requirements

Information about the occurrence of faults, components, inventory, different demands and different orders continuously needs to be shared among smart devices/processes resulting in improved efficiency, tracking, capacity use, quality of production and development in industries. From IIoT perspective; smart cities, smart factories, and smart products are important IIoT beneficial examples. The basic three-layer architecture of IoT as discussed in Section 2.3, need to be evolved according to the requirements of specific IIoT applications.

3.1. IIoT-Applications

From the identification of faults to solutions via communication and networking technologies, every step needs to be optimized and has research potential. Possessing attributes of intelligence, reliability, safety, sustainability, privacy, and efficiency; these applications are called *smart* in the literature. This revolution results in the development of new infrastructure. According to [37], smart city, smart factory and smart product are the main applications of industry 4.0.

3.1.1. Smart City Applications

About 6 billion people are expected to be part of cities of the earth in 2050 [38]. With the advent of technology infrastructure, this increase in population will result in more data origination and demand for services. This big data origination is called *Big Data*. To develop future smart city, there are many domains that need to be intelligent, such as the smart home, smart office, smart institution, smart health-care centers, smart agriculture, and smart transportation. All these domains have different IIoT applications with different requirements. The development policy of a smart city has six factors namely, smart economy, smart mobility, smart environment, smart people, smart living, and smart governance [2]. Numerous research has been carried out in the domain of smart cities. For instance, [39] presents a framework with which, smart cities can overcome current limitations. This smart city transformation will take time.

3.1.2. Smart Factory Applications

Smart industry comprises distributed automated systems and robotics. This future smart factory floor is possible using ML algorithms and AI technology. These automated devices are integrated with sensors, actuators, microchips, autonomous systems, and controllers. Relying upon CPS and IoT technologies for evolution industrial processes required ultra-reliable and low-latency communication (URLLC). For monitoring, managing and controlling such environment, IoT nodes can handle bounded latency of millisecond scale. These applications are characterized using latency, jitter, energy consumption, workload parameters. M2M and D2D are emerging supporting technologies for smart factory development. In the smart factory manufacturing process, machines will have high-level of automation and self-optimization attributes. These attributes will fulfill complex requirements of products. This has open issues regarding network communication technologies, number of devices, security, and cost.

3.1.3. Smart Product Applications

IoT, cloud computing, big data, cloud computing and production time are drivers of industry 4.0 development. The products in industry 4.0 are smart because they are integrated with sensors and microchips. Existing production systems need to be integrated with industry 4.0 architecture (IoT+CPS+WSN). This integration will allow communication interaction between human beings and products [40].

3.2. IIoT Application Design Parameters

To the increasing demand of customers and market requirements, the manufacturing industry is now facing problems in achieving desired goals. Industry 4.0 came up with an innovative idea of automation inside in the industry increasing the production process flexibly. Industry 4.0 key point is

M2M communication, in which machines communicate with each other over the Internet. Developing and manufacturing industry, communication of intelligent machines with each other using different technologies depending on the coverage area, give rise to high production and self-regulation of the manufacturing process in an industry. These are the goals of industry 4.0; all this is assured by IT systems as they provide enabling smart technologies. The integration of these heterogeneous devices in a network with other devices and existing communication technologies is also the main requirement for designers. Every IIoT application has critical design goal requirements to improve QoS in providing solutions. These design parameters are:

- **Energy & Long Battery Life:** Overall network energy should be preserved for better and efficient outcomes. Smart devices should have enough battery storage so that they can use for long time.
- **Latency:** Some IIoT applications are time-sensitive, a bound should be there to limit all types of delays including processing, propagation, transmission, and computation.
- **Throughput:** Amount of data for processing is different for different applications. It should satisfy the application requirement.
- **Network Topology:** How the number of servers (cloud, fog, e-node) and smart devices are placed in a network for better QoS requirements.
- **Reliability:** Solutions by IIoT applications demand reliable real-time connectivity.
- **Security, Safety & Privacy:** These are very demanding and major requirements for all IIoT applications. For example, inside a smart factory there should be privacy and security such that no one can access the private information. For healthcare applications, patient's information should be safe and not easily accessible and changeable. 3A's; Authentication, Access, and Authorization are steps involved in the strictly secure system. The demand of end to end communication in IIoT applications requires privacy of data as well. Sensors and actuators should be safe from intruders as well as environmental hazards.
- **Low Cost:** Smart devices used for IIoT applications should be low cost so that doesn't affect the CAPEX/OPEX. Deployment involved in industry 4.0 should not be so much that will cause loss in marketplace.
- **Long Coverage:** A device should be capable enough to cover the desired range.
- **Standardization:** So far, there is no such network standardization and is an open challenge for researchers.
- **Integration:** IIoT applications are composed of heterogeneous devices and hybrid networks, there are a lot of issues in integration.
- **Communication/Enabling Technology:** Communication technology for supporting IIoT application should provide assured performance services.
- **Device Maintenance:** Heterogeneous device in an industry 4.0 environment, require constant device management as devices are connected with each other and the Internet. Software Defined Networking (SDN) is used for such failure and changing maintenance issues of devices.
- **Monitoring Network:** Wireless, environmental and mobility nature may cause a change in network topology which requires the system to be monitored and managed frequently.
- **Configuration & Management of System:** Self-configurable, self-control, reconfiguration functionality in addition of new devices in network.
- **Traffic congestion & Overload:** Smart devices will be increased with time in any IIoT application. System should be able to adjust according to the traffic burden and data requirement.
- **Mobility:** IIoT applications, such as transportation, inside industry and healthcare devices, have the property of mobility from one place to another.
- **Scalability:** Scalability brings many issues, some are: How many numbers of smart devices are enough to support an industrial application environment? or how many devices are served by a server easily? how to optimally design a system under energy/spectrum issues?

- **Heterogeneity & Interoperability:** Heterogeneous smart IIoT devices have to communicate and collect information among themselves and the Internet. This integration is an issue to solve. Standardization is required for interoperability of IIoT devices.
- **Performance:** There is always a performance trade-off among these QoS requirements. There should be an optimized, supportive, and efficient trade-off among the factors affecting performance. Performance maintenance solutions are required for future automation.

3.3. IIoT Applications in Relation to Cloud, Fog and Edge Computing

All IIoT applications require critical QoS parameters in order to produce benefits in every field. Fog term used in computer science is an extension of cloud architecture which offers services of cloud to edge devices. It is seen as a new cloud or it will replace the cloud in future but it is just an architecture that complements cloud architecture in order to provide solutions for critical applications. Entry points in any network are called edge devices. Entry points are part of second and third layer, it has hardware devices named as switches, routers and WAN devices. Cloud is a centralized solution and fog brings solutions at a distributed level by bringing data storage and its computation near the edge of a network. It allows getting services in the proximity of IoT devices. Fog combines services provided by cloud and IIoT applications; or it enables IIoT applications.

Cloud architecture is efficient, beneficial and provides solutions. It offers services after storing data in remote centers from the Internet. These remote data centers face less delay and computation as compared to internet. It saves the cost of physical resources, makes connections more reliable and results in an increase in efficiency and performance. It is a flexible, innovative framework in the networking field. It helps in accessing resources anywhere anytime. Question is why fog architecture is needed when cloud is already the best solution to many issues? There are certain issues and requirements of IoT applications and fog can provide solutions. First and foremost, the low-latency requirement of IIoT applications in every field can be attained effectively. It stores data in the proximity of users. Propagation delay between cloud and users will be reduced; computation delay due to huge data traffic at cloud can be reduced. It supports time-sensitive tasks effectively. Second IoT challenge was using network bandwidth in an optimized manner, fog being in the center of cloud servers and end devices. It helps in less usage of bandwidth; data doesn't have to travel fog-to-cloud distance. Popular content is available at the network edge. This will also result in minimizing cost, lifespan of devices, energy consumption and complexity during every demand which goes on the users-to-cloud path. Security issues can be seen on the fog server; it can act as a proxy-server controller. Privacy and safety of data is another important IIoT devices requirement. Fog helps in monitoring such tasks.

Exponential increase in IIoT applications in every field making world a cutting edge technology paradigm. Figure 7 shows both these architectures; cloud computing in which IoT devices are directly connected to data centers and cloud server and fog computing in which fog server is in the middle of cloud and IoT devices. Fog works on network edge which improves speed, computing capabilities and provides distributed and better solutions. Fog has complemented the cloud architecture in many ways. Cloud is a centralized solution while fog can work in both a centralized and decentralized manner. Over large geographical area a group of fog nodes can be monitored and managed in a centralized way. Cloud size is large as compared to fog, as it has massive storage of data from the Internet. Fog size is a flexible parameter that can be altered according to the demand of users. For example, for a vehicle tiny sized fog would be enough while for an institution many small fogs can work in a form of network or may be a large-sized fog would be sufficient. Fog has fewer deployment complexities as compared to the cloud. Similarly, cloud management is tricky and more time taking as compared to a fog because of its flexibility. This flexibility will support mobility in networks.

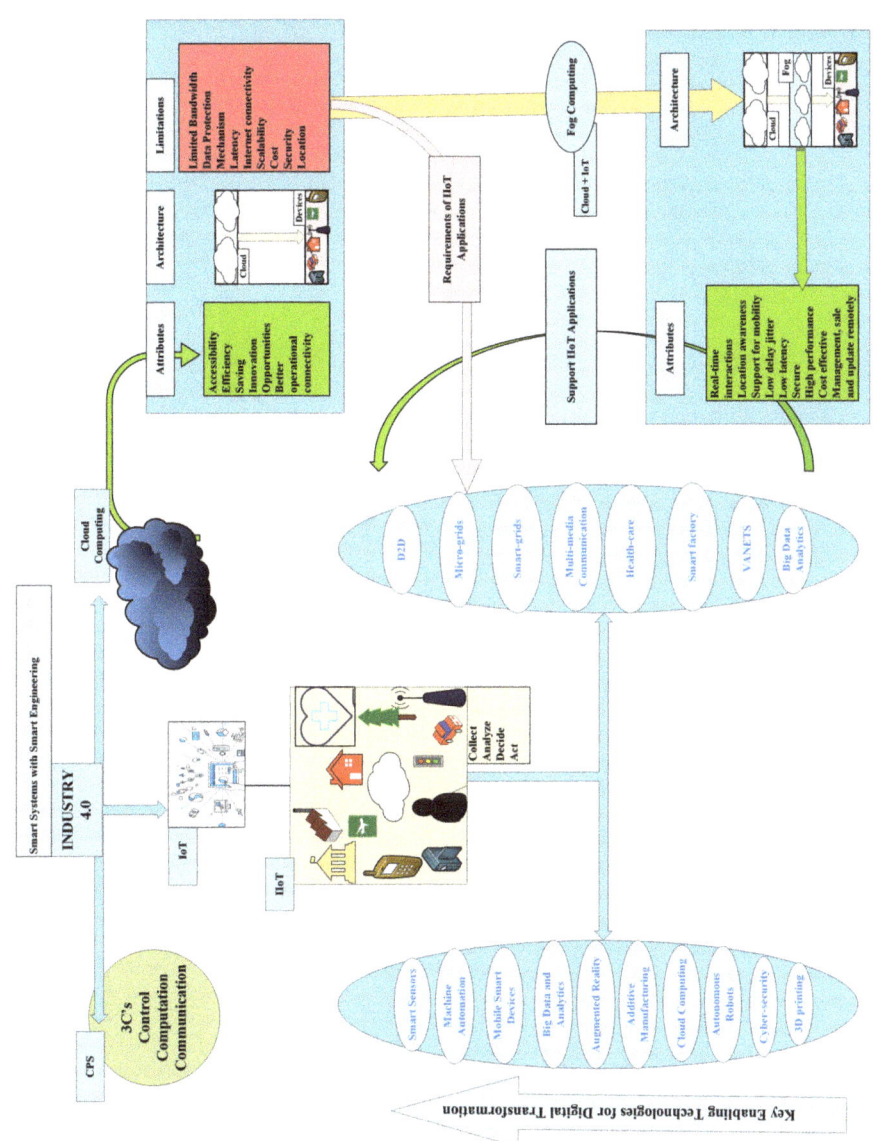

Figure 7. Industry 4.0, IIoT applications versus cloud and fog computing.

Because of architecture complexity, the cloud can only be operated by technical experts while fog can be managed and operated by little human effort. Large companies control cloud networks, while fog can be controlled by small as well as large companies. Low-latency critical IIoT applications will be handled by fog, while applications that can tolerate large delays leverage the cloud services. Internet connection between user IoT devices and a cloud server should be reliable for the entire time of connection, while in fog architecture connection with cloud is not necessarily required for the whole time. Bandwidth requirement increases for both frameworks as the number of IoT devices is increasing exponentially. It provides awareness property, which means it knows about the requirements of customers and will provide solutions accordingly. It can be placed anywhere between the cloud and user nodes according to the demands. Performance and efficiency parameters of IIoT application services can be enhanced using fog framework. In addition, the scalability issue can be handled using it. Data storage at network edge will result in minimization of service delays and supports real-time processing. The big amount of data produced from heterogeneous smart devices requires huge data storage and computation at the cloud server. IIoT applications requires data storage at their backbone. This big data is giving rise to edge computing for future. Edge computing is also called Mobile Edge Computing (MEC). This technology supports IIoT applications by building better operational connectivity. It brings cloud computing capabilities at the devices that are present on the edge of a network, these devices named as edge servers or edge devices. The edge devices are part of the Internet and participate in processing and computation near the data location. This technology is an industrial initiative by the European Telecommunication Standards Institute (ETSI) [41].

4. Protocols/Algorithms

Protocol is a basic set of rules that defines how communication happen between different devices in a network. Protocols have to be devised intelligently in order to achieve the defined goals. As the future is industrial automation in industry 4.0 revolutionized era, there is an exponential increase in smart, intelligent devices in IIoT applications. A major requirement is that the emerging network protocols must meet required goals in performance affecting parameters, such as energy efficiency, latency minimization, spectrum efficiency, cache memory maximization, and bandwidth use requirements. A summarized discussion on pre-existing protocols in the context of fog computing proposed by researchers is given below in following subsections.

4.1. Routing

Dong et al. [42] have introduced redundant fog loops for WSNs. The proposed fog loop-based scheme has two main steps. Creation of fogs using loop paths is the first step, while the second mechanism creates fog nodes in the source node areas along with many other interfering fogs within the network. This proposed scheme has helped in finding the exact location of the source node in terms of energy efficiency and privacy. Results were compared to the efficiency offered by the Phantom Routing Scheme (PRS). The proposed scheme gives improved efficiency by 4 folds and can also improve the privacy and security up to 8 folds.

Since fog computing lowers latency and offers energy saving, they are tailor-made for dealing with WSNs [43]. Sensors in WSNs are resource-constrained, therefore energy efficiency is an important issue. It needs to be addressed for the network to increase network life-time of operation and working efficiently for a prolonged period of time. Sensors in a form of clusters collected data and send to the base station using energy-efficient routing protocols. In this approach, using multi-hop communication, data is transferred to the sink/destination node. The nodes acting as a cluster head are used for multi-hop communications. For networking, apart from the routing problem, another issue that needs to be catered is increasing network lifetime. Network lifetime can be effectively increased by optimizing energy and power consumption at nodes. Some examples of these routing protocols are Low-Energy-Adaptive-Clustering-Hierarchy (LEACH) and Stable-Election-Protocol (SEP). LEACH protocol involves Deterministic Cluster Selection Head and abbreviated as LEACH-DCHS [44].

Handy et al. [45] have proposed LEACH in which rotation of the cluster heads in a randomized manner is used for energy distribution to the nodes evenly. These algorithms haven't been used for fog computing yet, but can be used in future incorporating fog computing. SEP was proposed by Smaragdakis et al. [46], prolongs the period of stability of wireless sensor networks. A modified version of SEP (M-SEP) proposed by Singh et al. [47] is suitable for heterogeneous WSNs. Some of the nodes in this approach might have a greater chance of getting selected as a cluster head and have more energy. Moreover, in large WSNs, the data to be processed from heterogeneous devices is in large volumes. Processing time is significantly large, so an alternate approach is to use fog computing. The time required to process large amount of data gets considerably reduced using fog as the sensors in the network get interconnected with the Internet in order to make smart by making them autonomous in making decisions.

Another routing protocol that is energy-efficient named as new-SEP is proposed by Naranjo et al. [48], prolongs the stability of sensor networks supported by fog more than SEP. Optimal clustering nodes are elected by considering various sensor node features such as the ratio of heterogeneity in the network, residual energy, distance between cluster heads. Results were compared with baseline schemes of LEACH and SEP, the proposed scheme performs better in terms of energy-preservation and network lifetime. Considering increasing the network's stability periods, N-SEP performs better than LEACH (50 percent) and SEP (25 percent). An efficient route optimization algorithm was presented in [49] to address the mobility control issue in fog-based SDN networks. The proposed SDN-enabled fog computing architecture had three-layered structure namely, fog layer, network layer and application layer. Results showed that there is a great improvement in network performance. A three-layered plane architecture was proposed to generate efficient routing paths by the authors in [50] for data-center-based heterogeneous networks, using tensor decomposition methods. These three planes do three different tasks, edge plane considers the traffic, bandwidth and delay requirements; fog planes computes and controls the available paths and finally cloud plane do the routing.

4.2. Resource Allocation

One way to meet the growing IIoT application requirements is to use fog computing. Interactions in real time takes place in fog application rather than batch processing. Services supported by fog include mobility, heterogeneity, working with cloud to extend cloud services, user optimization, etc. Resource allocation or resource management in any network is one of the biggest challenges. This defines new protocols for networking and communication. Many researchers have done work in this context and already proposed some protocols. Resource allocation was done with a specific defined objective, it can be maximizing energy efficiency or throughput; minimizing latency or power consumption or network cost. A joint min-max optimization problem of resource allocation and offloading decision making was proposed by the authors in [51]. They have proposed Computation-Offloading-Decision-Making and Resource-Allocation Algorithm (CORA) to minimize the maximal network cost (delay & energy consumption). Results were explained using fractional programming theory and lagrangian dual decomposition. Fog Radio Access Networks (F-RAN), an extension of the Services provided by Cloud Radio Access Network (C-RAN), has gathered attention globally and several advantages can be taken by providing functions of baseband signal processing near the edge or making the edge devices cache-enabled. F-RAN is the best example in which we are using edge devices as well as network devices (access points) as fog nodes to achieve the best results. Moreover, due to the cooperative communication, benefits of C-RAN such as enhanced spectral and energy efficiency are also conserved in F-RAN. Resource allocation is a challenge for the upcoming F-RAN. Authors in [52] have proposed a Stackelberg equilibrium (SE) for a hierarchical problem of network slicing customization between global radio resource manager (GRRM) and local radio resource managers (LRRMs). A game is formulated to alleviate the burdens in GRRM and LRRM; GRRM assigned resources to each slice and then LRRM in every slice provides resources to UEs. UEs could be divided into clusters on basis of respective objective functions and accordingly get resources. Authors have provide two algorithms for LRRM 1 slice and LRRM 2 slice with an objective function of maximizing high data rate and minimizing latency, respectively.

To meet low latency, high throughput and connectivity requirements in future RANs, non-orthogonal multiple access (NOMA) is a promising technique. Zhang et al. in [53] have formulated a problem for maximizing the net utility under interference constraint for resource allocation problem for NOMA-based FRAN system model. Results were compared with the conventional orthogonal frequency division modulation (OFDM) technique. Madsen and Albeanu [54] discuss platforms involving fog computing in which there is communication going on between the smart devices, fog, and cloud. A model for internet applications in future is presented by Hong et al. [55] in which applications are delay-sensitive and distributed geographically. For improving the rendering performance of a webpage, Zu et al. [56] exploit information available only at the network's edge. Companies that want to deliver content such as Netflix uses fog computing to reach their geographically distributed customers. As indicated by [57], ensuring significantly large streamed data to be delivered in the proximity of the end-user (customer), is done using fog computing.

Minimizing the energy consumed in geographically distributed applications and resource allocation using fog computing is discussed in [58] for video services. Size of fog nodes deployment shows the application demand in that region. To maximize social welfare, some of the user data needs to be controlled. Optimization on large scale is also possible using proximal and distributed algorithms. The algorithm proposed gives a near-optimal solution.

4.3. Load Balancing

When one edge device has to do a lot of work e.g., computing tasks all by itself, it consumes resources. However, this can be reduced with the help of distributed architecture in which load is evenly distributed among the edge devices present in the network. This distribution of load is called load balancing. Load Balancing in combination with fog computing makes a formidable combination. A fog network can provide a platform for cooperation and coordination between edge devices in a load-balanced network. Authors have proposed a load balancing offloading algorithm for latency minimization in Vehicular fog computing (VFC). VFC is an integration of vehicular networks and fog computing technology, which is an efficient field to achieve real-time and location-aware for vehicles in a smart city. Zhaolong et al. in [59] has formulated a response time minimization problem using a three-layered decentralized network system model to balance the traffic among vehicles. These three layers are cloud layer with a high computing cloud data center, cloudlet layer processed the received data from vehicles before sending to the cloud center. Lower layer has vehicle clusters, clusters are made for traffic balancing. The fog-vehicle interface manages and alleviates the traffic load.

Traffic overhead results in inefficient resource management, which is also a challenge to be solved for healthcare applications. WSNs based health monitoring systems have become a convenient choice as elderly people can frequently require health services. The basic requirements for healthcare applications are energy efficiency, high-response time, low-latency and real-time connection. Fog computing can be considered to be an important enabling technology for such time-sensitive applications. Authors in [60] provide a very significant critical review of existing solutions provided by researchers in the healthcare domain using fog computing. The authors in [61] proposed a fog-cloud hybrid solution to load-balancing problem. If a client's requirement is more critical, it will be handled by cloud otherwise servicing is done by foglets. Results shows that network utility can be enhanced in terms of latency and load balancing. They used iFogSim tool for experiments. Forough et al. [62] proposed an Energy Balancing Algorithm (EBA) for Fog-IoT networks to reduce delay and energy consumption. The authors have proposed two optimization problems, first one is to find an optimal transmission rate and power for terminal nodes (TNs) and the second one is to find an efficient topology between TNs and fog nodes (FNs). The system model was designed under the constraint of channel conditions between TN and corresponding FN. A major challenge in edge computing is the inefficient deployment of resources as it reduces the overall efficiency of the network. This results in definite increase in power consumption. A low service block is preferred for maintaining lower latency, which will otherwise be detrimental to the high-performance requirements of edge computing. Unlike cloud computing, fog computing has a limitation of resources, which can be overcome if we allow cooperation between various data centers. A cooperative scheme by Beraldi et al. [63] is proposed in which data centers near the edge exchange processing requests and shares the load of highly loaded data centers. The request arriving at a busy data center is forwarded to any other data center having the request buffer partially filled. The proposed scheme maintains a threshold of requests with the help of Markov Chains to make sure that load is equally distributed among all the data centers. This leads to much-improved performance as lightly loaded data centers can absorb the burden of heavily loaded ones during peak hours.

Full use of edge resources cannot be done by cloud computing polymerization calculation [64]. The edge devices are not part of the cloud computing, which is undesirable for delay-sensitive requirements. Ningning et al. [65] have explored how fog computing can turn the nodes or edges into virtual machines using Cloud Atomization Technology to improve on this problem. The authors use graph partitioning for developing an efficient load balancing algorithm. Consequently, a flexible network can be built by making fog networking possible after atomizing the cloud and eventually reduce the cost of the system that were high before implementing the load balancing algorithm. Deng et al. [66] investigate power consumption and transmission delay trade-off. Problem of workload allocation is formulated to obtain the minimal power consumption subject to the constraint of service delay. Three sub-problems result after the decomposition of the primal problem. Based on the obtained results, it is shown that cloud performance is enhanced by using fog computing. Moreover, latency is reduced and bandwidth can be saved by sacrificing some of the computational resources. Computational clusters are formed on the basis of cooperation behavior between the Small Cells (SC) to share computational resources. However, the cooperation is dependent upon many factors such as resource availability, resource allocation, delay constraints of the application, distribution of computational load and size of the cluster. The joint distribution of resources for the mobile end-users and the cloud is the main objective problem of this framework.

All the data that is being frequently accessed by the edge devices is stored at the Radio Units (RU), which considerably decreases the overall delay in the network of Fog-Radio Access Networks (F-RAN). However, in such systems, the energy efficiency aspect has always been a matter of serious concern due to the addition of extra smart components in the system. In [67], the authors have proposed a novel scheme by designing a green network in which an efficient algorithm is incorporated to optimize the selection of RU. Furthermore, the algorithm also jointly optimizes the formation of clustering and beamforming while maintaining the QoS and balancing the load of each of active backhaul as a measure of its capacity. While copying data from the database, edge devices may interfere with each other. To avoid this mishap, data replication techniques are used for copying data electronically from the main database where the data of all the users is hosted. Uniform distribution of data and processing is crucial over the network which in turn helps in managing the large amount of data and workload with efficiency. Fog computing not only helps in achieving higher efficiency, it also helps in balancing load across distributed platforms and achieving higher energy efficiency due to fewer performance bottlenecks. Verma et al. [68] have focused on making a network that is less dependent on cloud computing and bringing the storage and processing capabilities near the edge devices. The results are simulated with the help of CloudSim by testing different geographically separated servers and their configurations and then make comparisons between cloud and fog computing for various attributes.

5. Challenges with Solutions

In general, there are many challenges towards industry 4.0 digital transformation. For M2M communication, reliable and stable connectivity with bounded delays is a mandatory requirement. Real-time communication is on the higher priority for this fourth revolution, which brings many technical challenges for future network development. A well-designed network architecture can increase the sustainability and performance of the entire system. The journey towards industry 4.0 is on its way and until now there are no such standards, regulations, and certification to follow them.Although fog becoming a developing key enabling technology for IoT architecture, it still faces some issues while integrating into the current architecture. Complex software applications and solutions are needed to achieve an efficient fog network. Fog network will be analyzed in terms of key performance measurements, which are bandwidth use, energy consumption, low latency, maximum throughput, and resource management. In this section, critical technical communication and networking challenges in the context of fog computing for IIoT applications are listed. All of them have potential for future work.

- **Power Consumption/Energy Efficiency:** Several smart devices supporting an IIoT application will consume a massive amount of energy on a different scale according to their requirements. Ensuring network QoS with minimum energy consumption of smart IoT devices, fog nodes and cloud in an optimized way is an open challenge for every upcoming future IIoT application.
- **Throughput/Rate/Capacity:** Throughput or network bandwidth, data rate and storage capacity depends on how much data is used and where data is stored in a fog network. This data placement on fog nodes or edge devices or cloud server has effects on cost, delays, bandwidth, and network coverage. The optimal placement of data on cloud server or fog cloudlet is one of the critical technical challenges for fog-IoT architecture.
- **Spectrum Use/Resource Allocation:** Geographically separated fog, cloud nodes and their interconnection makes the backbone of any network that relies on offloading services. Most of the cloud computing interconnection mechanisms are not enough for fog networking due to their limitations including relying on a centralized cloud which cannot fulfill the latency and location awareness requirements of distributed devices, etc. Fog computing must encompass features, such as multi-tenancy, scalability, heterogeneity and quick resource provisioning. An architecture including fog and cloud computing must meet all these requirements for which resource allocation/use is the most critical challenge for better network performance. It has effects on all other QoS parameters.
- **Latency:** IIoT applications requirement is real-time connectivity. All applications are time-sensitive and require real-time streaming rather than batch processing. Fog computing gives a better result for such decentralized solutions. It gives low latency with reliable connectivity and mobility. Optimized placement of data centers, resource allocation, network architecture, energy consumption of nodes, and storage capacity of nodes have impact on latency. Latency for a network is the sum of transmission, processing, propagation and queuing delays. To achieve the low-latency requirement, there is need mitigate all types of delays.
- **Cache Enabled Edge Devices:** Caching content locally, reduces the access delay time and increases the energy and spectral usage efficiency. Since the Internet has multiple bottlenecks while accessing data from across continents and oceans, caching does not have to be dependent on any of these bottlenecks and instead makes the same data available locally. Furthermore, caching incredibly reduces the load on backhaul links since they do not have to be used anymore for accessing data. Since all the users have to access the data from the same centralized location (internet/cloud server), a certain degree of fairness is needed to avoid inefficiency in accessing data. Since all the backhauls have certain capacity constraints, there is a need for an efficient load balancing mechanism to overcome this issue.

Further, we discuss existing solutions to these mentioned challenges.

5.1. Power Consumption/Energy Efficiency

Objects are extensively being connected together using the IoT technologies. Heterogeneous smart objects, in the context of hardware and software, can perform efficiently in the availability of memory and high computational power. Network's complexity increases day by day, because of scalability issues of smart objects supporting IIoT applications.

Virtualization has played a very crucial role in improving the overall efficiency of the data centers and now research has also been done on how virtualization can help in fog networks. Virtualization makes it easiest to deploy fog functionalities on an existing node (by isolating and securing fog services in a virtual machine or container). It also helps in conserving energy by efficiently consolidating the tasks on a single fog node. A comparison in terms of power consumption was studied in [69] between C-RAN and F-RAN using network function virtualization (NFV) technology. The authors have formulated a mixed integer linear programming (MILP) problem, results for F-RAN are 30% more improved in terms of power-saving as compared to C-RAN.

Roca et al. [70] have developed a platform using Fog-Function Virtualization, which works and builds on the concept of Network Function Virtualization (NFV) for multiple IoT applications. This, further with the help of node constellation creation, helps in easy deployment and reduces the cost considerably as less energy is required for running the system due to efficiently performing virtualization. Task scheduling is necessary as it helps with the load balancing aspect of networking and can provide services to multiple users. Cooperative games between the containers and brokers are studied for energy-efficient task selection algorithm. Kaur et al. [71] have achieved efficiency with the help of container-as-a-service (CaaS). Lightweight containers have been used which considerably reduce the energy consumption by a container migration techniques. This kind of virtualization is more cost-efficient for distributed architecture, where a large number of devices with different running applications/processes can be allocated to resources efficiently. Results achieved by the authors prove that the system is more energy-efficient.

Graph-Based Heuristic algorithms were proposed by the authors in [72]. Type of problem is the integer linear programming (ILP) problem, the objective is to increase the energy efficiency under association and capacity threshold constraints for hybrid cloud-fog RAN (CF-RAN) architecture. Minimization of latency and power consumption is also addressed in the proposed system model. Fog computing along with NFV gives excellent outcomes in terms of reducing latency and power consumption. Energy can be saved by incorporating techniques, such as Message Queue Telemetry Transport (MQTT) in a fog-based environment [22]. In this scheme, the number of transmissions is reduced to save energy of the end devices. Using energy-efficient routing protocols is imperative to achieve energy efficiency. MQTT supports sensor data in real time due to its many-to-many communication nature. The concept of MQTT focuses on introducing another layer between the fog and cloud with lower complexity. MQTT broker place is at the fog layer. The intermediate layer is responsible for predicting the future measurements, and acts as a gateway for the upper layer. It helps to offload the computationally expensive tasks from the cloud to save in the storage memory of the fog server. This results in a reduced number of transmissions as the update only occurs in case of a mismatch.

5.2. Throughput/Rate/Capacity

For network designing, a new paradigm known as Socially-Aware-Networking (SAN) has been of major interest [73]. To achieve efficient performance, SAN brings the human behavior and CPS together via intelligently designing of a network. This design should be adaptable as well for all environments. The resources available to mobile devices differ depending on the models and specifications. This resource availability results in a group of mobile devices that might be sufficient in terms of processing and storage parameters. Group of some might not be self-sufficient. The best solution to this problem is given by SAN and Fog-Radio Access Networking (F-RAN).

D2D communication comprises the direct sharing of contents among mobile devices. Direct sharing is a key feature supporting D2D communication. For achieving efficient performance results, an imperative design of network embedded with all technologies is required. Research has shown that the system performance in terms of utility, throughput and energy efficiency is maximized using the download mode with the help of branch and bound algorithm [74]. Klas et al. [75] and Cau et al. [76] worked on improving network efficiency Mobile Edge Computing (MEC). MEC is a domain under fog and focuses on providing cloud computing capabilities near the edge of network. Efforts and improvements have been made in MEC in order for it to support 5G communications. Another target of MEC is to achieve access to information provided by the radio network for application development and distribution of the content. It is predicted that data traffic over mobile devices will increase manifold in the next few years. Such increasing needs must be satisfied using efficient mechanisms.

A reliable service with an extremely low latency (URLLC) and high capacity is the foremost requirement of IIoT networks. Throughput maximization of the F-RAN system is compared with three different back-haul strategies for Small Cell Networks (SCN) in [77]. All these strategies namely, decode and forward, direct transmission and C-RAN were studied under delay threshold, rate constraint, and backhaul and fronthaul links. The authors proposed iterative algorithms for all strategies. For F-RAN, Pontois et al. in [78] formulated a non-convex optimization problem under fronthaul constraints. A hybrid semi-distributed resource allocation algorithm was proposed by the authors for the proposed weighted sum-rate maximization problem. Results show that there is a trade-off between maximum throughput and system latency. A multi-objective optimization problem was proposed by the authors in [79]. They have proposed three parallel algorithms to improve latency, throughput and resource management. A queuing model was studied under task buffering, offloading and resource allocation algorithms. Lyapunov drift was used by the authors to design the resource allocation policy. In results for better system performance, trade-off between latency and throughput is observed. [80] end-to-end performance is guaranteed after composition of problem as Multi-Constrained Optimal Path (MCOP). The authors propose a solution from the network architecture's perspective and cloud service relation. The proposed algorithm provides better results in terms of efficiency and effectiveness. QoS parameters (capacity, delay and cost) gets improved by the proposed network-cloud service provisioning system model.

5.3. Spectrum Use/Resource Allocation

Fog networking can provide a solution for resource management issue in future 5G networks. With a growing number of smart devices, the major issue is of spectrum use and resource allocation. Spectrum pricing and allocation scheme (SPAS) was proposed in [81] for F-RAN framework. Three areas were defined for the whole network, and double game theory was applied to give the solution for efficient spectrum use. Proposed two algorithms are named as game-model (GM-SPAS) and multiple-spectrum-reuse-technologies (MSRT-SPAS). Their presented results are effective in terms of revenue and efficiency. The problem of resource allocation with utility maximization objective under QoS constraints was formulated in [82]. Analytic Hierarchy Process (AHP) was proposed by the authors for finding the QoS requirements of IoT devices. Afterward, the matching theory was applied for user association.

In [83], the authors have used the calculus of variation to find out the optimal spatial density of nodes such that the distribution of nodes can support the whole network. A Parallel Imperialistic Competitive Algorithm (PICA) is used to determine the initial positions of the access points. WSNs with immense scalability, immobility factor and low-deployment cost properties have limited processing and are resource-constrained. Smart Mobile Access Point (SMAP), which has been proposed by Majd et al. [84] aims at achieving higher resource use and energy efficiency with the adoption of hierarchical placement of SMAP in WSNs using fog computing. Moreno-Vozmediano et al. [85] have presented a framework for interconnection of fog and cloud computing, Hybrid Fog and Cloud (HFC), that provides effective, simple and productive resource provisioning. It also automates the process of configuring various virtual networks that are part of the network for interconnecting important components. The proposed architecture covers the salient features, such as security and scalability along with fog to fog, and fog to cloud communication mechanism.

Shojafar et al. [86] propose a resource scheduler for Networked Fog Centers (NetFCs) that not only provides services to the vehicular client but is also energy efficient. NetFCs operate from vehicular network's edge via Infrastructure to Vehicle (I2V) mobile links to the vehicular clients that are being served. Not only does the overall computation and communication energy efficiency get maximized but also the network performs better in terms of QoS requirements. Moreover, hard QoS requirements induced by the application are met, such as reducing transmission rates, delay, and jitter. Resource scheduler is responsible for admission control, dispatching the allowed traffic using minimum energy and adaptively controlling the traffic that is being injected into the mobile connections. Few of the important characteristics of the scheduler include providing QoS guarantees induced by the application, implantation of the scheduler is both scalable and distributive.

5.4. Latency

Fog networking is a key paradigm to provide solutions to latency-sensitive future IIoT applications. Many researchers have contributed to cater latency minimization problem. Online Fog Network Formation Algorithm was proposed by Gilsoo et al. in [87] for minimizing the overall maximum latency (communication and processing) of a fog network. The objective problem constitutes of the sum of two types of delays i.e., fog network formation and task distribution. A joint energy and latency optimization (JELO) scheme for F-RAN was proposed in [88]. The joint optimization of energy consumption and latency was formulated as an integer-programming (IP) problem subject to user association, capacity and latency threshold. The proposed complex problem was divided into two sub-problems of knapsack and semi-assignment problem. The proposed algorithm gives better results as compared to the existing techniques in the literature. A latency minimization problem was formulated for IoT-fog network under user association, workload and latency threshold constraints. The authors have proposed a matching-game theory for the proposed resource management problem for network latency minimization [89].

Over the last few years, the Internet of Vehicles (IoV) has been a matter of growing interest. Cloud computing provides high performing IoT services to IoV, still, there are many shortcomings when it comes to mobility support, latency and location awareness. Xiuli Hi et al. [90] have integrated fog computing into SDN. While fog Computing helps with the latency of the network, SDN provides flexibility in the centralized control and providing the complete global knowledge of the network. Vehicle to Vehicle (V2V), Vehicle to Infrastructure (V2I) and Vehicle to Base station (V2B) communication can be supported by a weighted undirected graph having Road Side Units (RSUs). The simulation results given by the aforementioned authors decrease the latency and achieve a higher QoS. The scheme proposed by Skarlat et al. [91] tackles an optimization problem, providing a solution that there is no trade-off between communication and using resources and energy for computation. The applied system model demonstrates that in optimization scenarios, reduced average round-trip time (RTT) and delays up to 39 percent can be achieved using fog. The scheme also considers an independent cooperative number of working nodes.

For the smart city, Fairness Cooperation Algorithm (FCA) was proposed by Dong et al. [92], for the joint optimization problem. The authors have formulated a convex-non-linear programming problem of minimizing the total system cost (delay and energy consumption), subject to power, workload and computation capacity threshold constraints. QoE and fairness of users under FCA was compared with baseline algorithm (BA) and distributed optimization algorithm (DOA).

5.5. Cache Enabled Edge Devices

Fog networking is a vast field, which uses both smart objects and the already deployed network infrastructure. System can ensure QoS either by using already deployed network devices or by optimally dividing the tasks among edge devices or using both at the same time. Cache placement in edge devices/fog nodes in a network improves the efficiency of a system in terms of low power consumption, low-latency, high throughput, and efficient spectrum use. To deal with the congestion problem at the backhaul link of F-RAN, authors have proposed a projection gradient method in [93]. To increase the throughput, they have used stochastic geometry to derive a close-form of successful transmission probability (STP). Afterward, the optimal placement of cache over the network was done. To fulfill the delay requirements, authors in [94] have proposed a decentralized asynchronous coding caching scheme. The result shows that the proposed algorithm is more efficient than existing algorithms present in the literature. A convex optimization problem of minimizing the worst-case fronthaul delay was formulated by the authors. Depending on application delay requirements, the proposed coded scheme gives synchronous and asynchronous transmission methods. The problem of user association with a fog node on the basis of lower latency is formulated in [95]. The authors have formulated the problem using game theory, for which they have used a proactive caching scheme and Boltzmann-Gibbs learning algorithm for solution. Latency of proposed fog network is the sum of computing and queuing delays.

Different researchers have worked on the F-RAN design considering various aspects. For a green system, as discussed by Chen et al. [67,96], minimization of energy consumption is done by considering an F-RAN that is cache-enabled for the selection of Remote Radio Head (RRHs). For balancing front-haul traffic, Park et al. [97] have discussed a scheme to deliver data from Base Band Unit (BBU) and RRHs. Content placement problem involving caching has been discussed in X Peng et al. [98] and Dai et al. [99]. F-RANs accommodate caching in the road-side units. Di Chen et al. [100] have worked on maximizing the Signal to Interference Noise Ratio (SINR) to ensure fairness while jointly optimizing cluster formation and multicast beamforming. The objective function in this work has non-convex constraints and is collectively an NP-hard problem. Sensor-cloud system gives solutions to many applications in a smart city. The system was developed by the integration of CPS and cloud computing. Besides many benefits, a major problem is coupling resource management, which was discussed in [101]. They have introduced a fog layer between sensor and cloud layer, which emphasizes the services. Firstly, authors have proposed an algorithm for caching at fog layer, afterwards, Hungarian algorithm was extended to deal with the optimal use of resources on basis of maximum matching. The proposed algorithms result in the minimization of latency for sustainable services.

6. Open Research IIoT Application Domains, Fog Computing as an Enabler

In Industry 4.0, the digital transformation of the industry requires research development in all fields (smart city, D2D communication, transportation, healthcare, etc.). These all IIoT domains have same critical issues in communication and networking. Even though, fog computing has umpteenth applications in multiple research areas, few notable applications in respective of IIoT domain have been listed in Table 1 to gain an idea about the diverse use of fog computing. The main objective is achieving maximum benefits/solutions using fog computing for this industrial revolution at efficient optimized QoS measurements. Many authors have proposed solutions in past years to networking and communication challenges in order to leverage benefits using fog computing in all domains of IIoT. Some has started to provide prototypes for supporting their research, Table 2 has listed some case studies for various areas of open research. In this section, some of the past research works are summarized for readers to get an idea about the integration of fog computing with different pre-existing network architectures. This integration brings solutions to support IIoT applications, yet there are many open research areas in all fields that need to be solved in the coming future towards the industrial development.

Table 1. Literature Review: R.A=Resource Allocation, L=Latency, E=Energy, T/R/C=Throughput/Rate/Capacity, Cc=Cache, P=Power, H=Handover, B=Bandwidth, S=Security, T.L.=Transmission Link.

Ref. No	IIoT Application Domain	R.A	L	E	T/R/C	Cc	P	H	B	S	T.L	Architecture
[49]	Mobility	✓						✓			Routing	SDN
[50]	Big Data Analytics	✓	✓								Routing	HetNets
[51]	Smart IoT devices	✓	✓	✓							Downlink	Cloud Computing
[52]	Smart IoT devices	✓					✓				Downlink	RANs
[53]	Big Data Analytics	✓					✓				Downlink	NOMA+RANs
[59]	VANETS		✓								Downlink	Cloud Computing
[61]	Healthcare	✓	✓						✓		Downlink+Uplink	Cloud Computing
[62]	Smart IoT devices		✓				✓				Downlink	Fog-IoT
[69]	5G network			✓			✓				Downlink	NFV+RANs
[72]	Virtualized Passive Optical Networks (VPON)/5G						✓				Downlink+Uplink	RANs + Cloud Computing
[77]	Small Cell Networks (SCNs)/5G	✓			✓		✓				Uplink	RANs
[78]	5G network	✓			✓						Downlink	RANs
[79]	5G network	✓					✓				Downlink	Cloud Computing
[81]	Heterogeneous IoT applications	✓	✓				✓				Downlink	RANs
[82]	Heterogeneous IoT applications	✓			✓				✓		Downlink	HetNets
[102]	VANETs	✓			✓						Downlink	VFC
[84]	Smart monitoring systems	✓	✓								Downlink	WSN+CPS
[85]	Security	✓			✓						Routing	VN+Cloud Computing
[87]	Heterogeneous IoT applications		✓				✓				Downlink	Cloud Computing
[88]	Time-sensitive IoT applications		✓	✓							Uplink	RANs
[89]	Heterogeneous IoT applications	✓	✓		✓						Downlink	Cloud Computing
[93]	Wireless network		✓			✓	✓				Downlink	RANs
[94]	5G network		✓			✓					Downlink	RANs
[95]	Microgrid	✓				✓					Downlink	Fog-IoT
[103]	Security+Microgrids		✓				✓				Downlink	VM+Cloud Computing
[104]	Microgrid		✓				✓				Downlink	Cloud Computing
[105]	Smart city		✓	✓		✓					Downlink	Cloud Computing
[101]	Smart city	✓		✓							Routing	CPS+Cloud Computing
[92]	Multimedia	✓	✓				✓				Downlink	Fog-IoT
[106]	Secure and time saving multimedia		✓							✓	Downlink	Cloud Computing
[107]	Secure IoT applications		✓							✓	Routing	ICN
[108]	VANETs		✓								Downlink	D2D
[109]	5G mobile network+V2G services		✓								Downlink	D2D+RANs
[110]	VANETs		✓								Routing	V2G
[111]	VANETs									✓	Routing	IoT+ITS

Table 1. Cont.

Ref. No	IIoT Application Domain	R.A	L	E	T/R/C	Cc	P	H	B	S	T.L	Architecture
[112]	Mobility+VANETs		✓	✓							Downlink	Cloud Computing
[113]	Mobility+VANETs	✓	✓				✓				Downlink	Fog-Ues
[114]	Mobility+Smart city		✓			✓						RANs+Cloud Comptinig
[115]	VANETs			✓					✓		Routing	SDN
[116]	Heterogeneous IoT applications			✓						✓	Routing	SDN+Blockchain
[117]	e-Healthcare									✓	Downlink	Blockchain
[118]	Cooperative+secure healthcare									✓	Routing	Fog+IoT
[119]	Big-Data Analytics+ security											Cloud Computing
[120]	Smart home										a case study	Cloud Computing
[121]	Smart city video applications	✓								✓	Routing	Cloud Computing

Table 2. Case studies, Fog Computing as an enabler.

Ref. No	IIoT Application Domain	Case Study: Key Focus
[122]	smart city	Smart city solutions have been deployed in cities, such as Barcelona and Venice, to make further advancements in e-governance
[123]	smart traffic control and health monitoring	To increase flexibility in a fog computing in the context of Complex event processing (CEP), a case study is presented. The methodology, called "mechanism transitions", is used to study how and where a query should be processed and how this decision affects the performance.
[124]	smart city	Fog Computing Architecture Network (FOCAN) is presented to give low-latency and energy-efficient solution for smart city applications. It manages different application's requirements by categorizing the traffic type and its flow.
[125]	city, factory, building, home	Using an open-source platform Distributed Node-RED (DNR), authors have presented how applications can be decomposed and deployed. They build prototype for scalability and dynamic nature solutions using the network simulator Omnet++.
[126]	smart pipeline monitoring	A sequential machine learning algorithm on every layer of fog-cloud architecture, sensors and Markov model are used to monitor, control and detection of hazardous events of a pipeline system. A working prototype was constructed to observe 12 distinct events. This prototype could be used for future city-wide pipeline safety measurements
[127]	smart transportation	The extended policy management to support secure travel to user's is presented by the authors. Four different route guiding scenarios are explained; namely depending on traffic condition, emergency connected vehicles (ECV), connected vehicle (CV) and probable collision detection.
[128]	smart transportation	Smart transportation framework is proposed for Vehicle to Vehicle (V2V) communication by the authors, on basis of the current traffic situation (road and vehicle's condition, capacity).
[129]	big data	Case study named as "Streamcloud" is presented to provide real-time energy-efficient solution.
[130]	healthcare	Personalized missing data resilient decision-making approach is validated on a real human subject trial on maternity health. Data missing in critical applications is a very crucial challenge, that needs to be solved.
[131]	healthcare	Table 2 in the mentioned paper gives some projects for healthcare monitoring supported by fog computing, cloud computing, and IoT.
[132]	healthcare	A demo test-bed is developed on edge-IoT architecture for e-healthcare applications. Proposed EH-IoT gives better results towards bandwidth and latency requirements. The article also presents the benefits leveraging from IoT and edge computing from an industrial perspective.
[133]	cardiac diseases	A case study using Electrocardiogram (ECG) feature is discussed in the article to monitor health in real time.

6.1. Micro-Grids (MGs)

Multiple loads and distributed renewable energy resources combine to form an electrical system named as MG. This energy is stored in the storage devices, though, it can have significant power losses when power is exchanged between different MGs. It increases reliability and efficiency of the system. In [104] authors implemented their proposed framework on a test MG system. Performance was observed for the proposed three-layer fog computing system. A convex linear programming problem was formulated with an objective of minimizing the total cost in terms of power consumption. The constraints are total load threshold which was calculated using the power balance equation. The equation has three types of powers with threshold range namely, dispatchable, non-dispatchable and grid power. Graph theory was applied to the proposed system and a fast consensus-based algorithm by taking advantage of fog computing was proposed. An optimization problem of power demand problem was discussed in [105] for hybrid fog-cloud system. The high volume of data by the number of smart devices results computation delays at cloud also it involves more power consumption. This power consumption can be optimized after load balancing among fog nodes and cloud server. Jalali et al. [134] have discussed various ways for the deployment of IoT nodes in an energy-efficient way. The authors have used MGs and fog computing as an enabler with an aim to reduce the energy consumption by the IoT specific applications. Energy consumption of various types has been discussed i.e., energy consumed during computation and balancing traffic on fog as well as the cloud. Renewable energy is stored in MGs, which can be used for further processing. Dynamic decisions such as weather forecasting or renewable energy availability causes dependent energy-saving processes.

6.2. Smart-Grids (SGs)

Due to an increase in the number of devices required by a user at home, residential usage of electricity has been increasing with time. The proposed modern solutions, such as SGs, greatly improve the reliability, efficiency, and sustainability of the system in an automated way. Using communication technologies, SGs works on gathered information about consumer's and supplier's behavior. As SGs are distributed systems implemented on a very large scale, fog computing is ideal to deal with such a scenario. For this purpose, Foteini Beligianni et al. [135] have discussed how IoTs and fog computing can be integrated into the power systems without compromising user privacy. The proposed architecture works on demand-response service to ensure privacy by using standard protocols and open-source resources. For devising a power plan for all the devices, a load scheduler is placed in the system while data management performs all the necessary actions that are required to process the data. The software architecture of the proposed solution is based on Lambda architecture which embodies edge computing. To meet the demands of real time processing, collecting information, computing and storing the data generated by multiple smart meters, the aforementioned researchers have proposed a fog-based solution. Fog networking acts as a bridge between the cloud and the SG. With fog networking, geographically distributed smart meters can be employed. Latency is reduced while improving location awareness and privacy for SG. Furthermore, Rao et al. [136] have worked on how to maintain the QoS by minimizing the expense of electricity using coordination between data centers.

6.3. Multimedia

Multimedia communication involves a large amount of audio and video data, being produced by smart IIoT devices. This ever-increasing amount of multimedia content brings new challenges. Low-latency and energy-efficient solutions are required to provide services. A direct consequence of increasing multimedia traffic is overburdening of the already strained mobile access network channels. The essential components of a multimedia communication system can be visualized with the help of Figure 1. The authors in [106], have formulated a cost minimization optimization problem, this cost depends on scheduler decisions. Scheduler takes decisions regarding distribution and states (open or close) of fog nodes. This cost minimization problem alternatively converted into multimedia user's (MMU) response time minimization problem under capacity, coverage zone area and association constraints. The authors have introduced a fog node to resolve the resource management and latency issues between cloud and MMUs. Using Stackelberg game, an online resource allocation scheme was proposed. The enormous amount of data production for smart applications, need security and privacy as well. The authors have proposed a chaotic cryptographic method to ensure security along with low-latency requirements for Information-Centric Multimedia Network (ICMN) [107]. The speed of encryption and decryption of multimedia streams in the cryptographic method, was enhanced using fog computing.

6.4. Device to Device (D2D) Communication

Researchers have to find a new way in which devices can independently work without using the existing cellular infrastructure. The demand for new infrastructure development is due to the increase in the number of smart devices that causes scarcity in spectrum and resources. This idea gives rise to D2D communication technology in which devices are part of the Internet and can communicate with each other without the involvement of internet infrastructure. This direct communication will result in less usage of the wireless spectrum. A device is itself constrained-bounded in terms of parameters, such as energy or resources, hence cloud and fog computing technologies support them. These emerging technologies enhance processing capabilities and result in efficient system formation. Cloud services for D2D communications have performance bottleneck as devices use direct communication, hence an alternative fog service is required. Fog with the same set of services, provide better performance. Fog networking act as an alternative of cloud, as it provides desired requirements in the vicinity of edge devices. Researchers have proposed D2D fogging to address energy-efficient task offloading in D2D communication [137]. The proposed method includes a framework for task offloading and uses assistance from the network for D2D. Mobile users use the communication and computation resources of each other dynamically. After discussing several security issues in fog computing, authors have proposed three lightweight anonymous authentication protocols (LAAPs) [108].

The proposed scheme with the aid of D2D communication, is feasible for IoT devices which are resource-limited. Li et al. [109] proposed a F-community architecture for F-RAN followed by a data caching scheme. The caching scheme for UEs in D2D aided F-RAN helps in reduction of delays. Nodes in the system with higher chances of being selected as the central nodes store the most popular content in their cache. For access to any data, a user receives data from the cache stored previously. Kaur et al. [138] have proposed cachinMobile, which, compared to the cloud, can meet the low latency and energy efficiency requirements by using the elastic services provided by the nodes at the edge. Energy efficiency and low latency requirements is a major issue in this technique as the resources at the edge and the devices that are mobile. This approach has proposed to use D2D communication to carry out communication at a short distance and save network resources. CachinMobile not only improves energy efficiency but also maintains QoS.

6.5. Vehicular Ad-hoc Networks (VANETs)

This area is named as Intelligent Transportation System (ITS), which is an emerging area with many open issues that need to be solved. A transportation service and automobile service management involves the controlling and monitoring of the transportation network. ITS is designed in such a way that this system can satisfy the required QoS parameters of the transportation network. These QoS requirements involve reliable connection, efficient performance, safety, and privacy requirement, mobility and scalability requirement. A transportation network system aided with ITS technology has components that can be optimized. These are Global Positing System (GPS), RFID sensor tags and readers, road-side equipment for example traffic lights, signals, road bank cameras, cars. These ITS subsystems integrated with IoT technology elements will help in monitoring and managing of transportation environment; distribution of vehicles according to the scenario; manufacturing of ITS; shipping, tracking and monitoring of physical objects.

Enormous number of such physical objects embedded with data processing capabilities, integrated with RFID sensors along with networking technologies will promote IIoT applications. These IIoT applications will help in monitoring the exact original location as well as destination location of vehicle or aero-plane or ships along with the followed path traffic condition or environment effects or any emergency road situation. Many authors have done research in designing ITS supportive transportation networks optimizing wireless communication technologies, RFID tags or antennas. The use of IoT technology in the transportation industry and systems results in new research domains such as vehicular ad-hoc networks (VANETs), internet of vehicles (IoV) and vehicle to grid (V2G). These domains are very promising areas of research. In recent times the integration of VANETs with fog computing results in better efficient system designing. Services of automotive connectivity architecture can be improved using fog computing as an enabler for smart transportation (ITS), such as the one shown in Figure 1 (Transportation). Efficiency is improved in terms of minimizing the latency for time-sensitive applications using fog computing. Being an integral part of the ITS, VANETs have many applications. Computation and communication demands are hard to meet in the cloud architecture. Incorporating fog networking not only fulfills the demands but also improves latency, location awareness and energy efficiency. For power management, there is a need for new technologies. For this problem Vehicle to Grid (V2G) is a recent concept with open issues, that uses renewable energy. When renewable energy is available it is used to charge plug-in electric vehicles. Otherwise, these electric vehicles are used as the source of energy. The use of alternative energy source will reduce the power burden on the grid during peak hours. Owners of these plug-in electric vehicles got paid by electric companies using metering systems. It provides some relief to electric companies in terms of payment, traffic load, and energy consumption during peak hours. In smart grids, a distributed architecture with storage and processing capabilities is need to be deployed, as there is a high factor of mobility in V2G. To implement the V2G services in the 5G network, a hybrid fog and cloud architecture was proposed by researchers in [110]. Open issues, such as energy efficiency, resource management, security, and privacy are needed to be addressed in the future to improve system efficiency. Security is a big challenge in VANETS, fog computing integration is smart transportation brings solution to this challenge. Ma et al. [111] have designed a new authenticated key agreement (AKA) protocol for fog-based VANETs.

The authors in [112] compared fog and cloud computing performance in a real VANET environment. Results shows that fog computing gives better services for real-time scenario applications, namely traffic detection and time estimation. Fog computing performs well because it supports the main attributes of VANETs that are location-awareness, mobility and real-time communication. Integration of the Internet of Things with VANETs gives rise to Internet of Vehicles (IoV) is a matter of growing interest over the last few years [139].

6.6. Big-Data Analytics

As the next revolutionized era comprises enormous smart devices that supports IIoT applications, this causes the generation of a significant amount of data. Due to the widespread acceptance of fog computing, significant research has been carried out in the domain of big-data analytics. Recent advancements show an indication that fog networking has the potential to provide solutions in this field. Fog computing, as an extension of cloud computing, gives solutions to problems such as location-awareness, mobility, big-data analytics, and cyber threats. Author in [119] have designed a three-layer architecture of IoT-Fog-Cloud that supports big-data analytics and security applications. Author summarized cyber attack types and compared the existing security solutions. The authors in [140] provide a review article on fog computing challenges in the context of big IoT analytics.

6.7. Software Defined Networking (SDN)

SDN is an evolving concept to remotely control the entire network from a centralized location. Distributed networks can be integrated with SDN architecture for more granularity of control in the network. Since edge computing is done near the edge devices, there is a high probability that the edge devices are mobile due to the exponential evolution of smartphones and other handheld devices. Similarly, the ever-changing requirements of resources for each user brings a certain degree of dynamism in the network which needs to be catered to ensure efficient operation. Since fog is a new concept and it cannot completely replace the existing cloud architecture, both fog and cloud work hand in hand for smooth operation of the network. The interplay between fog networking and cloud computing is always there when both work in an integrated fashion. The four-layer architecture was studied for end-to-end delay, energy consumption and packet loss ratio in [115]. The authors formulate mixed integer programming (MIP) problem of minimizing the energy consumption under data rate, bandwidth and delay threshold constraints. A new routing protocol was proposed for VANET applications using SDN and fog computing, named as Energy Efficient Multicast routing protocol (EEMSFV). SDN controller, OpenFlow switches and fog computing works under two algorithms, priority based scheduling algorithm for classifying the traffic message type (emergency or safety applications) and a classification algorithm to schedule the requests.

7. Conclusions

Industry 4.0, a revolutionized era which will have a massive number of smart devices that will support IIoT applications in every field. This deployment of smart devices will change all domains of human life's perspective. IIoT applications will provide solutions to all fields, such as transportation, healthcare, food supply chain, education, and industry. These IIoT applications will provide efficient, effective solutions for future networks. There are challenges in communication and networking in terms of latency, bandwidth, resource allocation, and storage.

All these advanced IIoT applications will create a huge amount of data, causing a burden on the cloud. Even though cloud computing provides services to the edge devices, it incurs huge latency, resource allocation challenges, caching placement problems, energy consumption. These issues are detrimental to the QoS aspect of a network. Fog as an extension of cloud provides a platform to compute, control, store and manage these IIoT devices. In the future, it will reshape all sectors involving IIoT applications, with the integration of important existing communication technologies namely, CPS, SDN, NFV, 5G, D2D. This brings computation, resource management, and storage challenges. In this paper, first, we gave an overview of IIoT applications and its enabling technologies used for new revolutionized era. The pre existing protocols and solutions to challenges related to fog computing are summarized. In the end, we have mentioned open research IIoT domains, in which fog computing can act as an enabler. We have previewed the work carried out by numerous researchers incorporating fog computing to provide services to IIoT edge devices leveraging towards Industry 4.0 way. Critical review of some existing work is summarized in the table, which can be used to find open research challenges. Towards the development of this industrial transformative epoch, most of the research work is still uncertain and waiting. This is an interesting era to discover what fog computing may contribute to the world of automation in the coming future.

Author Contributions: R.B. is the main author for this survey article, who wrote the original draft. M.A. (Mudassar Ali), S.Q., M.A. (Monther Aldwairi), M.I.A., and A.M. contributed in terms of conceptualization, organization and validation of the article. A.M., M.G. and M.A. (Monther Aldwairi) also contributed to acquire funds.

Funding: This work was supported by the Swedish Knowledge Foundation under Grant 20180178.

Acknowledgments: This work was supported by Zayed University Research Office, Research Cluster Award # R17079.

Conflicts of Interest: The authors declare no conflict of interest.

References

1. Rivera, J.; Goasduff, L. Gartner says a thirty-fold increase in internet-connected physical devices by 2020 will significantly alter how the supply chain operates. *Gartner* **2014**. Available online: https://www.gartner.com/en/newsroom/press-releases/2014-03-24-gartner-says-a-thirty-fold-increase-in-internet-connected-physical-devices-by-2020-will-significantly-alter-how-the-supply-chain-operates (accessed on 19 July 2019).
2. Roblek, V.; Meško, M.; Krapež, A. A complex view of industry 4.0. *Sage Open* **2016**, *6*. [CrossRef]
3. Thames, L.; Schaefer, D. Software-defined cloud manufacturing for industry 4.0. *Procedia CIRP* **2016**, *52*, 12–17. [CrossRef]
4. Varghese, A.; Tandur, D. Wireless requirements and challenges in Industry 4.0. In Proceedings of the 2014 International Conference on Contemporary Computing and Informatics (IC3I), Mysore, India, 27–29 November 2014; pp. 634–638.
5. Akpakwu, G.A.; Silva, B.J.; Hancke, G.P.; Abu-Mahfouz, A.M. A survey on 5G networks for the Internet of Things: Communication technologies and challenges. *IEEE Access* **2018**, *6*, 3619–3647. [CrossRef]

6. Vangelista, L.; Zanella, A.; Zorzi, M. Long-range IoT technologies: The dawn of LoRa™. In *Future Access Enablers of Ubiquitous and Intelligent Infrastructures*; Springer: Cham, Switzerland, 2015; pp. 51–58.
7. Osseiran, A.; Boccardi, F.; Braun, V.; Kusume, K.; Marsch, P.; Maternia, M.; Queseth, O.; Schellmann, M.; Schotten, H.; Taoka, H.; et al. Scenarios for 5G mobile and wireless communications: The vision of the METIS project. *IEEE Commun. Mag.* **2014**, *52*, 26–35. [CrossRef]
8. Christensen, J.H. Using RESTful web-services and cloud computing to create next generation mobile applications. In Proceedings of the 24th ACM SIGPLAN Conference Companion on Object Oriented Programming Systems Languages and Applications, Orlando, FL, USA, 25–29 October 2009; pp. 627–634.
9. Buyya, R.; Yeo, C.S.; Venugopal, S.; Broberg, J.; Brandic, I. Cloud computing and emerging IT platforms: Vision, hype, and reality for delivering computing as the 5th utility. *Future Gener. Comput. Syst.* **2009**, *25*, 599–616. [CrossRef]
10. Atlam, H.F.; Alenezi, A.; Alharthi, A.; Walters, R.J.; Wills, G.B. Integration of cloud computing with internet of things: Challenges and open issues. In Proceedings of the 2017 IEEE International Conference on Internet of Things (iThings) and IEEE Green Computing and Communications (GreenCom) and IEEE Cyber, Physical and Social Computing (CPSCom) and IEEE Smart Data (SmartData), Exeter, UK, 21–23 June 2017; pp. 670–675.
11. Ai, Y.; Peng, M.; Zhang, K. Edge computing technologies for Internet of Things: A primer. *Digit. Commun. Netw.* **2018**, *4*, 77–86. [CrossRef]
12. Peter, N. Fog computing and its real time applications. *Int. J. Emerg. Technol. Adv. Eng.* **2015**, *5*, 266–269.
13. Aazam, M.; Huh, E.N. Fog computing and smart gateway based communication for cloud of things. In Proceedings of the 2014 International Conference on Future Internet of Things and Cloud, Barcelona, Spain, 24–29 August 2014; pp. 464–470.
14. Atlam, H.F.; Alenezi, A.; Walters, R.J.; Wills, G.B.; Daniel, J. Developing an adaptive Risk-based access control model for the Internet of Things. In Proceedings of the 2017 IEEE International Conference on Internet of Things (iThings) and IEEE Green Computing and Communications (GreenCom) and IEEE Cyber, Physical and Social Computing (CPSCom) and IEEE Smart Data (SmartData), Exeter, UK, 21–23 June 2017; pp. 655–661.
15. Bonomi, F.; Milito, R.; Zhu, J.; Addepalli, S. Fog computing and its role in the Internet of things. In Proceedings of the First Edition of the MCC Workshop on Mobile Cloud Computing, Helsinki, Finland, 13–17 August 2012; pp. 13–16.
16. Wen, Z.; Yang, R.; Garraghan, P.; Lin, T.; Xu, J.; Rovatsos, M. Fog orchestration for internet of things services. *IEEE Internet Comput.* **2017**, *21*, 16–24. [CrossRef]
17. Verma, M.; Bhardwaj, N.; Yadav, A.K. Real time efficient scheduling algorithm for load balancing in fog computing environment. *Int. J. Inf. Technol. Comput. Sci* **2016**, *8*, 1–10. [CrossRef]
18. Fog Computing and the Internet of Things: Extend the Cloud to Where the Things Are. White Paper. 2016. Available online: https://www.cisco.com/c/dam/en_us/solutions/trends/iot/docs/computing-overview.pdf (accessed on 19 September 2019).
19. Atlam, H.; Walters, R.; Wills, G. Fog computing and the Internet of things: A review. *Big Data Cogn. Comput.* **2018**, *2*, 10. [CrossRef]
20. Chiang, M.; Zhang, T. Fog and IoT: An overview of research opportunities. *IEEE Internet Things J.* **2016**, *3*, 854–864. [CrossRef]
21. Liu, Y.; Fieldsend, J.E.; Min, G. A framework of fog computing: Architecture, challenges, and optimization. *IEEE Access* **2017**, *5*, 25445–25454. [CrossRef]
22. Peralta, G.; Iglesias-Urkia, M.; Barcelo, M.; Gomez, R.; Moran, A.; Bilbao, J. Fog computing based efficient IoT scheme for the Industry 4.0. In Proceedings of the 2017 IEEE International Workshop of Electronics, Control, Measurement, Signals and their Application to Mechatronics (ECMSM), Donostia-San Sebastian, Spain, 24–26 May 2017; pp. 1–6.
23. Bonomi, F.; Milito, R.; Natarajan, P.; Zhu, J. Fog computing: A platform for internet of things and analytics. In *Big Data and Internet of Things: A Roadmap for Smart Environments*; Springer: New York, NY, USA, 2014; pp. 169–186.
24. Agarwal, S.; Yadav, S.; Yadav, A.K. An efficient architecture and algorithm for resource provisioning in fog computing. *Int. J. Inf. Eng. Electron. Bus.* **2016**, *8*, 48. [CrossRef]

25. Ketel, M. Fog-cloud services for iot. In Proceedings of the of the SouthEast Conference, Kennesaw, GA, USA, 13–15 April 2017; pp. 262–264.
26. Georgakopoulos, D.; Jayaraman, P.P.; Fazia, M.; Villari, M.; Ranjan, R. Internet of Things and edge cloud computing roadmap for manufacturing. *IEEE Cloud Comput.* **2016**, *3*, 66–73. [CrossRef]
27. Lu, Y. Industry 4.0: A survey on technologies, applications and open research issues. *J. Ind. Inf. Integr.* **2017**, *6*, 1–10. [CrossRef]
28. Camarillo, A.; Ríos, J.; Althoff, K.D. Product Lifecycle Management as Data Repository for Manufacturing Problem Solving. *Materials* **2018**, *11*, 1469. [CrossRef]
29. Obst, M.; Holm, T.; Urbas, L.; Fay, A.; Kreft, S.; Hempen, U.; Albers, T. Semantic description of process modules. In Proceedings of the 2015 IEEE 20th Conference on Emerging Technologies & Factory Automation (ETFA), Luxembourg, 8–11 September 2015; pp. 1–8.
30. Li, B.; Zhao, Z.; Guan, Y.; Ai, N.; Dong, X.; Wu, B. Task Placement Across Multiple Public Clouds With Deadline Constraints for Smart Factory. *IEEE Access* **2018**, *6*, 1560–1564. [CrossRef]
31. Chen, H.; Abbas, R.; Cheng, P.; Shirvanimoghaddam, M.; Hardjawana, W.; Bao, W.; Li, Y.; Vucetic, B. Ultra-reliable low latency cellular networks: Use cases, challenges and approaches. *IEEE Commun. Mag.* **2018**, *56*, 119–125. [CrossRef]
32. Ladiges, J.; Fay, A.; Holm, T.; Hempen, U.; Urbas, L.; Obst, M.; Albers, T. Integration of modular process units into process control systems. *IEEE Trans. Ind. Appl.* **2018**, *54*, 1870–1880. [CrossRef]
33. Vogel-Heuser, B.; Diedrich, C.; Pantförder, D.; Göhner, P. Coupling heterogeneous production systems by a multi-agent based cyber-physical production system. In Proceedings of the 2014 12th IEEE International Conference on Industrial Informatics (INDIN), Porto Alegre, Brazil, 27–30 July 2014; pp. 713–719.
34. Wan, J.; Tang, S.; Shu, Z.; Li, D.; Wang, S.; Imran, M.; Vasilakos, A.V. Software-defined industrial internet of things in the context of industry 4.0. *IEEE Sens. J.* **2016**, *16*, 7373–7380. [CrossRef]
35. Gruber, F.E. Industry 4.0: A best practice project of the automotive industry. In Proceedings of the IFIP International Conference on Digital Product and Process Development Systems, Dresden, Germany, 10–11 October 2013; pp. 36–40.
36. Vaquero, L.M.; Rodero-Merino, L. Finding your way in the fog: Towards a comprehensive definition of fog computing. *ACM SIGCOMM Comput. Commun. Rev.* **2014**, *44*, 27–32. [CrossRef]
37. Stock, T.; Seliger, G. Opportunities of sustainable manufacturing in industry 4.0. *Procedia Cirp* **2016**, *40*, 536–541. [CrossRef]
38. Elliott, J.A. *An Introduction to Sustainable Development*; Routledge: Abingdon, UK, 2012.
39. Bibri, S.E.; Krogstie, J. Smart sustainable cities of the future: An extensive interdisciplinary literature review. *Sustain. Cities Soc.* **2017**, *31*, 183–212. [CrossRef]
40. Schlechtendahl, J.; Keinert, M.; Kretschmer, F.; Lechler, A.; Verl, A. Making existing production systems Industry 4.0-ready. *Prod. Eng.* **2015**, *9*, 143–148. [CrossRef]
41. Mouradian, C.; Naboulsi, D.; Yangui, S.; Glitho, R.H.; Morrow, M.J.; Polakos, P.A. A comprehensive survey on fog computing: State-of-the-art and research challenges. *IEEE Commun. Surv. Tutor.* **2017**, *20*, 416–464. [CrossRef]
42. Dong, M.; Ota, K.; Liu, A. Preserving source-location privacy through redundant fog loop for wireless sensor networks. In Proceedings of the 2015 IEEE International Conference on Computer and Information Technology; Ubiquitous Computing and Communications; Dependable, Autonomic and Secure Computing; Pervasive Intelligence and Computing (CIT/IUCC/DASC/PICOM), Liverpool, UK, 26–28 October 2015; pp. 1835–1842.
43. Huang, L.; Li, G.; Wu, J.; Li, L.; Li, J.; Morello, R. Software-defined QoS provisioning for fog computing advanced wireless sensor networks. In Proceedings of the 2016 IEEE SENSORS, Orlando, FL, USA, 30 October–3 November 2016; pp. 1–3.
44. Liu, Y.; Gao, J.; Jia, Y.; Zhu, L. A cluster maintenance algorithm based on LEACH-DCHS protoclol. In Proceedings of the International Conference on Networking, Architecture, and Storage, 2008. NAS'08, Chongqing, China, 12–14 June 2008; pp. 165–166.
45. Handy, M.; Haase, M.; Timmermann, D. Low energy adaptive clustering hierarchy with deterministic cluster-head selection. In Proceedings of the 4th International Workshop on Mobile and Wireless Communications Network, Stockholm, Sweden, 9–11 September 2002; pp. 368–372.

46. Smaragdakis, G.; Matta, I.; Bestavros, A. *Sep: A Stable Election Protocol for Clustered Heterogeneous Wireless Sensor Networks*; Technical Report; Boston University Computer Science Department: Boston, MA, USA, 2004.
47. Singh, D.; Panda, C.K. Performance analysis of modified stable election protocol in heterogeneous wsn. In Proceedings of the 2015 International Conference on Electrical, Electronics, Signals, Communication and Optimization (EESCO), Visakhapatnam, India, 24–25 January 2015; pp. 1–5.
48. Naranjo, P.G.V.; Shojafar, M.; Abraham, A.; Baccarelli, E. A new stable election-based routing algorithm to preserve aliveness and energy in fog-supported wireless sensor networks. In Proceedings of the 2016 IEEE International Conference on Systems, Man, and Cybernetics (SMC), Budapest, Hungary, 9–12 October 2016; pp. 2413–2418.
49. Bi, Y.; Han, G.; Lin, C.; Deng, Q.; Guo, L.; Li, F. Mobility support for fog computing: An SDN approach. *IEEE Commun. Mag.* **2018**, *56*, 53–59. [CrossRef]
50. Wang, X.; Yang, L.T.; Kuang, L.; Liu, X.; Zhang, Q.; Deen, M.J. A tensor-based big-data-driven routing recommendation approach for heterogeneous networks. *IEEE Netw.* **2019**, *33*, 64–69. [CrossRef]
51. Du, J.; Zhao, L.; Feng, J.; Chu, X. Computation offloading and resource allocation in mixed fog/cloud computing systems with min-max fairness guarantee. *IEEE Trans. Commun.* **2018**, *66*, 1594–1608. [CrossRef]
52. Sun, Y.; Peng, M.; Mao, S.; Yan, S. Hierarchical radio resource allocation for network slicing in fog radio access networks. *IEEE Trans. Veh. Technol.* **2019**, *68*, 3866–3881. [CrossRef]
53. Zhang, H.; Qiu, Y.; Long, K.; Karagiannidis, G.K.; Wang, X.; Nallanathan, A. Resource allocation in NOMA-based fog radio access networks. *IEEE Wirel. Commun.* **2018**, *25*, 110–115. [CrossRef]
54. Madsen, H.; Burtschy, B.; Albeanu, G.; Popentiu-Vladicescu, F. Reliability in the utility computing era: Towards reliable fog computing. In Proceedings of the 2013 20th International Conference on Systems, Signals and Image Processing (IWSSIP), Bucharest, Romania, 7–9 July 2013; pp. 43–46.
55. Hong, K.; Lillethun, D.; Ramachandran, U.; Ottenwälder, B.; Koldehofe, B. Mobile fog: A programming model for large-scale applications on the internet of things. In Proceedings of the Second ACM SIGCOMM Workshop on Mobile cloud Computing, Hong Kong, China, 12–16 August 2013; pp. 15–20.
56. Zhu, J.; Chan, D.S.; Prabhu, M.S.; Natarajan, P.; Hu, H.; Bonomi, F. Improving web sites performance using edge servers in fog computing architecture. In Proceedings of the 2013 IEEE Seventh International Symposium on Service-Oriented System Engineering, Redwood City, CA, USA, 25–28 March 2013; pp. 320–323.
57. Rudenko, E. Fog Computing Is a New Concept of Data Distribution. Available online: https://xcluesiv.com/fog-computing-is-a-new-concept-of-data-distribution/ (accessed on 19 July 2019).
58. Do, C.T.; Tran, N.H.; Pham, C.; Alam, M.G.R.; Son, J.H.; Hong, C.S. A proximal algorithm for joint resource allocation and minimizing carbon footprint in geo-distributed fog computing. In Proceedings of the 2015 International Conference on Information Networking (ICOIN), Siem Reap, Cambodia, 12–14 January 2015; pp. 324–329.
59. Ning, Z.; Huang, J.; Wang, X. Vehicular fog computing: Enabling real-time traffic management for smart cities. *IEEE Wirel. Commun.* **2019**, *26*, 87–93. [CrossRef]
60. Mutlag, A.A.; Ghani, M.K.A.; Arunkumar, N.a.; Mohamed, M.A.; Mohd, O. Enabling technologies for fog computing in healthcare IoT systems. *Future Gener. Comput. Syst.* **2019**, *90*, 62–78. [CrossRef]
61. Khattak, H.A.; Arshad, H.; ul Islam, S.; Ahmed, G.; Jabbar, S.; Sharif, A.M.; Khalid, S. Utilization and load balancing in fog servers for health applications. *EURASIP J. Wirel. Commun. Netw.* **2019**, *2019*, 91. [CrossRef]
62. Abkenar, F.S.; Jamalipour, A. EBA: Energy Balancing Algorithm for Fog-IoT Networks. *IEEE Internet Things J.* **2019**, *6*, 6843–6849. [CrossRef]
63. Beraldi, R.; Mtibaa, A.; Alnuweiri, H. Cooperative load balancing scheme for edge computing resources. In Proceedings of the 2017 Second International Conference on Fog and Mobile Edge Computing (FMEC), Valencia, Spain, 8–11 May 2017; pp. 94–100.
64. Armbrust, M.; Fox, A.; Griffith, R.; Joseph, A.D.; Katz, R.; Konwinski, A.; Lee, G.; Patterson, D.; Rabkin, A.; Stoica, I.; et al. A view of cloud computing. *Commun. ACM* **2010**, *53*, 50–58. [CrossRef]
65. Ningning, S.; Chao, G.; Xingshuo, A.; Qiang, Z. Fog computing dynamic load balancing mechanism based on graph repartitioning. *China Commun.* **2016**, *13*, 156–164. [CrossRef]
66. Deng, R.; Lu, R.; Lai, C.; Luan, T.H.; Liang, H. Optimal workload allocation in fog-cloud computing toward balanced delay and power consumption. *IEEE Internet Things J.* **2016**, *3*, 1171–1181. [CrossRef]

67. Chen, D.; Kuehn, V. Adaptive radio unit selection and load balancing in the downlink of Fog radio access network. In Proceedings of the 2016 IEEE Global Communications Conference (GLOBECOM), Washington, DC, USA, 4–8 December 2016; pp. 1–7.
68. Verma, S.; Yadav, A.K.; Motwani, D.; Raw, R.; Singh, H.K. An efficient data replication and load balancing technique for fog computing environment. In Proceedings of the 2016 3rd International Conference on Computing for Sustainable Global Development (INDIACom), New Delhi, India, 16–18 March 2016; pp. 2888–2895.
69. Zeng, Y.; Al-Quzweeni, A.; Elgorashi, T.E.; Elmirghani, J.M. Energy Efficient virtualization framework for 5G F-RAN. *arXiv* **2019**, arXiv:1904.02481.
70. Roca, D.; Quiroga, J.V.; Valero, M.; Nemirovsky, M. Fog function virtualization: A flexible solution for iot applications. In Proceedings of the 2017 Second International Conference on Fog and Mobile Edge Computing (FMEC), Valencia, Spain, 8–11 May 2017; pp. 74–80.
71. Kaur, K.; Dhand, T.; Kumar, N.; Zeadally, S. Container-as-a-service at the edge: Trade-off between energy efficiency and service availability at fog nano data centers. *IEEE Wirel. Commun.* **2017**, *24*, 48–56. [CrossRef]
72. Tinini, R.I.; Batista, D.M.; Figueiredo, G.B.; Tornatore, M.; Mukherjee, B. Low-latency and energy-efficient BBU placement and VPON formation in virtualized cloud-fog RAN. *IEEE/OSA J. Opt. Commun. Netw.* **2019**, *11*, B37–B48. [CrossRef]
73. Xia, F.; Liu, L.; Li, J.; Ma, J.; Vasilakos, A.V. Socially aware networking: A survey. *IEEE Syst. J.* **2015**, *9*, 904–921. [CrossRef]
74. Zhang, C.; Sun, Y.; Mo, Y.; Zhang, Y.; Bu, S. Social-aware content downloading for fog radio access networks supported device-to-device communications. In Proceedings of the 2016 IEEE International Conference on Ubiquitous Wireless Broadband (ICUWB), Nanjing, China, 16–19 October 2016; pp. 1–4.
75. Klas, G.I. Fog Computing and Mobile Edge cloud Gain Momentum Open Fog Consortium, Etsi Mec and Cloudlets. 2015. Available online: https://yucianga.info/wp-content/uploads/2015/11/15_11_22-_Fog_computing_and_mobile_edge_cloud_gain_momentum_Open_Fog_Consortium-ETSI_MEC-Cloudlets_v1_1.pdf (accessed on 19 September 2019).
76. Cau, E.; Corici, M.; Bellavista, P.; Foschini, L.; Carella, G.; Edmonds, A.; Bohnert, T.M. Efficient exploitation of mobile edge computing for virtualized 5G in EPC architectures. In Proceedings of the 2016 4th IEEE International Conference on Mobile Cloud Computing, Services, and Engineering (MobileCloud), Oxford, UK, 29 March–1 April 2016; pp. 100–109.
77. Li, Z.; Sichitiu, M.L.; Qiu, X. Fog Radio Access Network: A New Wireless Backhaul Architecture for Small Cell Networks. *IEEE Access* **2018**, *7*, 14150–14161. [CrossRef]
78. Pontois, N.; Kaneko, M.; Dinh, T.H.L.; Boukhatem, L. User pre-scheduling and beamforming with outdated CSI in 5G fog radio access networks. In Proceedings of the 2018 IEEE Global Communications Conference (GLOBECOM), Abu Dhabi, UAE, 9–13 December 2018; pp. 1–6.
79. Li, L.; Guan, Q.; Jin, L.; Guo, M. Resource allocation and task offloading for heterogeneous real-time tasks with uncertain duration time in a fog queueing system. *IEEE Access* **2019**, *7*, 9912–9925. [CrossRef]
80. Huang, J.; Duan, Q.; Guo, S.; Yan, Y.; Yu, S. Converged network-cloud service composition with end-to-end performance guarantee. *IEEE Trans. Cloud Comput.* **2015**, *6*, 545–557. [CrossRef]
81. Yu, Y.; Liu, S.; Tian, Z.; Wang, S. A dynamic distributed spectrum allocation mechanism based on game model in fog radio access networks. *China Commun.* **2019**, *16*, 12–21.
82. Abedin, S.F.; Alam, M.G.R.; Kazmi, S.A.; Tran, N.H.; Niyato, D.; Hong, C.S. Resource allocation for ultra-reliable and enhanced mobile broadband IoT applications in fog network. *IEEE Trans. Commun.* **2018**, *67*, 489–502. [CrossRef]
83. Majd, A.; Sahebi, G.; Daneshtalab, M.; Plosila, J.; Lotfi, S.; Tenhunen, H. Parallel imperialist competitive algorithms. *Concurr. Comput. Pract. Exp.* **2018**, *30*, e4393. [CrossRef]
84. Majd, A.; Sahebi, G.; Daneshtalab, M.; Plosila, J.; Tenhunen, H. Hierarchal Placement of Smart Mobile Access Points in Wireless Sensor Networks Using Fog Computing. In Proceedings of the 2017 25th Euromicro International Conference on Parallel, Distributed and Network-based Processing (PDP), St. Petersburg, Russia, 6–8 March 2017; pp. 176–180.
85. Moreno-Vozmediano, R.; Montero, R.S.; Huedo, E.; Llorente, I.M. Cross-site virtual network in cloud and fog computing. *IEEE Cloud Comput.* **2017**, *4*, 46–53. [CrossRef]

86. Shojafar, M.; Cordeschi, N.; Baccarelli, E. Energy-efficient adaptive resource management for real-time vehicular cloud services. *IEEE Trans. Cloud Comput.* **2016**, *7*, 196–209. [CrossRef]
87. Lee, G.; Saad, W.; Bennis, M. An online optimization framework for distributed fog network formation with minimal latency. *IEEE Trans. Wirel. Commun.* **2019**, *18*, 2244–2258. [CrossRef]
88. Vu, D.N.; Dao, N.N.; Jang, Y.; Na, W.; Kwon, Y.B.; Kang, H.; Jung, J.J.; Cho, S. Joint energy and latency optimization for upstream IoT offloading services in fog radio access networks. *Trans. Emerg. Telecommun. Technol.* **2019**, *30*, e3497. [CrossRef]
89. Ali, M.; Riaz, N.; Ashraf, M.I.; Qaisar, S.; Naeem, M. Joint cloudlet selection and latency minimization in fog networks. *IEEE Trans. Ind. Inform.* **2018**, *14*, 4055–4063. [CrossRef]
90. He, X.; Ren, Z.; Shi, C.; Fang, J. A novel load balancing strategy of software-defined cloud/fog networking in the Internet of vehicles. *China Commun.* **2016**, *13*, 140–149. [CrossRef]
91. Skarlat, O.; Schulte, S.; Borkowski, M.; Leitner, P. Resource provisioning for IoT services in the fog. In Proceedings of the 2016 IEEE 9th Conference on Service-Oriented Computing and Applications (SOCA), Macau, China, 4–6 November 2016; pp. 32–39.
92. Dong, Y.; Guo, S.; Liu, J.; Yang, Y. Energy-Efficient Fair Cooperation Fog Computing in Mobile Edge Networks for Smart City. *IEEE Internet Things J.* **2019**, *6*, 7543–7554. [CrossRef]
93. Wang, R.; Li, R.; Wang, P.; Liu, E. Analysis and Optimization of Caching in Fog Radio Access Networks. *IEEE Trans. Veh. Technol.* **2019**, *68*, 8279–8283. [CrossRef]
94. Jiang, Y.; Huang, W.; Bennis, M.; Zheng, F. Decentralized Asynchronous Coded Caching Design and Performance Analysis in Fog Radio Access Networks. *IEEE Trans. Mob. Comput.* **2019**. [CrossRef]
95. Abouaomar, A.; Elmachkour, M.; Kobbane, A.; Tembine, H.; Ayaida, M. Users-Fogs association within a cache context in 5G networks: Coalition game model. In Proceedings of the 2018 IEEE Symposium on Computers and Communications (ISCC), Natal, Brazil, 25–28 June 2018; pp. 14–19.
96. Chen, D.; Schedler, S.; Kuehn, V. Backhaul traffic balancing and dynamic content-centric clustering for the downlink of Fog Radio Access Network. In Proceedings of the 2016 IEEE 17th International Workshop on Signal Processing Advances in Wireless Communications (SPAWC), Edinburgh, UK, 3–6 July 2016; pp. 1–5.
97. Park, S.H.; Simeone, O.; Shamai, S. Joint optimization of cloud and edge processing for fog radio access networks. In Proceedings of the 2016 IEEE International Symposium on Information Theory (ISIT), Barcelona, Spain, 10–15 July 2016; pp. 315–319.
98. Peng, X.; Shen, J.C.; Zhang, J.; Letaief, K.B. Joint data assignment and beamforming for backhaul limited caching networks. In Proceedings of the 2014 IEEE 25th Annual International Symposium on Personal, Indoor, and Mobile Radio Communication (PIMRC), Washington, DC, USA, 2–5 September 2014; pp. 1370–1374.
99. Dai, B.; Yu, W. Joint user association and content placement for cache-enabled wireless access networks. In Proceedings of the 2016 IEEE International Conference on Acoustics, Speech and Signal Processing (ICASSP), Shanghai, China, 20–25 March 2016; pp. 3521–3525.
100. Chen, D.; Kuehn, V. Weighted max-min fairness oriented load-balancing and clustering for multicast cache-enabled F-RAN. In Proceedings of the 2016 9th International Symposium on Turbo Codes and Iterative Information Processing (ISTC), Brest, France, 5–9 September 2016; pp. 395–399.
101. Wang, T.; Liang, Y.; Jia, W.; Arif, M.; Liu, A.; Xie, M. Coupling resource management based on fog computing in smart city systems. *J. Netw. Comput. Appl.* **2019**, *135*, 11–19. [CrossRef]
102. Zhou, Z.; Liu, P.; Feng, J.; Zhang, Y.; Mumtaz, S.; Rodriguez, J. Computation resource allocation and task assignment optimization in vehicular fog computing: A contract-matching approach. *IEEE Trans. Veh. Technol.* **2019**, *68*, 3113–3125. [CrossRef]
103. Javaid, S.; Javaid, N.; Saba, T.; Wadud, Z.; Rehman, A.; Haseeb, A. Intelligent resource allocation in residential buildings using consumer to fog to cloud based framework. *Energies* **2019**, *12*, 815. [CrossRef]
104. Tajalli, S.Z.; Tajalli, S.A.M.; Kavousi-Fard, A.; Niknam, T.; Dabbaghjamanesh, M.; Mehraeen, S. A Secure Distributed Cloud-Fog Based Framework for Economic Operation of Microgrids. In Proceedings of the 2019 IEEE Texas Power and Energy Conference (TPEC), College Station, TX, USA, 7–8 February 2019; pp. 1–6.
105. Barros, E.B.C.; Dionísio Machado Filho, L.; Batista, B.G.; Kuehne, B.T.; Peixoto, M.L.M. Fog Computing Model to Orchestrate the Consumption and Production of Energy in Microgrids. *Sensors* **2019**, *19*, 2642. [CrossRef] [PubMed]

106. Jie, Y.; Li, M.; Guo, C.; Chen, L. Game-theoretic online resource allocation scheme on fog computing for mobile multimedia users. *China Commun.* **2019**, *16*, 22–31.
107. Zhou, Y.; Shen, Q.; Dong, M.; Ota, K.; Wu, J. Chaos-Based Delay-Constrained Green Security Communications for Fog-Enabled Information-Centric Multimedia Network. In Proceedings of the 2019 IEEE 89th Vehicular Technology Conference (VTC2019-Spring), Kuala Lumpur, Malaysia, 28 April–1 May 2019; pp. 1–6.
108. Gope, P. LAAP: Lightweight Anonymous Authentication Protocol for D2D-Aided Fog Computing Paradigm. *Comput. Secur.* **2019**, *86*, 223–237. [CrossRef]
109. Li, Z.; Chen, J.; Zhang, Z. Socially Aware Caching in D2D Enabled Fog Radio Access Networks. *IEEE Access* **2019**, *7*, 84293–84303. [CrossRef]
110. Tao, M.; Ota, K.; Dong, M. Foud: Integrating fog and cloud for 5G-enabled V2G networks. *IEEE Netw.* **2017**, *31*, 8–13. [CrossRef]
111. Ma, M.; He, D.; Wang, H.; Kumar, N.; Choo, K.K.R. An Efficient and Provably-Secure Authenticated Key Agreement Protocol for Fog-Based Vehicular Ad-Hoc Networks. *IEEE Internet Things J.* **2019**, *6*, 8065–8075. [CrossRef]
112. Pereira, J.; Ricardo, L.; Luís, M.; Senna, C.; Sargento, S. Assessing the reliability of fog computing for smart mobility applications in VANETs. *Future Gener. Comput. Syst.* **2019**, *94*, 317–332. [CrossRef]
113. Wang, D.; Liu, Z.; Wang, X.; Lan, Y. Mobility-Aware Task Offloading and Migration Schemes in Fog Computing Networks. *IEEE Access* **2019**, *7*, 43356–43368. [CrossRef]
114. Chen, Y.S.; Tsai, Y.T. A mobility management using follow-me cloud-cloudlet in fog-computing-based RANs for smart cities. *Sensors* **2018**, *18*, 489. [CrossRef]
115. Kadhim, A.J.; Seno, S.A.H. Energy-efficient multicast routing protocol based on SDN and fog computing for vehicular networks. *Ad Hoc Netw.* **2019**, *84*, 68–81. [CrossRef]
116. Muthanna, A.; Ateya, A.A.; Khakimov, A.; Gudkova, I.; Abuarqoub, A.; Samouylov, K.; Koucheryavy, A. Secure IoT Network Structure Based on Distributed Fog Computing, with SDN/Blockchain. *J. Sens. Actuator Netw.* **2019**, *8*, 15. [CrossRef]
117. Islam, N.; Faheem, Y.; Din, I.U.; Talha, M.; Guizani, M.; Khalil, M. A blockchain-based fog computing framework for activity recognition as an application to e-Healthcare services. *Future Gener. Comput. Syst.* **2019**, *100*, 569–578. [CrossRef]
118. Tang, W.; Zhang, K.; Zhang, D.; Ren, J.; Zhang, Y.; Shen, X.S. Fog-Enabled Smart Health: Toward Cooperative and Secure Healthcare Service Provision. *IEEE Commun. Mag.* **2019**, *57*, 42–48. [CrossRef]
119. Moustafa, N. A Systemic IoT-Fog-Cloud Architecture for Big-Data Analytics and Cyber Security Systems: A Review of Fog Computing. *arXiv* **2019**, arXiv:1906.01055.
120. Yassine, A.; Singh, S.; Hossain, M.S.; Muhammad, G. IoT big data analytics for smart homes with fog and cloud computing. *Future Gener. Comput. Syst.* **2019**, *91*, 563–573. [CrossRef]
121. Nasir, M.; Muhammad, K.; Lloret, J.; Sangaiah, A.K.; Sajjad, M. Fog computing enabled cost-effective distributed summarization of surveillance videos for smart cities. *J. Parallel Distrib. Comput.* **2019**, *126*, 161–170. [CrossRef]
122. Mora, L.; Bolici, R.; Deakin, M. The first two decades of smart-city research: A bibliometric analysis. *J. Urban Technol.* **2017**, *24*, 3–27. [CrossRef]
123. Luthra, M.; Koldehofe, B.; Steinmetz, R. Transitions for Increased Flexibility in Fog Computing: A Case Study on Complex Event Processing. *Inform. Spektrum* **2019**, *42*, 244–255. [CrossRef]
124. Naranjo, P.G.V.; Pooranian, Z.; Shojafar, M.; Conti, M.; Buyya, R. FOCAN: A Fog-supported smart city network architecture for management of applications in the Internet of Everything environments. *J. Parallel Distrib. Comput.* **2019**, *132*, 274–283. [CrossRef]
125. Giang, N.K.; Lea, R.; Leung, V.C. Developing applications in large scale, dynamic fog computing: A case study. *Softw. Pract. Exp.* **2019**, doi: 10.1002/spe.2695. [CrossRef]
126. Tang, B.; Chen, Z.; Hefferman, G.; Wei, T.; He, H.; Yang, Q. A hierarchical distributed fog computing architecture for big data analysis in smart cities. In Proceedings of the ASE BigData & SocialInformatics 2015, Kaohsiung, Taiwan , 7–9 October 2015; p. 28.
127. Dsouza, C.; Ahn, G.J.; Taguinod, M. Policy-driven security management for fog computing: Preliminary framework and a case study. In Proceedings of the 2014 IEEE 15th International Conference on Information Reuse and Integration (IEEE IRI 2014), Redwood City, CA, USA, 13–15 August 2014; pp. 16–23.

128. Aamir, M.; Masroor, S.; Ali, Z.A.; Ting, B.T. Sustainable Framework for Smart Transportation System: A Case Study of Karachi. *Wirel. Pers. Commun.* **2019**, *106*, 27–40. [CrossRef]
129. Baccarelli, E.; Cordeschi, N.; Mei, A.; Panella, M.; Shojafar, M.; Stefa, J. Energy-efficient dynamic traffic offloading and reconfiguration of networked data centers for big data stream mobile computing: Review, challenges, and a case study. *IEEE Netw.* **2016**, *30*, 54–61. [CrossRef]
130. Azimi, I.; Pahikkala, T.; Rahmani, A.M.; Niela-Vilén, H.; Axelin, A.; Liljeberg, P. Missing data resilient decision-making for healthcare IoT through personalization: A case study on maternal health. *Future Gener. Comput. Syst.* **2019**, *96*, 297–308. [CrossRef]
131. Kumari, A.; Tanwar, S.; Tyagi, S.; Kumar, N. Fog computing for Healthcare 4.0 environment: Opportunities and challenges. *Comput. Electr. Eng.* **2018**, *72*, 1–13. [CrossRef]
132. Ray, P.P.; Dash, D.; De, D. Edge computing for Internet of Things: A survey, e-healthcare case study and future direction. *J. Netw. Comput. Appl.* **2019**, *140*, 1–22. [CrossRef]
133. Gia, T.N.; Jiang, M.; Rahmani, A.M.; Westerlund, T.; Liljeberg, P.; Tenhunen, H. Fog computing in healthcare internet of things: A case study on ecg feature extraction. In Proceedings of the 2015 IEEE International Conference on Computer and Information Technology; Ubiquitous Computing and Communications; Dependable, Autonomic and Secure Computing; Pervasive Intelligence and Computing, Liverpool, UK, 26–28 October 2015; pp. 356–363.
134. Jalali, F.; Vishwanath, A.; De Hoog, J.; Suits, F. Interconnecting Fog computing and microgrids for greening IoT. In Proceedings of the 2016 IEEE Innovative Smart Grid Technologies-Asia (ISGT-Asia), Melbourne, VIC, Australia, 28 November–1 December 2016; pp. 693–698.
135. Beligianni, F.; Alamaniotis, M.; Fevgas, A.; Tsompanopoulou, P.; Bozanis, P.; Tsoukalas, L.H. An internet of things architecture for preserving privacy of energy consumption. In Proceedings of the Mediterranean Conference on Power Generation, Transmission, Distribution and Energy Conversion (MedPower 2016), Belgrade, Serbia, 6–9 November 2016.
136. Rao, L.; Liu, X.; Ilic, M.D.; Liu, J. Distributed coordination of internet data centers under multiregional electricity markets. *Proc. IEEE* **2012**, *100*, 269–282.
137. Pu, L.; Chen, X.; Xu, J.; Fu, X. D2D fogging: An energy-efficient and incentive-aware task offloading framework via network-assisted D2D collaboration. *IEEE J. Sel. Areas Commun.* **2016**, *34*, 3887–3901. [CrossRef]
138. Wang, S.; Huang, X.; Liu, Y.; Yu, R. CachinMobile: An energy-efficient users caching scheme for fog computing. In Proceedings of the 2016 IEEE/CIC international conference on communications in China (ICCC), Chengdu, China, 27–29 July 2016; pp. 1–6.
139. Yang, F.; Wang, S.; Li, J.; Liu, Z.; Sun, Q. An overview of internet of vehicles. *China Commun.* **2014**, *11*, 1–15. [CrossRef]
140. Anawar, M.R.; Wang, S.; Azam Zia, M.; Jadoon, A.K.; Akram, U.; Raza, S. Fog computing: An overview of big IoT data analytics. *Wirel. Commun. Mob. Comput.* **2018**, *2018*, 7157192. [CrossRef]

© 2019 by the authors. Licensee MDPI, Basel, Switzerland. This article is an open access article distributed under the terms and conditions of the Creative Commons Attribution (CC BY) license (http://creativecommons.org/licenses/by/4.0/).

Article

Narrowband Internet of Things (NB-IoT): From Physical (PHY) and Media Access Control (MAC) Layers Perspectives

Collins Burton Mwakwata [1,*], Hassan Malik [1], Muhammad Mahtab Alam [1], Yannick Le Moullec [1], Sven Parand [2] and Shahid Mumtaz [3]

[1] Thomas Johann Seebeck Department of Electronics, Tallinn University of Technology (TalTech), Ehitajate tee-5, 19086 Tallinn, Estonia; hassan.malik@taltech.ee (H.M.); muhammad.alam@taltech.ee (M.M.A.); yannick.lemoullec@taltech.ee (Y.L.M.)
[2] Telia Estonia Ltd., 10616 Tallinn, Estonia; sven.parand@telia.ee
[3] Instituto de Telecomunicações, 1049-001 Aveiro, Portugal; smumtaz@av.it.pt
* Correspondence: collins.burton@taltech.ee; Tel.: +372-58-663569

Received: 27 April 2019; Accepted: 5 June 2019; Published: 8 June 2019

Abstract: Narrowband internet of things (NB-IoT) is a recent cellular radio access technology based on Long-Term Evolution (LTE) introduced by Third-Generation Partnership Project (3GPP) for Low-Power Wide-Area Networks (LPWAN). The main aim of NB-IoT is to support massive machine-type communication (mMTC) and enable low-power, low-cost, and low-data-rate communication. NB-IoT is based on LTE design with some changes to meet the mMTC requirements. For example, in the physical (PHY) layer only single-antenna and low-order modulations are supported, and in the Medium Access Control (MAC) layers only one physical resource block is allocated for resource scheduling. The aim of this survey is to provide a comprehensive overview of the design changes brought in the NB-IoT standardization along with the detailed research developments from the perspectives of Physical and MAC layers. The survey also includes an overview of Evolved Packet Core (EPC) changes to support the Service Capability Exposure Function (SCEF) to manage both IP and non-IP data packets through Control Plane (CP) and User Plane (UP), the possible deployment scenarios of NB-IoT in future Heterogeneous Wireless Networks (HetNet). Finally, existing and emerging research challenges in this direction are presented to motivate future research activities.

Keywords: narrowband; IoT; PHY; NB-IoT; MAC; deployment; survey; mMTC; 5G

1. Introduction

According to Information Handling Services (IHS) technology forecast, the Internet of Things (IoT) market is expected to grow to billions of devices by 2020 [1]. Massive connections are expected to respond to different IoT use cases such as smart city, smart wearables, smart home, etc. [2]. For these applications, latency-insensitive devices can be positioned in hard-to-reach areas and do not require high throughput or frequent reporting. Therefore, to cope with such tremendous IoT trends, the Third-Generation Partnership Project (3GPP) introduced the Narrowband Internet of Things (NB-IoT) standard as a communication technology enabler. NB-IoT is categorized as one of the licensed Low-Power Wide-Area Networks (LPWAN) cellular technologies based on Long-Term Evolution (LTE) with long range and low cost. In the LPWAN category, there exist other licensed technologies, i.e., Long-Term Evolution Category M1 (LTE-M), and unlicensed technologies, i.e., Long Range (LoRa), SigFox, Ingenu, etc. [3–7], but they are not the focus of the current work since they are not based on cellular technology.

The term Narrowband refers to NB-IoT's bandwidth of maximum 200 kHz thanks to which it can coexist either in the Global System for Mobile Communications (GSM) spectrum or by occupying one of the legacy LTE Physical Resource Blocks (PRBs) as in-band or as guard-band. Since it coexists in the LTE spectrum, NB-IoT follows the legacy LTE numerologies as it uses Orthogonal Frequency Division Multiplexing (OFDM) and Single-Carrier Frequency Division Multiple Access (SC-FDMA) in the downlink and uplink transmission schemes, respectively. Some modifications in the physical (PHY) and medium access control (MAC) layers are implemented to support the long-range massive machine-type (mMTC) connections with low power, low data rates, low complexity, and hence low cost. However, despite its low complexity, this new radio access technology (RAT) delivers better performance in terms of the supported number of devices, and coverage enhancements for latency-insensitive applications with maximum coupling loss (MCL) of about 20 dB higher than LTE (i.e., 164 dB) [5–11].

With flexible deployment as well as the possibility to implement over-the-air (OTA) firmware upgrades, many telecommunication operators across the globe (as shown in Figure 1) deployed NB-IoT to test its practical feasibility on diverse use cases with real-life trials such as connected sheep in Norway [12], smart metering and tracking in Brazil [13], NB-IoT at sea in Norway [14], smart city in Las Vegas, USA [15], etc. The trials are enabled by different NB-IoT software and hardware solutions from different chip or module vendors such as Skyworks [16], Media tek [17], Neul (Huawei) [18], Quectel [19], Nordic Semiconductors [20], Intel [21], Sequans [22], Qualcomm [23], Siera wireless [24], Samsung [25], Altair [26], U-Blox [27], and so on.

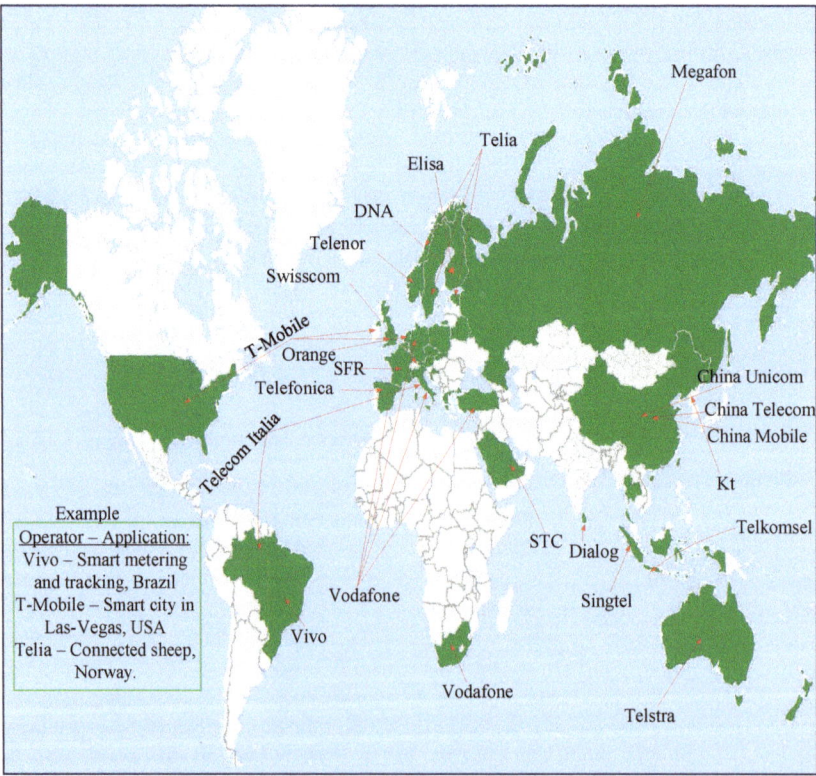

Figure 1. The geographical representation of countries with the ongoing NB-IoT real-life deployments for diverse use cases (May 2019).

The availability of such commercial off-the-shelf solutions speeds up the adoption of NB-IoT. For this reason, numerous studies addressing segmented enhancement criteria including survey articles emerged to analyze NB-IoT performance and implementation. Table 1 presents, in a nutshell, the main differences and similarities between this survey and the other existing ones by displaying the key focus features.

Table 1. Summarized comparison of this survey's contribution with respect to the existing surveys.

Survey	The Third Generation Partnership Project				Layers		Deployment Strategies
[Ref]	Rel 13	Rel 14	Rel 15	Rel 16	Physical	Media Access Control	
[28] 2017	✓						
[29] 2017	✓	✓					
[30] 2017		✓					
[31] 2017	✓						
[32] 2018	✓						
[33] 2019	✓	✓					
This survey	✓	✓	✓	✓	✓	✓	✓

For example, in [32], the authors surveyed the development path of MTC and elaborated the NB-IoT evolution in Release 13. Similarly, in [28], the authors discussed the Release 13 features and compared its performance with respect to other communication technologies such as LTE-M, SigFox, Lora and Wireless-Fidelity (WiFi), etc. In [29,30], the authors gave an overview of NB-IoT Release 14; however, in [30], the authors elaborated more on the expectations for NB-IoT Release 15 agenda. In [31], the authors presented a survey on the NB-IoT downlink scheduling issues by highlighting the associated scheduling process in terms of offset index selection. In [33], the authors surveyed the uplink and downlink performance evaluation of NB-IoT systems by analyzing the main causes of latency, trade-off between throughput and free resources, channel occupancy etc. with respect to Release 13 and Release 14 updates.

In contrast to the above surveys, this paper presents:

- A comprehensive survey of NB-IoT, from Release 13 to the ongoing Release 16 prospects.
- An all-inclusive overview of the state of the art of PHY and MAC layers by addressing the key improvement concerns in terms of challenges and the corresponding potential solutions.
- The possible NB-IoT deployment strategies for synchronous and asynchronous network structures in HetNet scenarios to foster the NB-IoT coexistence with legacy technologies as well as with the fifth generation (5G) networks.
- Discussion on the open research challenges to motivate future research directions.

To the best of the authors' knowledge, this is the first survey that covers broadly these above-mentioned contributions and hence will facilitate the reader's knowledge related to NB-IoT from standardization, ongoing research, and its practical implementation.

The rest of this paper is organized as follows: Section 2 discusses NB-IoT standards by elaborating the key design changes and the related ongoing enhancements. Section 3 presents the state of the art of NB-IoT protocol stack by detailing the PHY layer and MAC layer features. Section 4 discusses the open research questions and their potential solutions, and the conclusion is drawn in Section 5.

2. Narrowband-IoT Standard and Releases

Early in 2014, the LPWAN market rapidly developed thanks to the emergence of IoT. Realizing the need and potential for new communication ways, 3GPP started a feasibility study on cellular system support for an ultra-low complexity and low throughput IoT solution referred to as cellular IoT. In May 2014, Huawei and Vodafone proposed the Narrowband Machine to Machine (NB-M2M) to 3GPP as a study item to cope with the IoT market needs. Additional telecom industrial players got interested and later the same year Qualcomm proposed narrowband orthogonal frequency division multiplexing (NB-OFDM). In May 2015, 3GPP merged the two proposals (i.e., NB-M2M and NB-OFDM) and formed the Narrowband Cellular IoT (NB-CIoT). Eight months later, Ericsson proposed the Narrowband Long-Term Evolution NB-LTE. In September 2015, 3GPP included all proposals as a work item for Release 13. The key difference between NB-CIoT and NB-LTE was the number of the reused legacy LTE network resources to support interoperability. In June 2016 NB-IoT was recognized as a new clean slate RAT. Only further improvement changes were allowed and implemented thereafter.

In this regard, this section presents the main NB-IoT design changes from Release 13 until today that enabled the massive IoT connections with the corresponding solutions to respond to the adopted NB-IoT objectives. The enhancement features are classified following the objectives that are presented in the releases which would make it easier for the readers to refer back to the official 3GPP documents [8,9,34–38].

2.1. Release 13

3GPP introduced the following techniques in NB-IoT Release 13 to enable cellular massive IoT deployment for diverse use cases with low power, low complexity, and hence low cost. The introduced features and their corresponding objectives are as follows.

2.1.1. Mode of Operation

With the limited bandwidth requirement, NB-IoT can be deployed in three different modes i.e., standalone, in-band, and guard-band, as depicted in Figure 2. In in-band and guard-band modes, NB-IoT occupies one PRBs of 180 KHz in LTE spectrum both in the downlink and uplink. It can also be allocated as standalone where it occupies the 200 KHz bandwidth by "refarming" the GSM spectrum. These flexible deployment possibilities enable fast integration and coexistence with legacy LTE and GSM systems.

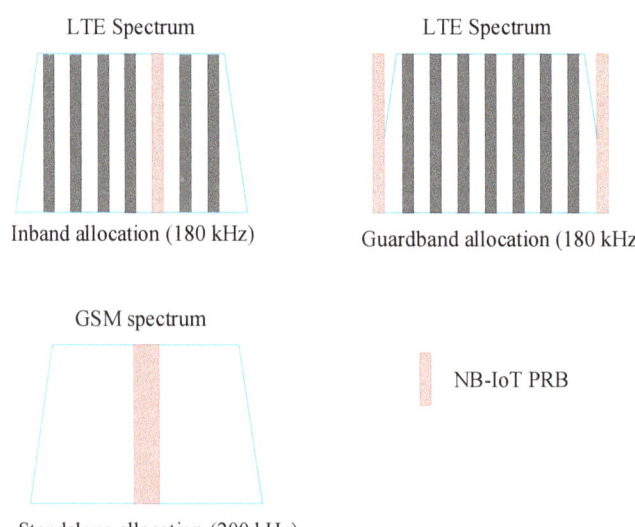

Figure 2. Narrow band Interet of Things (NB-IoT) Flexible Allocation inside Long-Term Evolution (LTE) spectrum (in-band and guard-band) and when refarming the Global System for Mobile Communications (GSM) spectrum (standalone).

2.1.2. Multi-Tone Transmission Support

To reach the massive device deployment objective, NB-IoT introduces the allocation of Resource Units (RU) to multiple User Equipment (UE) contrary to LTE where the whole resource block is allocated to a single UE in the uplink. In this regard, tones (frequency domain) with different duration are allocated to UEs. For the uplink transmission, each tone may either occupy 3.75 kHz or 15 kHz of transmission bandwidth based on the SC-FDMA scheme; for downlink NB-IoT uses 15 kHz of transmission bandwidth with OFDM scheme as LTE. With 15 kHz spacing, NB-IoT can dedicate either single-tone (8 ms) or multi-tone (3 tones, 6 tones, and 12 tones) to different UEs with the duration of 4 ms, 2 ms, and 1 ms, respectively. On the other hand, the 3.75 kHz spacing supports only single-tone allocation to different users with 48 subcariers of 32 ms duration [11,39,40].

2.1.3. Complexity and Cost Reduction Techniques

NB-IoT is required to have low complexity to reach the low-cost objective to facilitate massive connections. The features that were implemented to reach this objective include relaxed base-band processing, low memory storage, and reduced radio-frequency (RF) components. In this regard, the system bandwidth is set as narrow as 180 kHz with reduced frequency and time synchronization requirement. Also, NB-IoT uses the restricted BPSK and QPSK modulation schemes with only one antenna support both in uplink and downlink transmission.

2.1.4. Power Reduction Method

NB-IoT devices are intended to have a 10 years battery life to support massive deployment with limited human intervention. In this regard, two features i.e., Power Saving Mode (PSM), (from Release 12), and extended Discontinuous Reception (eDRx) (new feature from Release 13) were supported. These features are intended to extend the UE's battery longevity as follows:

In PSM, the NB-IoT device is configured to completely sleep while remaining registered online but cannot be reached by the base station signaling. In Release 13, the device can be in PSM mode for approximately up to about 413 days. In eDRX, the device is in an inactive mode for a few minutes to a few hours only.

In both cases, the partial or complete inability to receiving and sending different signals enhance the battery life longevity; however, choosing either PSM, eDRX or both depends on the corresponding use-case requirement. In this regard, the device can be synchronized to wake up from these modes by either Real-Time Clock (RTC), triggering from sensors, or both.

2.1.5. Physical Channels and Signals

NB-IoT adopts the same frame structure as LTE, with 1024 hyper frames, consisting of 1024 frames that contain 10 subframes of two slots with a duration of 0.5 ms each in the time domain. Similarly, in the frequency domain, NB-IoT contains 12 subcarriers of 7 OFDM symbols mapped in each slot. In addition to that, when NB-IoT uses the 3.75 kHz spacing on the uplink, 48 subcarriers are used with a slot duration of 2 ms.

The following channels and signals are used in the uplink:

- Narrowband Physical Random Access Channel (NPRACH).
- Narrowband Physical Uplink Shared Channel (NPUSCH).
- Demodulation Reference Signal (DMRS).

And the following are in the downlink frame:

- Narrowband Physical Downlink Shared Channel (NPDSCH).
- Narrowband Physical Downlink Control Channel (NPDCCH).
- Narrowband Reference Signal (NRS).
- Narrowband Primary Synchronization Signal (NPSS).
- Narrowband Secondary Synchronization Signal (NSSS).
- Narrowband Physical Broadcast Channel (NPBCH).

In general, NPRACH is used by UEs to perform initial access to the network, to request transmission resources, and to reconnect to the base station after a link failure. NPDSCH and NPUSCH are used to carry the downlink and uplink data packets transmissions, respectively. DMRS is used for uplink channel estimation accuracy. The UE acquires Master Information Block (MIB) from NPBCH and System Information Block (SIBs) from the NPDCCH. The defined MIB and SIB are broadcasted once during 640 ms and 2560 ms intervals, respectively. The timing of the remaining SIBs is configured in SIB1-NB. NRS is used for cell search and initial system acquisition. NPSS and NSSS are used by the UE for its frequency and timing synchronization with the base station. Due to overhead scheduling gaps in NPDCCH, the downlink and uplink peak data rates are ~250 kb/s and ~2267 kb/s, respectively, [34,40–43].

2.1.6. Coverage Enhancement Method

NB-IoT is designed to enhance coverage for the applications that are in hard-to-reach areas such as deep indoors and basements. In this regard, NB-IoT delivers an additional coverage of 20 dB as compared to the legacy LTE system. This corresponds to 164 dB of MCL. To enhance its coverage, NB-IoT uses up to 128 and 2048 retransmissions in uplink and downlink, respectively. Hence, this makes NB-IoT suitable for use cases that are latency insensitive as it can tolerate up to 10 seconds transmission delay.

2.2. Release 14 Enhancements

After the implementation of Release 13 features, studies erupted along with field trials that revealed the need for further enhancements to improve the quality of service as well as user experience. In this regard, 3GPP introduced further enhancement features to NB-IoT.

The enhancements features in Release 14 include positioning update, multicast services, and a new UE output power class in which the NB-IoT system throughput, mobility, service continuity and non-anchor carrier operation are improved [29,30].

2.2.1. Improved Positioning Technique

3GPP Release 14 introduces an indoor advanced positioning method of observed time difference of arrival (OTDOA) for NB-IoT to enhance UE position measurement of cell identity (CID). In OTDOA method, the UE measures the times of arrival (ToAs) of positioning reference signals (PRSs) received from different transmitters with respect to a reference node's PRS transmission to form the reference signal time difference (RSTD) measurements. In enhanced CID, the measurement requirements include the base station receive (Rx) and transmit (Tx) time difference, reference signal received power (RSRP), and reference signal received quality (RSRQ).

2.2.2. Multicast Services

The main objective of this mechanism is to optimize resources as well as transmission latency by addressing the data to a group of UEs at the same time rather than sending it multiple times to separate devices.

Therefore in Release 14, Multimedia Broadcast Multicast Services (MBMS) is supported through single-cell point-to-multipoint (SC-PTM). In general, SC-PTM is an efficient dynamic mechanism for optimal radio resource usage as it allows broadcast or multicast services to a specific group based on real-time traffic load and user requirement. SC-PTM uses NPDSCH by mapping Single-cell MBMS Control CHannel (SC-MCCH) and Single-Cell MBMS Traffic CHannel (SC-MTCH) that carry control and data traffic to the physical layer scheduled by using the downlink control information (DCI).

2.2.3. New Power Class for Narrowband-IoT User Equipment

Instead of the two power classes of Release 13 (i.e., 20 dBm and 23 dBm), in Release 14, the maximum allowed device's output power is reduced to 14 dBm. This has led to coverage relaxation of 9 dB that corresponds to 155 dB MCL as compared to 164 dB MCL and hence reduces the drained current. Technically, the use of the new power class facilitates the use of small coin-cell batteries and hence can be suitable for limited-size devices and applications that need a small battery. The compensation of the reduced NB-IoT power is achieved by increasing the NB-IoT transmission time to maintain the same energy per bit as the UE in Release 13 achieves. The newly introduced power class allows the serving base station to acquire the device power class during the establishment of the connection.

2.2.4. New Transport-Block-Size Support

Contrary to Release 13 where NB-IoT supports relatively low data rates (~250 kb/s and ~226.7 kb/s in downlink and uplink, respectively), 3GPP Release 14 introduces a new NB-IoT device category which supports the improved data rates by enhancing the Transport Block Size (TBS) to 2536 bits. These data rates can be reached thanks to the ability to support a second Hybrid Automatic Repeat Request (HARQ) process. This second HARQ is useful for enhancing the reliability of the link for the UEs that experience favorable channel conditions. Implementation of this optional second HARQ process results in throughput gain as it reduces the overhead caused by NPDCCH scheduling gaps.

2.2.5. Multicarrier Operation

To enable the massive NB-IoT deployment, in Release 14, NB-IoT can monitor paging and perform random access on non-anchor carriers. With this feature, one or more non-anchor carriers are added to the anchor carrier to carry out the synchronization and mobility measurements by using the NRS. Non-anchor carriers should also perform random access or paging when needed. Therefore, paging occasions and hence paging load will be spread over the anchor and non-anchor carriers and all carriers can then monitor paging.

2.2.6. User Equipment Mobility Enhancement

For the use cases that involve mobility, the temporary loss of radio interface impacts the system to a degree that can degrade link performance in terms of transmission errors. In this regard, 3GPP Release 14 introduces the possibility of Radio Resource Control (RRC) re-establishment for NB-IoT UE that supports data transfer via the control plane, i.e., the UE will try to re-establish the connection on that cell and resume the data transfer. This new RRC re-establishment feature hides the temporary loss of the radio interface to the upper layers.

2.3. Release 15 Enhancements

On top of all the enhancements that were introduced in Releases 13 and 14, the following improvements were introduced in Release 15 to satisfy the fast adoption of massive deployment with further improved quality of service.

2.3.1. Latency Reduction

In Release 15, NB-IoT supports new features to further reduce the transmission delay as well as to further reduce the power consumption dissipated during long transmission requirements.

In this regard, the NB-IoT UE is now able to support the physical layer Scheduling Request (SR) which is a special physical layer message to request the network to send the access grant (DCI format 0) so that the UE can transmit the uplink data. Also, NB-IoT uses a wake-up (Wu) signal to wake up the main receiver. This signal is transmitted in idle mode only when the UE is required to decode the physical downlink control channel in paging occasions. Therefore, power consumption reduction with the wake-up signal technique is larger when the UE wakes up from deep sleep more frequently (i.e., for shorter DRX/eDRX cycles). Also, significant power consumption reduction is achieved even when a common wake-up signal is used for a group of UEs. Quick RRC release and early data transmission during random access channel (RACH) procedure are supported to reduce the UE transmission latency and hence power consumption.

2.3.2. Semi-Persistent Scheduling

To enable better support of voice messages for the corresponding use cases, in Release 15, Semi-Persistent Scheduling (SPS) feature is introduced. In general, SPS is comprised of persistent scheduling for initial transmissions and dynamic scheduling for retransmissions. The base station assigns specific resource units to be used for NB-IoT UE voice messages with specific interval to save control plane overhead and hence optimize the radio resource usage. By principle, the base station preconfigures the UE with the Radio Network Temporary Identifier (SPS-RNTI) which is used to specifically differentiate one NB-IoT UE from another, or one radio channel from another. This SPS enables the NB-IoT data reception at a regular configured periodicity.

2.3.3. Small Cell Support

To further improve the capacity as well as coverage, in Release 15, NB-IoT supports small cell deployments. The downlink power to be reused for NB-IoT small cells is specified in section 16.2.2 of TS 36.213 [44]. In general, NB-IoT UE is not allowed to transmit more power than the configured maximum power, even if the configured power is lower than UE's maximum capability. This is done to avoid interference.

On the other hand, to extend the IoT connectivity especially in remote and rural areas for use cases such as agriculture, logistics, and environmental monitoring, NB-IoT is now able to support up to 100 km range. According to Ericsson, this could be achieved with a software upgrade only, without any changes in the existing NB-IoT hardware [45].

2.3.4. Enhanced User Equipment Measurements

Like in legacy LTE systems, UE measurements are critical since the corresponding reporting is mainly used to characterize the reference signal of a given bandwidth.

In Release 15, UE measurements are improved in a way that only NSSS additionally to NRS is defined for radio resource management measurement enhancement. This means that NRS is determined by the resource elements that carry NSSS in the NSSS occasions that the UE measures, through which the cell search and initial cell acquisition are improved.

2.3.5. Time Division Duplex (TDD) Support

In Release 15, a new feature of TDD support is introduced with a new TDD frame structure (type 2). For both 3.75 kHz and 15 kHz spacing, some specified restrictions are introduced i.e., only a normal cyclic prefix is supported for NB-IoT transmission. To support some of the TDD configurations with few downlink subframes, some of the system information (SI) can be transmitted on non-anchor carriers. In this way, the UE will have reduced system information acquisition and search time, and hence reduced UE differentiation and access control [30,46,47].

2.4. Release 16 Enhancement Prospects

3GPP and many industrial players are involved in ongoing discussions for Release 16 enhancements. The agenda includes the following objectives with their corresponding solutions.

2.4.1. Grant-Free Access

Most of the power consumption takes place during the NB-IoT UE active time, i.e., during Tx and Rx. In Release 16, the UE will be expected to transmit during RRC-Idle mode through Msg3 (RRC connection request) without access grant. A UE in RRC connected mode can transmit data without grant or with the simplified control-less grant. A further enhancement is on reducing NB-IoT signaling overhead while guaranteeing the needed quality of service. These features will reduce both power consumption and latency. In Release 16, it is also proposed to further study other signal waveforms (i.e., FDMA) that require less orthogonality with more relaxed timing advance (TA) alignment as compared to SC-FDMA.

2.4.2. Simultaneous Multi-User Transmission

The introduction of new schemes will enable simultaneous multi-user transmissions by using a shared resource in the time and frequency domains, such as Code division multiplexing (CDM), and multi-user multiple inputs multiple outputs (MU-MIMO), without increasing the number of antennae at the UE. In this regard, more dynamic access can also be achieved through enhanced base station receiver for detection of multiple users that are using the same resource unit as cluster and hence be able to schedule them effectively. This is because, for the last releases, NB-IoT UE uses the static or semi-static configuration of more resources for the unexpected application traffic handling. Similarly, the introduction of NB-IoT transmission without grant will cause a collision of data packets so dynamic handling of multiplexing is necessary.

2.4.3. Enhanced Group Message Mechanism

In Release 16, there should be more enhancements to support downlink command between user groups and group RNTIs. This is because MBMS which was proposed in Release 14 is only efficient for large size downlink command message transmission and requires many UEs to be deployed. For example, the application layer common message can be very small but sent to many UEs under a small group of UEs hence making MBMS not efficient for such applications.

2.4.4. Inter-RAT Idle-Mode Mobility

For applications such as smart tracking of logistics that involve mobility, the NB-IoT UE may still need to be accessible even when moved to the area served by other base station.

In this regard, 3GPP should introduce the new feature for NB-IoT UE support for inter-RAT mobility during idle mode. The mentioned feature is introduced along with optional handover support during connected mode through procedure simplification i.e., without dedicated signaling for measurement control and report. This is because handover helps to reduce system information reading time.

2.4.5. Network Management Tool Enhancement to Improve UE Differentiation

NB-IoT UE is expected to be able to perform differentiation according to maximal tolerable delay per service to optimize the radio resource usage. This is because, in the last release, the UE can be differentiated according to traffic model (periodic communication indicator, periodic time, scheduled communication time, traffic profile) and battery indication.

Section 2 has presented the NB-IoT standard and the corresponding enhancements from Release 13 until today. It has highlighted the main design changes and the corresponding further enhancements, i.e., deployment flexibility, physical channels and signals, positioning, multicast, new power classes, improved data rates, multicarrier operations, mobility support, improved scheduling, NB-IoT small cell support etc.

3. Narrowband-IoT: Protocol Stack

This section presents the NB-IoT protocol stack based on state of the art of the PHY and MAC layers to identify the knowledge gap and define future research directions. NB-IoT adopts the same protocol stack as the legacy LTE. However, some design changes in both PHY and MAC layers were introduced to support the massive long-range connections with up to additional 20 dB MCL than in legacy technologies such as LTE, GSM, and GPRS. Those changes are described in what follows.

3.1. Physical Layer

On the physical layer, NB-IoT adopts the same numerologies as legacy LTE along with OFDM and SC-FDMA signal waveforms in downlink and uplink, respectively. However, the resource scheduling unit in NB-IoT is the subcarrier (or tone) instead of PRB, to foster the network scalability by serving multiple UEs in a 180 kHz bandwidth. The downlink and uplink frame structures are as depicted in Figures 3 and 4, respectively.

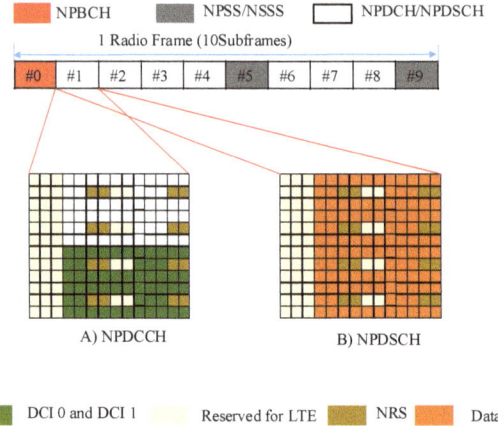

Figure 3. NB-IoT Downlink Frame Structure: subframe number 0 carries the Narrowband Physical Broadcast Channel (NPBCH), 1 to 4, and 6 to 8 carry the Narrowband Physical Downlink Control Channel (NPDCCH)/Narrowband Physical Downlink Shared Channel (NPDSCH), and 5 and 9 carry the Narrowband Primary Synchronization Signal (NPSS)/Narrowband Secondary Synchronization Signal (NSSS) (**A**) When the subframe is carrying control channels and (**B**) when the subframe is carrying data.

In general, the base station uses DCI to specify the scheduling information for a downlink/uplink transmission in NB-IoT. Then NB-IoT UE learns the deployment mode (standalone, in-band, or guard-band) as well as the cell identity through its initial acquisition, and it figures out which resource elements are already used by LTE. This is the way by which the UE can map NPDCCH and NPDSCH symbols to available resource elements. For example, in the downlink, NPDCCH is transmitted by aggregating the narrowband control elements (element 0 and element 1) where element 0 is occupied in subcarrier 0 to 5 and element 1 occupies subcarrier 6 to 11 in a subframe. The elements are determined by the type of DCI which is carried by NPDCCH to deliver scheduling command. Either two DCIs can be multiplexed in one subframe, or one DCI can be mapped in one subframe, corresponding to the aggregation level used [48]. However, NPDCCH, NPDSCH, and NRS cannot be mapped to the already occupied resource elements for LTE signals such as cell-specific reference symbols (CRS) and LTE physical downlink control channel (PDCCH). When NB-IoT UE receives NPDCCH which carries DCI, it decodes it and uses the device's scheduling feature (k0) to know the delay over which it will start to receive NPDSCH. The scheduling information is used to identify the allocated resources over NPDSCH and NPUSCH, respectively. In each NPDCCH, a maximum of two DCIs can be transported, and each UE can receive up to one DCI. The time interval between two successive NPDCCH opportunities is referred to as an NPDCCH period (PP) [48].

In the state of the art, different works have proposed solutions to the challenges that occur in PHY layer features, such as initial cell acquisition and synchronization, random access, channel estimation, error correction, and co-channel interference, as summarized in Table 2.

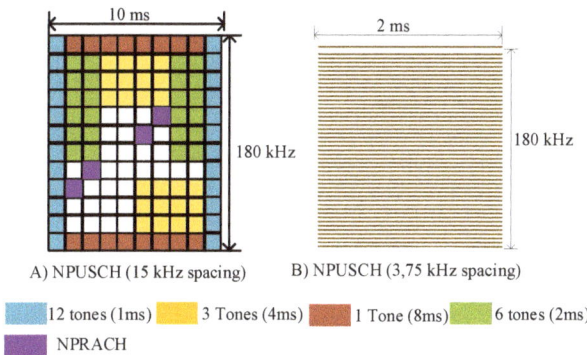

Figure 4. NB-IoT Uplink Frame Structure, (**A**) when 15 kHz spacing is used with different tone-allocation possibilities with slot duration of 0.5 ms and (**B**) when 3.75 kHz is used only single-tone allocation is supported with 4 times longer slot duration (2 ms).

Table 2. Articles on the proposed PHY layer enhancement techniques.

Feature	Article	Technique Used	Enhancement Criteria	Limitation
Cell Acquisition	[49]	Maximum-Likelihood (ML) NPSS detector	Average latency reduction for timing synchronization	It is a computationally complex detection method
	[50]	Cell search and initial synchronization algorithm	Time and frequency synchronization by using NPSS and NSSS with two-stage time domain NPSS correlation	mobility and new NB-IoT transmit power are not considered which have a direct impact on inter-RAT camping and the detected SNR, respectively
	[51]	Non-orthogonal spectral efficient frequency division multiplexing (SEFDM) waveform and an overlapped sphere decoding (OSD) detector	Resource optimization by the use of less bandwidth with better data rates compared to OFDM signal waveform	The proposed method would lead to sampling rate mismatch, carrier frequency offset and also will need to raise the computation complexity to NB-IoT UE
	[52]	New synchronization signal structure with Zadoff-Chu conjugates	Minimization of timing errors due to low-complexity NB-IoT frequency offset	If the same model is used for uplink synchronization it might lead to estimation errors if mobility is involved in NB-IoT
	[53]	NPRACH detection and time-of-arrival estimation for NB-IoT system	Enhancement on cell acquisition and channel estimation accuracy	The algorithm might not work for multi-tone allocation. Also, frequency hopping may raise power consumption as well as device complexity
	[54]	Receiver algorithm for NPRACH timing advance estimation and detection	Modeling the detection threshold to satisfy the NPRACH performance by lowering the probabilities of false alarm	The paper did not explain how receiver sensitivity can affect the NPRACH detection

Table 2. Cont.

Feature	Article	Technique Used	Enhancement Criteria	Limitation
Cell Acquisition	[55]	Mathematical modeling of NB-IoT performance	Throughput enhancement and NPRACH optimization by the use of repetition number, NPRACH preamble transmission per second and intersite distance	The work did not include some parameters such as the impact of mobility and how the achieved MCL for different coverage classes can impact the repetition assignment
	[56]	NPSS and NSSS frequency diversity reception	Time and frequency synchronization for cell search improvement	Alternative switching of NPSS and NSSS may require additional control commands which may lead to higher device complexity
Random Access	[57]	Configurable signal propagation model	System performance analysis in terms of number of supported devices, BER performance, preamble retransmissions, etc.	The impact of preamble retransmission on the overall transmission latency is not considered
	[58]	Mathematical evaluation of RACH preamble transmission	Analysis of NB-IoT transmission delay by using periodicity, start time, number of repetitions, number of preamble attempts and random access response window	Their model used minimum, intermediate, and maximum values for simulation which is so deterministic. However, it could be better to use random distribution to characterize NB-IoT realistic channel variations
	[59]	Random Access with differentiated barring (RADB) algorithm	Minimization of random access collision	Not resource efficient method since it does not include the impact of scheduling in different tone configurations
	[60]	New frequency hopping pattern of NPRACH preamble	Time-of-arrival estimation by the use of all the hopping distances	It only used a small cell scenario, if applied in dense NB-IoT network, estimation by considering all hopping distances may lead to system overhead and possible interference
Channel estimation	[61]	Frequency tracking algorithm	Frequency synchronization, as well as channel estimation for NB-IoT systems	More pilot signals, are used. This increases the overhead and hence can degrade the spectral efficiency
	[62]	Timing advance (TA) adjustment	Preamble sequence decoding by means of round trip estimation for coverage enhancements (on the sea)	It might not work for applications that do not involve a direct line of sight such as in dense urban environment
	[63]	MCS and coverage level optimization	Mobility effect on different coverage levels and how MCS affect paging performance	The channel model does not include other factors such as the effect of repetition, multipath, different Tx power for NB-IoT UEs as well as carrier frequency offset and inter-RAT operability
	[64]	New iterative algorithm for NB-IoT transmission scheme	NB-IoT error correction by using cryptographic redundancy and error correcting code	The channel estimation model to characterize NB-IoT transmission is not good, because some errors might be due to intersymbol interference and others due to intercarrier interference however the model does not explain

Table 2. *Cont.*

Feature	Article	Technique Used	Enhancement Criteria	Limitation
Interference mitigation	[65]	Channel Equalization algorithm	Intersymbol Interference mitigation by the phase-shifted channel frequency responses (CFR) to conquer the sampling mismatch between NB-IoT and base station	The proposed model did not consider the NPSS and NSSS impact ON time and frequency synchronization
	[66]	Mathematical model for sample duration in LTE and NB-IoT system	Interference and close-form interference analysis due to sampling mismatch between NB-IoT and base station	The model is computational complex when implemented in NB-IoT systems

3.1.1. Cell Acquisition and Synchronization

NB-IoT UE goes through the same process as LTE UE where to camp on a cell, it goes through frequency and timing synchronization to obtain the center carrier frequency as well as the allocated slot and frame timing used for the cell acquisition. In general, if MIB and SIB are properly decoded, cell ID, a subframe number, scheduling information, and system bandwidth can be detected successfully. In NB-IoT, the low complexity of devices may lead to poor synchronization and cell acquisition performance, especially due to carrier frequency offsets and poor channel estimation capacity. The following are the papers that have proposed different solutions to optimize the initial cell acquisition and initial synchronization procedure.

In [49], the authors presented a Maximum-Likelihood (ML) NPSS detector which is based on frequency domain cross-correlation metrics by using an overlap-save method. Their method achieves an average timing synchronization latency of 140 ms for the in-band deployed mode with SNR of −12.6 dB. Their proposed method showed a 34% reduction of the energy that is required for NPSS detection. However, their work showed only how much energy could be reduced with respect to the autocorrelation NPSS detection methods. It could be better to show how much of the total device's energy is consumed by their proposed computationally complex detector, i.e., it could be more realistic to include analysis in terms of reduction with respect to the energy consumed during time synchronization but also in terms of energy optimization over the total device consumption.

In [50], the authors presented an algorithm for initial synchronization and cell search. The proposed algorithm uses NPSS for timing acquisition and initial Carrier Frequency Offset (CFO) estimation called the two-stage time domain NPSS correlation. They also used NSSS sequences for the cell ID and frame timing. Their proposed algorithm showed that under extremely low SNR and different fading conditions, NB-IoT could provide the required performance and could also quickly camp on the cell, if any. However, practical experiments are still needed to prove the feasibility of these simulations especially on how the newly introduced NB-IoT power class and actual channel variations could have an impact on the detected SNR at the base station.

In [51], the authors presented an NB-IoT framework by using an advanced signal waveform called non-orthogonal spectral efficient frequency division multiplexing (SEFDM). This waveform uses less bandwidth as compared to OFDM waveform. The designed signal could improve the data rate without the need for more bandwidth. At the base station, the minimum Euclidian norm search detector is used for better error correction. The simulation results reveal that the proposed advanced signal waveform could achieve 25% improvement on data rate as compared to the OFDM signal waveform. The work also proposed an overlapped sphere decoding (OSD) detector which reduces the computation complexity as compared to the single sphere decoding detector while guaranteeing the needed performance. However, the model does not explain the impact of CFO due to the non-orthogonality of the subcarriers on the received signal.

In [52], the work investigated the downlink synchronization signal design and proposed the novel general synchronization signal structure with a couple of Zadoff-Chu (ZC) conjugated sequences in order to remove the potential timing errors caused by large frequency offsets. Their new synchronization signal structure demonstrated better functionality with the frequency offset tolerance of up to 40 kHz. However, the model does not explain the number of samples per symbol involved in synchronization operation and decision.

In [53], the random access preamble is discussed based on the design of NPRACH for single-tone frequency hopping only. It introduces the new single-tone frequency hopping random access signal used by NPRACH in NB-IoT systems. It further explains the design rationale and proposes some possible receiver algorithms for NPRACH detection and ToA estimation. The simulation results show that NPRACH performance is improved. However, the paper did not discuss the impact of massive interference which may result in lower received Signal Interference plus Noise Ratio (SINR) at the base station such that the lower the SINR the lower the detection probability of NPRACH. In addition to that the higher the number of devices the higher probability of NPRACH preamble collision. So lower SINR detected at base station and higher collision probability both affect NPRACH detection negatively.

In [54], the authors described a NPRACH design as specified in 3GPP in standard in Release 13. They proposed a receiver algorithm for the NPRACH timing advance estimation as well as detection. The simulation results for the NPRACH detection shows that if one preamble sequence is transmitted, the detection threshold should be set between 55% to 70% of the average value to satisfy the desired NPRACH performance at the lowest SNR. The results also showed that at 5 and 11 preamble sequence transmission, the detection threshold should be 50% and 35% of the average value, respectively. It is noted that increasing the detection threshold lowers the false alarm probability, which leads to an increased likelihood of misdetection.

In [55], the authors provided a mathematical model of an NB-IoT network in order to predict the optimum performance with a specific configuration of some design parameters (i.e., repetition, number of the preamble in NPRACH per second, coverage classes and intersite distance). The paper analyzes the effects of parameter choice in outdoor, indoor, and deep indoor. The work finally proposes how to choose the optimal configuration i.e., by providing the highest throughput, as well as success probability higher than minimal success probability with minimal one being of 90%. The work showed that even though the success probability has a maximum limit, it can still be altered by modifying the number of repetitions to enhance the coverage or the system capacity in terms of throughput.

In [56], the authors presented the NB-IoT frequency diversity (FD) reception for NPSS as well as NSSS. In the reception mode, the NB-IoT UE alternatively receives the NPSS and NSSS in time domain radio frame by switching the received signals transmitted in different resource blocks in the frequency domain. Their simulation results show that using the proposed FD reception could improve the detection probability by 16% more than without applying the frequency diversity. Additionally, using FD with precoding vector switching (PVC) transmit diversity, achieves 90% of physical cell ID detection (PCID) probabilities at the average SNR of 0 dB with maximum carrier offset of 70 kHz. The method also achieves 97% of PCID detection probability without consideration of frequency carrier offset.

3.1.2. Random Access Procedure

Like in LTE, NB-IoT random access (RA) is intended for initial UE uplink synchronization through which the UE acquires its unique UE ID used for communication with the base station. RA is also used to regain the lost UE access due to the long state of inactivity which has led to the loss in uplink synchronization. In NB-IoT, RA faces several challenges as seen on the research discussions; some solutions to improve the RA performance have been proposed as described in what follows.

In [57], the authors presented the random access procedure (RAP) model and analyzed the system performance by taking into consideration the configurable signal propagation model, a number of supported users per cell, and the RAP configuration parameters. The paper used the contention-based random access with Msg3 collisions instead of Msg1 collision (as multipath transmission) for the random access procedure. The proposed model results show the impact of the parameters (Msg3 transmission mode, Msg3 modulation and coding scheme (MCS), power control schemes and power ramping step) in the single-tone and multi-tone transmissions Bit Error Rate (BER) performance. The results are presented in terms of the total number of preamble transmission success, preamble retransmissions and lost preamble attempts. The work concludes that Msg3 must be considered in the random access procedure analysis, the transmission mode as well as MCS, and for better system performance and fairness distribution of UEs in the cell, it is better to configure power control correctly.

In [58], the authors analyze NB-IoT transmission delay as well as mathematical evaluation of the probability of success for the random access procedure preamble transmission. The analysis is based on three scenarios; scenario one uses minimum values of parameters, scenario two uses the intermediate values, and scenario three uses maximum parameter values. The used parameters are NPRACH periodicity, start time, number of repetitions, number of preamble attempts, and random access response window size. The average delay analysis was performed such that k preamble sequences are mapped in n subcarriers. The preamble collision occurs when multiple UEs send preamble sequences in the same subcarrier. A successful preamble attempt occurs when only one UE sends the preamble to a given subcarrier.

In [59], the authors investigated a random access optimization algorithm and summarized the NPRACH feature and hence designed random access with differentiated barring (RADB) for NB-IoT system. It is observed that the RADB could solve the preamble request conflict caused by massive NB-IoT UEs and hence provide reliable random access for latency-sensitive devices. However, the authors did not consider the problems of channel resource distribution and resource use rate.

In [60], the authors designed a new frequency hopping pattern of NPRACH preamble which uses all feasible hopping distances for a given number of subcarriers. It is seen that their proposed pattern was compatible with standards that is keeping the same NPRACH structure with only very small changes (hopping in the standard is allowed only between the subcarriers of the same resource group). Their simulation where they adopted their first traffic model which deploys 3000 devices, 48 ms and 40 ms of NPRACH preamble and periodicity, respectively, show that the proposed hopping pattern could improve the ToA estimation without additional system overhead.

3.1.3. Channel Estimation and Error Correction

Like in LTE systems, NB-IoT system performance depends to some extent on the quality of the channel estimation. However, for NB-IoT systems massive deployment, the poor quality of channel estimates is highly influenced by the low complexity of the UEs that can lead to misdetection of some signals, frequency offset, phase noise, passive intermodulation (PIM) on the device level, etc. To address the challenges that affect the channel estimation as well as to improve the quality of error correction to ensure the required performance with the low complexity, several works have proposed some solutions, as summarized in the following paragraphs.

In [61], the authors presented an NPSS detection method whose timing metric is composed of symbol-wise autocorrelation and a dedicated normalization factor in an in-band downlink NB-IoT system. The authors proposed a novel low-power algorithm for frequency tracking by the use of more pilot signals as compared to the LTE system. Their algorithm is implemented to compensate for the accumulated frequency offset during the NB-IoT transmission of NB-IoT. Their proposed frequency tracking algorithm delivers high estimation efficiency in terms of Minimum Mean Square Error (MMSE), the probability of correct cell acquisition, etc. However, their study did not elaborate on what could be the impact of mobility and inter-RACT support in the cell search procedure for NB-IoT.

In [62], the authors presented a practical coverage test on the ocean; it is shown that the proposed solution (where the base station decides whether the compensated round trip delay is short or long enough to decode the preamble sequence) was done by considering NPRACH design and hence the authors proposed their solution which considers the TA adjustment. Their proposed solution proved that NB-IoT coverage could reach as far as 35 km. However, the paper does not elaborate on the solution feasibility in environments without line of sight.

In [63], the work provided optimization cases for NB-IoT downlink in terms of MCS. The work also provided the optimization cases of coverage level (CL) by taking into consideration the RACH success rate with different driving speeds of NB-IoT devices in a commercially deployed network. Their results show that the base station paging success rate is decreased as the adjacent cell interference increases. However, the decrease in MCS improves paging performance. Coverage level 0 is the best choice for NB-IoT use cases that involve mobility, whereas coverage level 1 and 2 are mostly for fixed location NB-IoT use cases.

In [64], the author presented an iterative algorithm for NB-IoT transmission procedure. The simulation results in terms of BER and blocks error rate (BLER) show that by use of concatenated error correcting codes or cryptographic redundancy and error correcting code, the algorithm improves the NB-IoT coverage and reduces the overall NB-IoT power consumption. The modification of additional correction of low reliable bits could demonstrate the error correction of the damaged messages by the noisy transmission and hence can reduce the repetition number. However, this work did not discuss how effective the algorithm is when taking into consideration different channel conditions, payload sizes, as well as different repetition numbers with respect to device signal quality.

In [67], the authors considered the presence of random phase noise of the received signals mainly caused by oscillators impairments in both the transmitting and receiving sides and how to lower the mean square error (MSE) estimates. They presented the sequential MMSE channel estimation method that could be implemented in NB-IoT systems. Their model shows that if random phase noise is considered during channel estimation, it is possible to improve the detected SNR by up to 1 dB. However, the model is assumed to be uniformly distributed hence does not present the real-time channel which is randomly changing over time.

3.1.4. Co-Channel Interference

NB-IoT being deployed in the existing LTE spectrum, co-channel interference may occur between NB-IoT and LTE UEs. This is due to several reasons such as sampling rate mismatch, inter-PRB interference due to power leaking between NB-IoT and LTE PRBs, etc. To mitigate the impact of co-channel interference in the NB-IoT/LTE coexistence scenario, the following works have addressed the problems and proposed potential solutions.

In [65], the authors proposed the design guidance for channel equalization that can be used for 5G networks. The proposal set some assumptions such that the currently most used algorithms in cyclic prefix—OFDM system for pilot design, channel estimation, equalization, synchronization, and system performance analysis may no longer be applicable to NB-IoT systems. Their mathematical modeling demonstrated that channel equalization coefficients for NB-IoT UE are a set of phase-shifted CFR combination and not a simple Fourier Transforms of the channel impulse responses. This is the consequence of sampling rate mismatch between NB-IoT user and base station.

In [66], the authors established a comprehensive system model for in-band and guard-band NB-IoT by considering sample duration. They derived the mathematical expressions of received LTE and NB-IoT signals and analyzed the close-form interference power on LTE signal from adjacent NB-IoT signal. It is observed that the sample duration of NB-IoT significantly impacts the desired signal as well as interference on LTE UE; this is due to mismatched sampling rate between NB-IoT UE and the base station. Their proposed system model and derivations match the simulations, hence can be used for coexistence analysis for NB-IoT system.

Summary: This subsection has addressed the state of the art of NB-IoT PHY layer protocol. The main focus was set on different approaches to improve cell acquisition process, random access process, channel estimation, and interference mitigation. The next subsection focuses on MAC layer features by addressing the corresponding challenges and potential solutions.

3.2. Media Access Control Layer

Handling retransmissions (HARQ), multiplexing, random access, timing advance, choice of transport block formats, priority management, and scheduling are the tasks executed by the MAC layer. The discussion on this part focuses on features such as radio resource management, link adaptation, coverage, and capacity improvement, power, and energy consumption reduction, as summarized in Table 3.

Table 3. Articles on proposed MAC layer enhancement techniques

Feature	Article	Technique Used	Enhancement	Limitation
Resource allocation	[68]	Resource blanking	Interference cancellation by resource blanking	The proposed technique may lead to performance degradation in terms of spectral efficiency, especially for NB-IoT massive deployment.
	[69]	Iterative algorithm by a cooperative approach	Radio resource management in terms of scheduling index, repetition number and interference	The proposed solution is sub-optimal hence it does not provide maximum achievable performance in terms of maximum rate and capacity
	[70]	Scheduling algorithm	Efficient resource allocation by reducing the NPDCCH periods	Mobility is not considered and reducing NPDCCH period could lower the channel estimation quality hence may degrade the performance by unrealistic channel estimation
	[71]	Resource allocation technique by extending the specific PRB for paging traffic offload	power consumption reduction for NB-IoT UE during paging loading and offloading	The use of specific PRB for paging offloading is not an efficient use of the existing resource blocks. Also, the model is not applicable in standalone mode.
	[72]	NB-IoT scheduling algorithm	Interference analysis for 15 kHz LTE coexistence with 3.75 kHz guard-band NB-IoT	Emptying the LTE resource is not efficient resource use. Also, the model is not applicable for the standalone mode of deployment
Link adaptation	[73]	NB-IoT basic scheduler algorithm	Optimal resource usage by considering an average device delay and processing time	The scheduler did not consider semi-persistent scheduling, especially for inter-RAT networks
	[31]	Offset index selection and UE specific and common search spaces for NB-IoT dense networks	Cell capacity enhancement by means of optimal scheduling	Did not consider the number of sessions that each device has to transmit with respect to different requirements and use cases
	[74]	Link adaptation algorithm by using the mathematical expression of Shannon theorem	Coverage enhancement by characterizing SNR, repetition number and NB-IoT supported bandwidth	The work did not consider the impact of channel state information on UE link adaptation
	[75]	Two-dimensional NB-IoT dynamic link adaptation algorithm	Optimization of repetition number by dynamically adjusting MCS to ensure better BLER and BER performance	the model does not encompass the effect of speed and the deployment of the optional HARQ process to ensure better channel modeling

Table 3. Cont.

Feature	Article	Technique Used	Enhancement	Limitation
Coverage and capacity	[11]		NB-IoT coverage comparisons in different scenarios for 15 kHz and 3.75 kHz spacing	The channel estimation impairments, carrier offset as well as mobility with respect to different configurations are not considered for the claimed 170 dB of achieved MCL of NB-IoT
	[76]	Preconfigured access scheme and the joint spatial and code domain scheme	capacity and spectral efficiency improvement	It can only be applicable in small cell configurations when NB-IoT is deployed in large scale, preconfiguring access for different require
	[77]	Control plane small data transmission scheme	Effective data transmission enhancement by transmitting small packets in RRC connection set up	This scheme may results in NB-IoT signaling overhead due to Radio Resource Control (RRC) connection setup process encompassed with small data
	[10]	UE coverage and capacity simulation measurement based on real operators network parameters	NB-IoT enhanced coverage measurements by the use of real network configuration parameters	Optimal repetition number for NB-IoT devices is not considered, with additional penetration loss, it does not explain the additional repetition requirement to enhance the coverage while guaranteeing the required performance
	[78]	Low Earth Orbit (LEO) satellite to extend NB-IoT coverage	NB-IoT Coverage extension beyond LTE achieved link budget	The work did not consider the impact of repetition number on extended coverage as well as time and frequency synchronization that can lead to sampling rate mismatch as well as carrier frequency offset for low-end NB-IoT modules
Power management	[79]	Practical power measurement	Power consumption analysis for NB-IoT by varying payloads and repetition numbers, I-eDRx and PSM	Using two devices is not representative massive NB-IoT devices in the because different chips have different power consumption depending on the enabled features such as inter-RAT support that can affect the overall device consumption
	[80]	Prediction-based energy-saving algorithm	Reduction of power consumption by reducing the scheduling request procedure	The solution is not optimal because it reduces scheduling request without considering the device requirement with respect to channel parameters
	[81]	Semi-Markov chain for energy evaluation	Energy consumption and delay requirement evaluation for NB-IoT systems by considering the four states, namely power saving mode, idle mode, RACH procedure, and transmission mode	The model does not include the energy consumption during transition between the four mentioned modes and it does not include the impact of repetition on the device power consumption

3.2.1. Radio Resource Allocation

In NB-IoT, resource allocation is the key feature to ensure the expected massive connections in a cell. Tone allocations, PRBs, repetition number options, power configurations, subframes, or time slots, etc. must be optimized to maximize performance with minimum possible resources. Since NB-IoT is intended for low rate, less frequent time insensitive applications but with the required performance metrics, better radio resource management will ensure the optimal resource usage for expected throughput, spectral efficiency, and coverage enhancement.

In [68], the authors discussed the impact of interference for partial deployment of NB-IoT such that if one PRB is used for NB-IoT in some of the cells, that same PRB could be used for LTE in other cells for the in-band mode of NB-IoT operation. In such a deployment, possible co-channel interference may appear between NB-IoT UE and LTE UE. The authors modeled the partial deployment in percentile such that 100%, 75%, 50%, 25% represent the percentage of cells where NB-IoT enabled. Their results were analyzed by means of cumulative density functions (CDF) of respective SINR detected and maximum coupling loss achieved. The work demonstrates possible NB-IoT interference between NB-IoT and another NB-IoT UEs from adjacent cells and between NB-IoT and LTE from the adjacent cell. The simulation is performed for the in-band mode of operation where both NB-IoT and LTE UEs share the same PRB. They proposed the PRB blanking i.e., blanking the resources that are used by NB-IoT to not be used by LTE, not even being used for CRS. Blanking of these resources will omit the interference from LTE UEs and will result in NB-IoT only access to this PRB. However, the paper did not consider the performance degradation due to reduced available radio resources after when resource blanking is applied.

In [69], the authors formulated an analytical model to characterize the maximum achievable data rate, then investigated the impact of intercell interference in a multicell environment (for in-band and standalone scenarios), and finally proposed an iterative algorithm which uses cooperative approach which takes into consideration the overhead of control channels, repetition number, intercell interference as well as time offset. The proposed sub-optimal solution ensured better radio resource allocation, which raised the data rate by 8% and reduced the overall device energy consumption by 17% with respect to the non-cooperative approach.

In [82], the authors presented preliminary results of RSSI and detected SNR by developing a DORM (integrateD cOmpact naRrowband platforM) node which was deployed on a university campus to test its practical feasibility in different indoor scenarios. Their SNR and RSSI values were observed to be in the range of 18 dB to 23 dB and −65 dBm to −70 dBm, respectively, which shows its suitability for indoor coverage. The RSSI and SNR values variations are considered to be due to different elevations that the nodes are, with respect to the serving base station. However, the paper does not explain the channel estimation and measurements quality and their impact on the achievable throughput, moreover their paper does not cover the outdoor deployment and the impact of repetition on the overall devices' energy consumption when devices are located in different indoor environments.

In [70], the authors introduced the NB-IoT radio access strategy in detail and studied the NB-IoT scheduling problem. Their primary objective is to lower the number of used radio resource while each device's data requirement can still be satisfied. They furthermore formulated the NB-IoT scheduling problem and proposed an efficient algorithm to overcome such a problem. Their simulation results show that they could minimize the number of NPDCCH periods (NPs) used to satisfy each device's data requirement.

However, the repetition number is given according to the distance between the base station and the device. It could be better to use real-time channel parameters or MCS or BLER value to schedule the respective downlink channels to the devices. This is because, within the same distances, devices may experience different signal attenuation due to different factors such as fading, non-line of sight, line of sight, indoor placement, outdoor placement, underground placement, etc. Therefore, there is still an open space for practical deployment to analyze the effectiveness of the different downlink and

NB-IoT scheduling schemes which considers the device's simplicity, modulation schemes, channel conditions, and delay requirement for specific use cases.

In [71], the paper proposed a new resource allocation technique by extending the paging resource that will be specific for paging traffic offload. The authors noted that the new paging PRB could lower power consumption which is mostly used to load and offload the paging load. Also, the work proposed the selection scheme based on UE identity (ID) that is used to balance the load between the paging resource blocks. The simulation results show that power consumption reduction and resource optimal usage are of 80% and 30.5%, respectively. This work considered adding other PRBs for paging monitoring; however, the authors do not demonstrate the trade-off between the newly introduced scheme and the UE complexity requirements.

In [83], the authors proposed an enhanced access reservation protocol (ARP) that allows the device to transmit a fraction of a preamble sequence by providing an analytical model that captures the performance of ARP in terms of the false alarm, misdetection, and collision probabilities. They mathematically analyze the trade-off between the misdetection and the collision probabilities. The drawback of this protocol is that with massive NB-IoT deployment, altering the configuration of the protocol may result in detection performance degradation which can lead to huge packet loss.

In [72], the authors analyzed the impact of interference when the 15 kHz LTE system coexists with a guard-band NB-IoT system with 3.75 kHz subcarrier separation. Their simulation results demonstrated that it is desirable that the scheduler of the LTE system empties the neighboring RBs of the NB-IoT system and allocates resources if possible. The authors then proposed an NB-IoT scheduling method for the LTE system to improve the performance of the studied NB-IoT system. Their results showed that if emptying is not done, at 10^3 BER there is 1 dB drop of SNR as compared to when emptying of RBs is done.

3.2.2. Link Adaptation

Like in LTE, NB-IoT link adaptation involves adaptive modulation and coding schemes as well as adaptive power allocation. However, the modulation schemes are limited to QPSK to enable low complexity and hence reduce the overall power consumption. To extend the coverage and increase the link reliability, a repetition number of up to 128 times is introduced. In the literature, it is seen that NB-IoT link adaptation has several issues; potential solutions are also proposed, as summarized below.

In [31], the author formulated the scheduling issue such that the resource assignment must be in a specific format taking into an account reserved signaling resources and capabilities of the NB-IoT UE. They proposed a solution that incorporates two parameters which are (i) offset index selection (k0) and (ii) UE specific and common search space configuration. The offset index selection was chosen because with the limited k0 and varying size of payloads, it is critical to adapt the scheduling process for high resource use to accommodate more devices at the same time. Additionally, UE specific and common search space configuration were chosen because it decides the timing of NPDCCH and NPDSCH for different UEs, hence it can consequently improve the overall scheduling efficiency.

In [73], the authors presented a basic NB-IoT scheduler for NB-IoT system and analyzed the enhancements on average delay, optimal resource usage, and processing time. The proposed algorithm demonstrated that shorter NPDCCH period selection may reduce the UE average delay and optimize the overall system resource usage. Also, the model shows that the scheduling delay (k0) should be determined before the allocation of subcarriers. However, the model does not elaborate the type of configuration used since the choice of configuration such as single tone or multi-tones have a direct impact on periodicity and transmission delay and hence can directly impact the system performance.

In [74], the authors analyzed NB-IoT repetition number and bandwidth allocation and proposed analytic expressions based on SNR, bandwidth, and energy per bit that can be derived from Shannon theorem in order to characterize the impact of the repetition number as well as bandwidth allocation to different UEs. Additionally, their work proposed an algorithm for link adaptation. The algorithm exploits resource unit number, repetition as well as bandwidth. Their results show that reducing bandwidth and performing repetitions could enhance the coverage. However, the work did not consider the actual impact of channel parameters as well as NB-IoT UE impairments such as CFO which may lead to transmission errors.

In [75], the authors proposed a new NB-IoT link adaptation scheme with the consideration of the repetition factor. They claim that their proposed two-dimensional scheme is composed of Inner Loop Link Adaptation that copes with BLER by periodically adjusting the repetition number and outer loop link adaptation which coordinates the MCS and repetition number. This is because 20 dB coverage enhancement beyond LTE can be achieved by the repetition of transmitted data. So, in this work, they proposed an algorithm that dynamically chooses MCS and repetition number based on estimated real-time channel state information (CSI). However, their algorithm does not elaborate on the different NB-IoT power classes and to which range their respective coverage could be enhanced.

3.2.3. Coverage and Capacity

NB-IoT support for extended coverage of up to 164 dB of maximum coupling loss is to enable the technology to be used for cellular IoT services, especially for applications that are in hard-to-reach areas. Its narrow bandwidth and support for repetition are the key features to enable the enhanced coverage.

In [10], the authors simulated and analyzed the NB-IoT wide-area rural deployment and deep indoor urban deployment by using the network parameters of one metropolitan operator. Their work showed that NB-IoT devices could still transmit and receive data at an MCL of 167dB, which is 3 dB higher than the 3GPP's 164 dB of MCL limit set. Furthermore, in different indoor scenarios, even with an addition of 30 dB as penetration loss, NB-IoT had better outage probabilities as compare to another LTE LPWAN technology (eMTC). For outdoor and light indoor conditions with an additional 10 dB penetration loss and an average intersite distance of 2.8 km, NB-IoT had less than 0.1% of outage probability. However, despite the varying additional penetration losses of 10 dB, 20 dB and 30 dB, their simulation does not consider the impact of features such as mobility, CFO, lower power class on the achieved MCL.

In [11], the authors showed that for the maximum number of repetitions (128 times), with 15 kHz and 3.75 kHz subcarrier spacing, coverage of up to 170.2 dB MCL and 174.2 dB MCL could be achieved, respectively. The work concluded that the evaluations show that NB-IoT could provide up to 20 dB coverage enhancement in various deployment scenarios as compared to legacy LTE. Similarly, the work did not study the impact of mobility and weak channel estimation quality to the achieved MCL.

In [76], the authors proposed two less complex scheduling schemes (compared to brute-force) that can be used NB-IoT. The first proposed scheme is called the preconfigured access scheme and the second is the joint spatial and code domain scheme. Their simulation performance results (spectrum efficiency, number of active devices as well as low collision rate achieved) by the two low complex schemes were found better when compared to the ones that can be achieved by the brute-force scheme.

In [78], the authors proposed a specific unidirectional system to study the coverage enhancement by using satellite network i.e., LEO constellation. Their proposed model with the mathematical derivations shows that NB-IoT could achieve the 20 dB more than LTE achieved MCL and could still operate according to Release 13 standards. From their results, it is seen that the packet error rate (PER) of the transmitted signal is distorted by Doppler spread. However, the model does not consider the clock synchronization between NB-IoT device and satellite, which can lead to performance degradation especially caused by the CFO or the sampling rate mismatch. Additionally, the work did not consider the maximum achievable throughput when their system is employed to comment on the effectiveness of the techniques as compared to terrestrial NB-IoT deployment.

3.2.4. Power and Energy Management

The NB-IoT reduced complexity is intended to reduce the power consumption in different modes. PSM and eDRX are the implemented features dedicated to foster the long-lasting battery life.

In [79], the authors presented the NB-IoT power measurement. Their measurements were set in such a way that NB-IoT transmission consumes 716 mW when at 23 dBm with a power efficiency of 37%. DL control and data signals consume 213 mW, idle-mode-eDRx and PSM consumes 21 mW and 13 µW, respectively. In general, according to their empirical measurements, it is shown that the power consumption is 10% lower than the 3GPP estimates. During measurements, parameters such as time domain repetition, I-eDRx, and PSM were taken into consideration. To characterize each component in the proposed model, several test cases such as Tx power, UL, and DL data rates, I-eDRx, and PSM were used and all parameters except one at a time were fixed. Their results showed that the NB-IoT devices power consumption is independent of the subcarrier spacing. However, the total ON time of the devices is in many cases defining the overall battery life. As their remark, the data rates do not directly impact the power consumption, but it has a major direct impact because it defines the overall device ON time. If the transmitting interval is 1 h, the device achieves only 2.5 weeks of battery life. Increasing the duration to 24 h, the lifetime of the device increases to 12.8 years in PSM.

In [80], the authors proposed a prediction-based energy-saving mechanism to reduce energy consumption by decreasing the number of scheduling request procedures. Their proposed scheme showed that it could reduce the NB-IoT active time from 5% to 16% for the medium and bad channel quality and achieve from 10% to 34% battery saving in different scenarios as compared to 3GPP consumption simulation specifications in [43].

In [81], the authors developed a semi-Markov chain with power saving mode, idle mode, random access, and transmission mode to study the energy requirement and delay performance for NB-IoT. It is noted that for massive synchronous connections, extra power is drained in random access and transmission states due to collisions. The paper further proposes an energy optimization model based on a priori method that takes into consideration the PSM duration as well as power consumption. The results demonstrate that for optimal energy and delay requirement, it is important to set the higher RACH transmission number to accommodate more delay on the UE. However, their optimization model did not consider the power consumption during the transition of different states, because when the UE is required to perform several sessions per day, it might go through several transitions that have a significant effect on power consumption. Furthermore, the mode does not include the small data transmission scheme during RRC connection as proposed in the updated standards. However, with the introduction of the new power class in NB-IoT Release 14, there is a need for practical experiments to evaluate the new coverage classes. With lower transmit power, the SNR detected at the base station becomes lower hence the device will need to perform more repetitions to enhance coverage.

3.3. Upper Layers

Although the focus of this paper is mainly on the features regarding PHY an MAC layers, it is still imperative to address some enhancements, challenges, and potential solutions to the upper layers. Especially the changes that are implemented in Evolved Packet Core (EPC) by adding the Service Capability Exposure Function (SCEF) to manage both IP and non-IP data packets [84].

Control and User Plane Optimization

To support massive end-to-end device connectivity with extremely low complexity and reduce the transmission signaling, NB-IoT implements new small data transmission procedures based on Cellular IoT (CIoT) Evolved Packet System on both Control Plane (CP) and User Plane (UP). These transmission procedures support small bursts of data efficiently while guaranteeing the long-range coverage as compared to legacy GPRS [85,86]. In this regard, NB-IoT is can support more than one data path in CP for the transmission of user data which is carried by the signaling messages managed by the Mobile Mobility Entity (MME) as shown in Figure 5. The procedures are optimized to efficiently support the small data transfer as follows:

- Mandatory CP CIoT EPS;
- Optional UP CIoT EPS.

CP CIoT EPS optimization encapsulates the data packets in Non-Access Stratum (NAS) by using control plane signaling messages. In this regard, this procedure is mandatory. Compared to the conventional SR procedure, the NB-IoT UE skips some steps required for each data transfer hence this optimization procedure best fits the short data transmission or reception.

On the other hand, UP CIoT EPS optimization requires the RRC connected mode to get the scheduled radio resources as well as Access Stratum (AS) between the UE and the network. This mode uses the newly introduced connection to Suspend and Resume procedures. Connection suspend procedure helps to retain the network context so that the UE can resume the connection when traffic is available. Retaining the context helps the UE and the network to skip the AS and RRC reconfiguration in each data transfer. Since it uses user plane, the UP CIoT EPS is suitable for both small and large transactions.

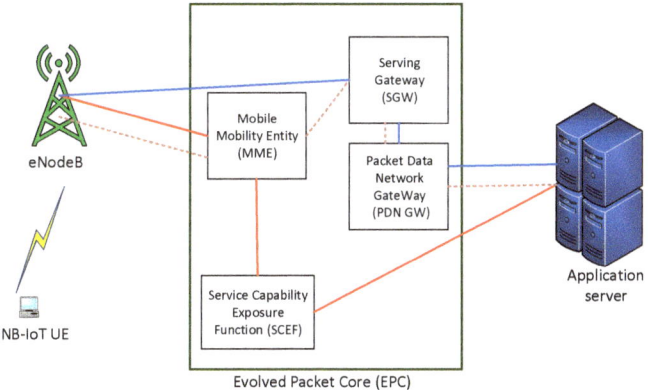

Figure 5. Representation of NB-IoT IP and Non-IP data path: Blue line displays the IP data path in UP mode (as Legacy LTE), Red line displays the non-IP data path in CP mode, and dashed-line displays the IP data path in CP mode.

Furthermore, the UE in Service Request procedure (an LTE procedure used by the UE and base station to transmit or receive data in RRC idle state) is required to be in a connected state in order for base station to allocate the radio resources. For NB-IoT this SR is optional; however, NB-IoT UE that supports UP optimization needs also to support SR. For example; if the NB-IoT UE wants to transmit the uplink data in idle state, it will send the random access preamble through which the base station and UE will establish RRC connection and UE will be allocated with the radio resources for data transfer. After a certain period of inactivity, the base station initiates the release procedure.

Similarly, for UE downlink data reception, if the UE is in DRX mode, the UE regularly listens to downlink signaling and if the UE notices the paging message, it will perform the SR procedure as described in uplink data transmission. Additionally, if the UE is in PSM mode, it will be completely inaccessible until it initiates the same SR procedure for the uplink grant or by using Tracking Area Update (TAU).

There are works that are addressing the upper layers such as [77], where the authors proposed an efficient small data transmission scheme by using CP procedure. The proposed scheme enables the devices to transmit data packets through the RRC connection setup procedure when the device is in idle mode. This process reduces the signaling overhead caused by the security setup process and data radio bearer setup process. However, a suggestion could be to analyze the power consumption during this small data transmission and compare its effectiveness to when the same data is transmitted during the UP procedure.

Summary: This section discussed PHY layer features, highlighting the corresponding enhancements on cell acquisition procedure, random access channel estimation, and interference mitigation. It then addressed the MAC layer enhancements regarding resource allocation, link adaptation, coverage and capacity, and power management. It further addressed the upper layers changes related to cellular IoT evolved packet system optimization through user and control planes to enhance the small data packets transmissions for end-to-end massive connectivity.

4. Narrowband-IoT Possible Deployment Strategies

This section proposes potential deployment strategies for NB-IoT massive deployments by considering the NB-IoT support for small cells in heterogeneous network scenarios.

HetNets are effective network deployment strategies in which small cells are incorporated in macrocells with the objective of improving performance in terms of capacity, coverage, and spectral efficiency. In general, the macrocells are characterized by higher transmit power and broader range as compared to small cells. When smalls cells are overlaid in macrocells, interference becomes a concern, especially to small cell edge users. Several techniques for interference cancellation, estimation and coordination that involve frequency hopping, frequency reuse, power control etc. have been proposed; however, the performance trade-offs for the proposed techniques for macrocells and small cells are still challenging [87–90].

Similarly, NB-IoT is expected to coexist with the currently deployed legacy LTE as well as the forthcoming 5G networks. This questions the existing interference management techniques i.e., are they applicable to the newly deployed technology since NB-IoT is expected to support different power classes while maintaining the low complexity which can severely affect the channel estimation quality and hence interference estimation quality. In NB-IoT coexistence with the legacy cellular networks, the possible deployment scenarios are as follows:

- Synchronous NB-IoT deployment in all small cells;
- Asynchronous NB-IoT deployment in all small cells;
- Synchronous NB-IoT deployment in small cells and Macrocells;
- Asynchronous NB-IoT deployment in small cells and LTE in macrocells.

These scenarios, as shown in Figure 6, are detailed in what follows.

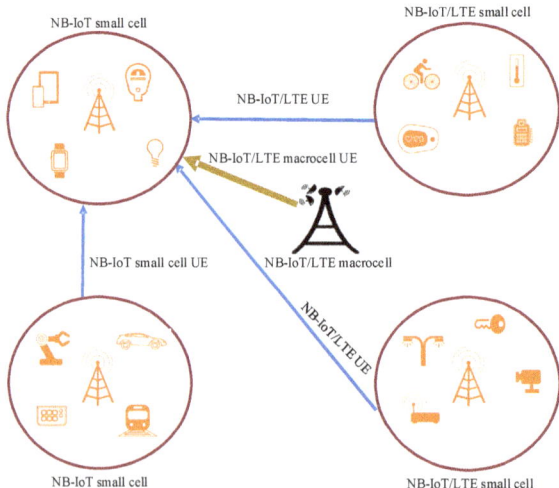

Figure 6. Summary of NB-IoT deployment strategies. For example, when NB-IoT is deployed in macrocell and LTE in small cell, when LTE is in macrocell and NB-IoT is in small cells, when NB-IoT is in macrocell and small cells support both NB-IoT and LTE, and when LTE is in macrocell and LTE/NB-IoT is in small cells

4.1. Synchronous NB-IoT Deployment in All Small Cells

This is the NB-IoT deployment strategy which is enabled in all the small cells by using the same physical resource blocks. All the small cells are synchronized in such a way that with the same PRBs, all the NB-IoT UEs are using the transmit power that is configured regardless of its maximum transmitting power capacity. This means that even though NB-IoT devices might support different power classes such as 14 dBm, 20 dBm, or 23 dBm, the NB-IoT devices will only be configured to use the minimum allowed transmit power in order to avoid causing the co-channel interference to other UEs using the same radio resources.In this strategy, power control may be the key feature to ensure the required performance. However, cell edge UEs may still suffer from the interference problem. This interference may highly be increased due to the low channel estimation quality of NB-IoT UEs associated by its reduced computational complexity.

4.2. Asynchronous NB-IoT Deployment in All Small Cells

This deployment strategy is employed in such a way that NB-IoT is enabled in all small cells by using different physical resource blocks. This implementation may avoid the interference between NB-IoT UEs from different small cells; however, this may result in co-channel interference between NB-IoT and LTE UEs that are using the same radio resources. When deploying under this strategy, it is imperative to implement proper frequency planning as well as proper power configuration for NB-IoT devices. As seen from the state of the art, some works have proposed blanking of the radio resources to the adjacent cells for the resources that are already occupied by NB-IoT even in the cells that NB-IoT is not enabled. However, blanking of the resources is a wastage of resources, so, there should be some other means such as frequency hopping to avoid wastage of resources (blanking) as well as to mitigate interference.

4.3. Synchronous NB-IoT Deployment in Small Cells and Macro Cells

In this strategy, NB-IoT is enabled in the small cells as well as in macro cell on the same PRBs. Macrocell UEs are configured to use higher transmit power as compared to small cell UEs while keeping the same PRBs for NB-IoT while others left for legacy LTE. Possible co-channel interference may occur in small cell edge UEs if the UEs are scheduled on the same resource units. The impact may further increase for UEs under mobility which might require the use of handover for smoothing the UEs transition from one serving cell to another. From our review, no work has addressed the interference cancellation mechanism for such a case. It is imperative to employ the existing geographical planning, frequency reuse, frequency hopping, and power control while considering the low complexity but high coverage range NB-IoT.

4.4. Asynchronous NB-IoT Deployment in Small Cells and LTE in Macrocells

In this strategy, NB-IoT uses separate PRBs between small cells and macro cells. This means that one or more PRBs are used for small cells and different PRB(s) for the macrocells. If the PRBs are not well planned, NB-IoT users from adjacent cells (using the same resource units) may suffer from interference. Also, LTE users that are using the same resource elements may interfere with small cell or macro cell UEs. Different transmit power control configurations may be used to control interference.

The choice of the deployment strategy depends on several factors such as use-case requirements, environmental conditions, equipment quality, etc. It is imperative to implement better interference estimation, mitigation or management techniques that will ensure better performance and spectral efficiency for the massive NB-IoT deployment in coexistence with other technologies.

Summary: This section has presented the possible NB-IoT deployment strategies by considering the NB-IoT support for small cells in coexistence with legacy LTE in HetNet scenario.

The following section presents the open research challenges to motivate future research directions.

5. Open Research Questions and Discussion

5.1. Battery Life

PSM and eDRx were introduced in NB-IoT Release 12 and 13 to lengthen the NB-IoT devices' battery life. Moreover, the most recent updates require the UE to be able to transmit during RRC-idle mode which will reduce the required ON time for data transmission. However, devices experiencing bad channel conditions due to hard-to-reach areas will require to perform several retransmissions per session, which will drain the device's energy and hence shortens the battery life. Similarly, devices that require a relatively large number of reporting sessions per day will consume more energy, which makes energy management a concern. As seen in Section 3, most of the proposed algorithms are power hungry because most of the power is consumed during transmission and reception. Therefore, energy harvesting alternatives such as solar, biogas, vibrations, etc. that will lengthen the NB-IoT device battery life should be introduced to complement or replace frequent battery charging.

5.2. Radio Resource Management

5.2.1. Tones Allocation.

As seen in the literature, most of the articles consider single-tone allocation for the simplicity in the simulation, thus, multi-tone allocation is not well studied. This causes a knowledge gap in the effectiveness of different tone-allocation possibilities. Moreover, for guard-band, in-band and standalone it is still not clear about the respective performance metrics that could be achieved in terms of throughput, coverage range, interference robustness etc. This restricts to a certain extent the optimal choice of deployment for a large number of devices with the required performance. Furthermore, different frame structures, especially for TDD configurations, are not discussed even though NB-IoT is required to support TDD. Therefore, optimal resource use techniques must be proposed that incorporate repetition, mobility, tones allocation, etc. for efficient spectrum usage.

5.2.2. Interference Mitigation

Interference prediction, estimation, cancellation, and coordination techniques for NB-IoT become a challenge. This is because of the sharing of spectrum resources between NB-IoT and legacy LTE. Similarly, with NB-IoT being deployed in a small cell or macrocell scenarios in heterogeneous networks, interference becomes a concern. Several works have tried to address this by means of resource blanking, power control, or better uplink and downlink scheduling schemes and frequency and timing synchronization, etc. However, it is still challenging to incorporate the NB-IoT features such as repetition, low complexity (which affects channel estimation quality), and mobility in deploying the already existing LTE interference management techniques. As seen in the possible NB-IoT deployment scenarios above, there is still a need for deploying effective schemes that will ensure better NB-IoT performance without degrading the LTE performance [91,92].

5.3. Mobility Management

As seen in Section 2, most of the simulation works have ignored the mobility impact of NB-IoT channel modeling. However, for use cases that involve movement, Doppler shift has to be taken into consideration during channel estimation, which might slightly increase the device complexity to support handover and other mobility features such as the support for inter-RAT mobility during idle mode [93,94]. The increase in NB-IoT UEs mobility makes the channel suffer from fast varying channel conditions, due to which adaptive transmission schemes that might involve channel estimation, error correction, etc. must be implemented.

Therefore, applying intelligent/adaptive algorithms that are low power and optimal for repetition number, yet mobility-aware, is of great importance. The algorithms could involve low-power frequent CSI reporting, early data transmission by using both user and control plane in either Msg 3 or Msg 4.

5.4. Latency

NB-IoT latency tolerance is set to 10 ms. This is due to its support for use cases of UEs that are in environments with bad channel conditions [95–97]. Initial cell acquisition, frequency, and timing requirements, RACH transmission, half duplex mode of transmission and several repetitions that are performed during transmission are some of the features that play part in the overall data transmission delay. Several works are trying to reduce the timing requirement so as to reduce transmission latency of devices; however, most of the works have not addressed delay by taking into consideration the massive congestion that is expected for the IoT networks, processing delays due to low complex devices, queuing delays, propagation delays especially with long-range feature, as well as errors and error recovery.

However, early data transmission schemes and the second NB-IoT HARQ process for devices that have good channel conditions are among the features that can be used to reduce the transmission latency and improving the transmission link performance. However, only a handful of research articles have discussed the effectiveness of these processes when applied in NB-IoT.

5.5. Semi-Persistent vs. Dynamic Scheduling

Most of the NB-IoT literature addresses dynamic uplink and downlink scheduling by studying the scheduling of logical channels and signals. There are still very few NB-IoT studies about the effectiveness of Semi-Persistent Scheduling schemes (SPS) even though SPS helps to reduce the NPDCCH overhead as compared to dynamic scheduling. It provides the NB-IoT UEs with longer allocated resources (more than one subframe) so that the NB-IoT device will not need the frequent downlink assignment as well as an uplink grant which is delivered by NPDDCH for each subframe. However, for applications that involve mobility or fast varying channel conditions, how is this scheduling scheme going to be effective knowing that NB-IoT has poor channel estimation capabilities as compared to LTE?

5.6. Random Access

Massive NB-IoT modules that try to request the radio channel resources at the same time for uplink data transmission may suffer from random access preamble collision. This is caused by several factors such as detection inaccuracy that may not satisfy the detection threshold, the high probability of false alarm, etc. Several works have proposed random access preamble detection algorithms (i.e., random access with differential barring etc.) and others have developed mathematical models to characterize the preamble transmissions in order to improve the NPRACH success rate and better time-of-arrival estimation and other NPRACH performance improvements. However, it is still unclear which scheme is effective for massive deployment, since most of the proposed schemes do not consider the heterogeneous network architecture, channel estimation impairments, or realistic channel conditions [98,99].

5.7. Timing Advance (TA)

When the base station responds to NB-IoT UEs about RRC connection request, it incorporates the TA command to be used for NB-IoT UE terminal data uplink transmission timing (i.e., to time-synchronize the UEs to the base station and help to compensate the propagation delays). However, for NB-IoT UE, the TA adjustment accuracy of the signaled timing advance with respect to the prior uplink transmission may highly be affected by the massive number of NB-IoT devices contending for the access. This is because the base station may need to correct some UE timing while for other NB-IoT UEs that had already transmitted NPRACH could receive the random access response which is not intended for them. Some works have addressed the receiver algorithms for NPRACH TA estimation as well as detection timing advance adjustment decoding schemes to improve the estimation but the NB-IoT receiver sensitivity and weak channel estimation quality still negatively affect the TA adjustment.

5.8. Cell Search and Initial Synchronization

NPSS and NSSS are two signals based on frequency domain Zadoff-Chu sequence that are used for NB-IoT time and frequency synchronization to the base station. According to NB-IoT standard, NPSS and NSSS may not be transmitted on the same antenna port hence NB-IoT initial synchronization may rely on NPSS only. The challenge is that the imperfect channel conditions may severely affect the cell camping procedure as a small CFO may result in a phase shift to a received frequency domain sequence which as a consequence may degrade the cell search and synchronization performance. To improve this, frequency diversity techniques should also be used for NPSS and NSSS reception improvement.

5.9. Unified NB-IoT Testing Tool

Since NB-IoT is a promising technology, there should be a unified testing tool used as a reference to verify if the produced products comply with the standards. Taking Bluetooth as an example, for better compatibility towards different available products from handsets to car kits, Profile Tuning Suite (PTS) software is used to automate the compliance testing to specific Bluetooth function. So, to support compliance with standards and hence backward compatibility and interoperability, what is the testing tool to validate if different available products will fit standards? Similarly, for simulation purposes, most of the works choose the parameters that can generate results easily. If there is a concrete simulation model that takes into account the major NB-IoT features and incorporating all the possibilities from repetition number allocation, mobility selection, modulation and coding scheme, real-time channel variations, etc. it would be easier to get realistic modeling for different scenarios.

5.10. Backward Compatibility and Interoperability

A ten-year telecommunication generation is characterized by different changes in releases and updates. In order to reach their lifespan as compared to what the standards stipulate, NB-IoT devices should operate for around ten years with a single battery charge. Whenever new releases or updates are introduced, backward compatibility and interoperability should be possible. Apparently, the device complexity is set as low as possible; will these simple devices (hardware) support hard and robust algorithms that will be implemented by over-the-air upgrades/updates to satisfy the demands of future NB-IoT use cases?

Summary: This section has presented the open research questions regarding battery life, radio resource allocation, cell search, and initial acquisition procedures, mobility management, latency, random access, etc., as summarized in Table 4, in order to motivate future research directions. The next section concludes the paper.

Table 4. Open Research Questions related to the physical layer, MAC layer, and standard.

Physical Layer	MAC Layer	Standard
Radio resource management	Timing advance adjustment	Support for small cell
Frequency and time synchronization	Dynamic scheduling and semi-persistent scheduling	TDD support
Random access	Latency	Antenna diversity
Channel estimation	Power management	Mobility and handover support
Error correction	Network throughput	More efficient group messages
Link adaptation	Control packet overhead	Multicarrier operation
Interference mitigation	Control plane small data transmission	Network management tool for UE differentiation

6. Conclusions

Due to the fact that most of the existing works are segmented and only consider one or two releases in their corresponding studies or simulations, this paper has presented a comprehensive overview of NB-IoT standard from Release 13 to Release 16 prospects to enhance and enable more realistic research. It further presented the detailed current state of the art of NB-IoT based on the ongoing discussion on NB-IoT protocol stack along with the related contributions and analyzed the knowledge gaps by using NB-IoT standard as a benchmark. It is observed that most of the articles focus on improving one or few features while neglecting others,it could be better to display the trade-offs between the improvement feature and the neglected ones, i.e., performance trade-off between PHY and MAC layer when one feature is changed in either of the layers, the impact of repetition on overall energy consumption, CFO on channel estimation quality etc. This paper also presented the NB-IoT deployment strategies to highlight the coexistence possibilities with other legacy technologies i.e., LTE, by considering the NB-IoT support for small cells in HetNet scenarios. Lastly, it discussed the open research challenges and the future common research focus on NB-IoT i.e., battery life, optimal resource usage, handover support during mobility, transmission latency, scheduling, etc. To the best of the author's knowledge, this is the first survey that covers broadly these mentioned contributions and hence this work will help the researchers get most of the needed information to accelerate their research by finding the relevant information and sources for deeper exploration of the research concepts as well as finding possible solutions.

Author Contributions: Conceptualization, C.B.M., H.M., M.M.A., Y.L.M., S.P. and S.M.; investigation, C.B.M., H.M., M.M.A., Y.L.M., S.P. and S.M.; resources, C.B.M., H.M., M.M.A., Y.L.M., S.P. and S.M.; writing—original draft preparation, C.B.M..; writing—review and editing, C.B.M., H.M.; formal analysis, C.B.M., H.M., M.M.A., Y.L.M., S.P. and S.M.; supervision, M.M.A., H.M., Y.L.M.

Funding: This research received no external funding.

Acknowledgments: This project has received funding partly from European Union's Horizon 2020 Research and Innovation Program under Grant 668995 and European Union Regional Development Fund in the framework of the Tallinn University of Technology Development Program 2016–2022. This material reflects only the authors' view and the EC Research Executive Agency is not responsible for any use that may be made of the information it contains.

Conflicts of Interest: The authors declare no conflict of interest.

References

1. Lucero, S. IoT Platforms: Enabling the Internet of Things. 2019. Available online: https://cdn.ihs.com/www/pdf/enabling-IOT.pdf (accessed on 12 January 2019).
2. Salman, L.; Salman, S.; Jahangirian, S.; Abraham, M.; German, F.; Blair, C.; Krenz, P. Energy efficient IoT-based smart home. In Proceedings of the IEEE 3rd World Forum on Internet of Things (WF-IoT), Reston, VA, USA, 12–14 December 2016.
3. Bardyn, J.; Melly, T.; Seller, O.; Sornin, N. IoT: The era of LPWAN is starting now. In Proceedings of the ESSCIRC Conference 2016: 42nd European Solid-State Circuits Conference, Lausanne, Switzerland, 12–15 September 2016; pp. 25–30. [CrossRef]
4. Devalal, S.; Karthikeyan, A. LoRa Technology—An Overview. In Proceedings of the Second International Conference on Electronics, Communication and Aerospace Technology (ICECA), Tamilnadu, India, 29–30 March 2018.
5. Mekki, K.; Bajic, E.; Chaxel, F.; Meyer, F. A comparative study of LPWAN technologies for large-scale IoT deployment. *ICT Express* **2019**, *5*, 1–7. [CrossRef]
6. Lauridsen, M.; Nguyen, H.; Vejlgaard, B.; Kovacs, I.Z.; Mogensen, P.; Sorensen, M. Coverage Comparison of GPRS, NB-IoT, LoRa, and SigFox in a 7800 square km Area. In Proceedings of the IEEE 85th Vehicular Technology Conference (VTC Spring), Sydney, Australia, 4–7 June 2017.
7. Sinha, R.S.; Wei, Y.; Hwang, S.H. A survey on LPWA technology: LoRa and NB-IoT. *ICT Express* **2017**, *3*, 14–21. [CrossRef]

8. Rico-Alvarino, A.; Vajapeyam, M.; Xu, H.; Wang, X.; Blankenship, Y.; Bergman, J.; Tirronen, T.; Yavuz, E. An overview of 3GPP enhancements on machine to machine communications. *IEEE Commun. Mag.* **2016**, *54*, 14–21. [CrossRef]
9. Gozalvez, J. New 3GPP Standard for IoT [Mobile Radio]. *IEEE Veh. Technol. Mag.* **2016**, *11*, 14–20. [CrossRef]
10. Lauridsen, M.; Kovacs, I.Z.; Mogensen, P.; Sorensen, M.; Holst, S. Coverage and Capacity Analysis of LTE-M and NB-IoT in a Rural Area. In Proceedings of the 2016 IEEE 84th Vehicular Technology Conference (VTC-Fall), Montreal, QC, Canada, 18–21 September 2016.
11. Adhikary, A.; Lin, X.; Wang, Y..E. Performance Evaluation of NB-IoT Coverage. In Proceedings of the IEEE 84th Vehicular Technology Conference (VTC-Fall), Montreal, QC, Canada, 18–21 September 2016.
12. Telia. Verdens Storste Iot Pilot. 2018. Available online: https://www.telia.no/magasinet/verdens-storste-iot-pilot/ (accessed on 23 December 2018).
13. U Blox. Press Releases. 2019. Available online: https://www.u-blox.com/en/press-releases/ (accessed on 26 January 2019).
14. Norway, T. Sub-Pump. 2018. Available online: https://www.teliacompany.com/en/news/news-articles/2017/sub-pump/ (accessed on 26 December 2018).
15. T Mobile. Smart Cities. 2019. Available online: https://iot.t-mobile.com/solutions/smart-cities/ (accessed on 10 February 2019).
16. Skyworks. Cellular IoT. 2019. Available online: http://www.skyworksinc.com/Products/1152/Cellular_IoT (accessed on 15 February 2019).
17. Mediatek. NB-IoT. 2019. Available online: https://www.mediatek.com/products/nbIot/mt2625 (accessed on 19 February 2019).
18. Huawei. 2019. Available online: https://e.huawei.com/my/solutions/technical/iot/nb-iot (accessed on 20 February 2019).
19. Quactel. LPWA IoT Module. 2019. Available online: https://www.quectel.com/product/list/LPWAIoTModule.htm (accessed on 15 March 2019).
20. Nordic Semiconductors. 2019. Available online: https://www.nordicsemi.com/News/2018/12/ (accessed on 10 February 2019).
21. Intel. Modem Solutions. 2019. Available online: https://www.intel.com/content/www/us/en/mobile/modem-solutions.html (accessed on 10 February 2019).
22. Sequans. 2019. Available online: http://www.sequans.com/products-solutions/ (accessed on 10 February 2019).
23. Qualcomm. IoT Modem. 2019. Available online: https://www.qualcomm.com/products/mdm9206-iot-modem (accessed on 10 February 2019).
24. Sierrawireless. Products and Solutions. 2019. Available online: https://www.sierrawireless.com/products-and-solutions/embedded-solutions/ (accessed on 10 February 2019).
25. Samsung. Samsung Exynos. 2019. Available online: https://news.samsung.com/global/samsungs-exynos-i-s111 (accessed on 10 February 2019).
26. Altair Semiconductors. 2018. Available online: https://altair-semi.com/ (accessed on 10 February 2019).
27. U blox. Cellular Modules. 2019. Available online: https://www.u-blox.com/en/cellular-modules (accessed on 10 February 2019).
28. Chen, M.; Miao, Y.; Hao, Y.; Hwang, K. Narrow Band Internet of Things. *IEEE Access* **2017**, *5*, 20557–20577. [CrossRef]
29. Hoglund, A.; Lin, X.; Liberg, O.; Behravan, A.; Yavuz, E.A.; Van Der Zee, M.; Sui, Y.; Tirronen, T.; Ratilainen, A.; Eriksson, D. Overview of 3GPP Release 14 Enhanced NB-IoT. *IEEE Netw.* **2017**, *31*, 16–22. [CrossRef]
30. Ratasuk, R.; Mangalvedhe, N.; Xiong, Z.; Robert, M.; Bhatoolaul, D. Enhancements of narrowband IoT in 3GPP Rel-14 and Rel-15. In Proceedings of the IEEE Conference on Standards for Communications and Networking (CSCN), Helsinki, Finland, 18–20 September 2017.
31. Boisguene, R.; Tseng, S.; Huang, C.; Lin, P. A survey on NB-IoT downlink scheduling: Issues and potential solutions. In Proceedings of the 13th International Wireless Communications and Mobile Computing Conference (IWCMC), Valencia, Spain, 26–30 June 2017.
32. Xu, J.; Yao, J.; Wang, L.; Ming, Z.; Wu, K.; Chen, L. Narrowband Internet of Things: Evolutions, Technologies, and Open Issues. *IEEE Internet Things J.* **2018**, *5*, 1449–1462. [CrossRef]

33. Feltrin, L.; Tsoukaneri, G.; Condoluci, M.; Buratti, C.; Mahmoodi, T.; Dohler, M.; Verdone, R. Narrowband IoT: A Survey on Downlink and Uplink Perspectives. *IEEE Wirel. Commun.* **2019**, *26*, 78–86. [CrossRef]
34. Zayas, A.D.; Merino, P. The 3GPP NB-IoT system architecture for the Internet of Things. In Proceedings of the IEEE International Conference on Communications Workshops (ICC Workshops), Paris, France, 21–25 May 2017.
35. ERICSSON. Scope of Release 16-NB-IoT, RP-181187. 2018. Available online: http://www.3gpp.org/ftp/TSG_RAN/TSG_RAN/TSGR_80/Report/ (accessed on 23 December 2018).
36. Huawei. On NB-IoT Evolution in Rel-16, RP-180877. 2018. Available online: http://www.3gpp.org/ftp/TSG_RAN/TSG_RAN/TSGR_80/Report/ (accessed on 23 December 2018).
37. 3GPP. Revision of WI on Enhancements of NB-IoT, RP-161901. 2018. Available online: https://portal.3gpp.org/ngppapp/CreateTdoc.aspx?mode=view&contributionId=730352 (accessed on 23 December 2018).
38. 3GPP. TS 36.300 V13.8.0 (2017-06). 2018. Available online: https://portal.3gpp.org/desktopmodules/Specifications/SpecificationDetails.aspx?specificationId=2430 (accessed on 23 December 2018).
39. Xu, T.; Darwazeh, I. Non-Orthogonal Narrowband Internet of Things: A Design for Saving Bandwidth and Doubling the Number of Connected Devices. *IEEE Internet Things J.* **2018**, *5*, 2120–2129. [CrossRef]
40. Beyene, Y.D.; Jantti, R.; Tirkkonen, O.; Ruttik, K.; Iraji, S.; Larmo, A.; Tirronen, T.; Torsner, J. NB-IoT Technology Overview and Experience from Cloud-RAN Implementation. *IEEE Wirel. Commun.* **2017**, *24*, 26–32. [CrossRef]
41. Ratasuk, R.; Mangalvedhe, N.; Kaikkonen, J.; Robert, M. Data Channel Design and Performance for LTE Narrowband IoT. In Proceedings of the IEEE 84th Vehicular Technology Conference (VTC-Fall), Montreal, QC, Canada, 18–21 September 2016.
42. Ratasuk, R.; Mangalvedhe, N.; Zhang, Y.; Robert, M.; Koskinen, J. Overview of narrowband IoT in LTE Rel-13. In Proceedings of the IEEE Conference on Standards for Communications and Networking (CSCN), Berlin, Germany, 31 October–2 November 2016.
43. 3GPP. Specs. 2019. Available online: http://www.3gpp.org/ftp/Specs/archive/45_series/45.820/45820-d10.zip (accessed on 24 April 2019).
44. 3GPP. TS 36.213. 2018. Available online: https://portal.3gpp.org/desktopmodules/Specifications/SpecificationDetails.aspx?specificationId=2427 (accessed on 29 December 2018).
45. Ericsson. Ground-Breaking Long-Range NB-IoT Connection. 2018. Available online: https://www.ericsson.com/en/press-releases/2018/9/ericsson-and-telstra-complete-ground-breaking-long-range-nb-iot-connection, (accessed on 25 December 2018).
46. Ericsson. Key Technology Choices for Optimal Massive IoT Devices. 2019. Available online: https://www.ericsson.com/en/ericsson-technology-review/archive/2019/key-technology-choices-for-optimal-massive-iot-devices (accessed on 20 March 2019).
47. Ericsson. Technology Overview. 2019. Available online: https://www.ericsson.com/en/ericsson-technology-review/archive/2019/ (accessed on 10 April 2019).
48. Wang, Y.E.; Lin, X.; Adhikary, A.; Grovlen, A.; Sui, Y.; Blankenship, Y.; Bergman, J.; Razaghi, H.S. A Primer on 3GPP Narrowband Internet of Things. *arXiv* **2016**, arXiv:1606.04171.
49. Kroll, H.; Korb, M.; Weber, B.; Willi, S.; Huang, Q. Maximum-Likelihood Detection for Energy-Efficient Timing Acquisition in NB-IoT. In Proceedings of the IEEE Wireless Communications and Networking Conference Workshops (WCNCW), San Francisco, CA, USA, 19–22 March 2017.
50. Ali, A.; Hamouda, W. On the Cell Search and Initial Synchronization for NB-IoT LTE Systems. *IEEE Commun. Lett.* **2017**, *21*, 1843–1846. [CrossRef]
51. Xu, T.; Darwazeh, I. Uplink Narrowband IoT Data Rate Improvement: Dense Modulation Formats or Non-Orthogonal Signal Waveforms? In Proceedings of the 2018 IEEE 29th Annual International Symposium on Personal, Indoor and Mobile Radio Communications (PIMRC), Bologna, Italy, 9–12 September 2018; pp. 142–146. [CrossRef]
52. Zou, J.; Xu, C. Frequency Offset Tolerant Synchronization Signal Design in NB-IoT. *Sensors* **2018**, *18*, 4077. [CrossRef]
53. Lin, X.; Adhikary, A.; Eriç Wang, Y. Random Access Preamble Design and Detection for 3GPP Narrowband IoT Systems. *IEEE Wirel. Commun. Lett.* **2016**, *5*, 640–643. [CrossRef]

54. Cho, S.; Kim, H.; Jo, G. Determination of Optimum Threshold Values for NPRACH Preamble Detection in NB-IoT Systems. In Proceedings of the 2018 Tenth International Conference on Ubiquitous and Future Networks (ICUFN), Prague, Czech Republic, 3–6 July 2018; pp. 616–618.
55. Feltrin, L.; Condoluci, M.; Mahmoodi, T.; Dohler, M.; Verdone, R. NB-IoT: Performance Estimation and Optimal Configuration. In Proceedings of the European Wireless 24th European Wireless Conference, Catania, Italy, 2–4 May 2018.
56. Shimura, A.; Sawahashi, M.; Nagata, S.; Kishiyama, Y. Physical Cell ID Detection Performance Applying Frequency Diversity Reception to NPSS and NSSS for NB-IoT. In Proceedings of the 24th Asia-Pacific Conference on Communications (APCC), Ningbo, China, 12–14 November 2018.
57. Martín, A.G.; Leal, R.P.; Armada, A.G.; Durán, A.F. NBIoT Random Access Procedure: System Simulation and Performance. In Proceedings of the Global Information Infrastructure and Networking Symposium (GIIS), Thessaloniki, Greece, 23–25 October 2018.
58. Baracat, G.H.; Brito, J.M.C. NB-IoT Random Access Procedure Analysis. In Proceedings of the 2018 IEEE 10th Latin-American Conference on Communications (LATINCOM), Guadalajara, Mexico, 14–16 November 2018; pp. 1–6. [CrossRef]
59. M, Y.; Cheng, Y.; Hossain, J.; Ahmed, M.S.G. RADB: Random Access with Differentiated Barring for Latency-Constrained Applications in NB-IoT Network. *Wirel. Commun. Mob. Comput.* **2018**, *2018*, 6210408.
60. Jeon, W.S.; Seo, S.B.; Jeong, D.G. Effective Frequency Hopping Pattern for ToA Estimation in NB-IoT Random Access. *IEEE Trans. Veh. Technol.* **2018**, *67*, 10150–10154. [CrossRef]
61. Li, Y.; Chen, S.; Ye, W.; Lin, F. A Joint Low-Power Cell Search and Frequency Tracking Scheme in NB-IoT Systems for Green Internet of Things. *Sensors* **2018**, *18*, 3274. [CrossRef] [PubMed]
62. Ha, S.; Seo, H.; Moon, Y.; Lee, D.; Jeong, J. A Novel Solution for NB-IoT Cell Coverage Expansion. In Proceedings of the Global Internet of Things Summit (GIoTS), Bilbao, Spain, 4–7 June 2018.
63. Chung, H.; Lee, S.; Jeong, J. NB-IoT Optimization on Paging MCS and Coverage Level. In Proceedings of the 15th International Symposium on Wireless Communication Systems (ISWCS), Lisbon, Portugal, 28–31 August 2018; pp. 1–5.
64. Živic, N. Improved Up-Link Repetition Procedure for Narrow Band Internet of Things. In Proceedings of the International Conference on Computational Science and Computational Intelligence (CSCI), Las Vegas, NV, USA, 13–15 December 2018.
65. Zhang, L.; Ijaz, A.; Xiao, P.; Tafazolli, R. Channel Equalization and Interference Analysis for Uplink Narrowband Internet of Things (NB-IoT). *IEEE Commun. Lett.* **2017**, *21*, 2206–2209. [CrossRef]
66. Yang, B.; Zhang, L.; Qiao, D.; zhao, G.; Imran, M. Narrowband Internet of Things (NB-IoT) and LTE Systems Co-existence Analysis. In Proceedings of the 2018 IEEE Global Communications Conference (GLOBECOM), Abu Dhabi, UAE, 9–13 December 2018.
67. Rusek, F.; Hu, S. Sequential channel estimation in the presence of random phase noise in NB-IoT systems. In Proceedings of the IEEE 28th Annual International Symposium on Personal, Indoor, and Mobile Radio Communications (PIMRC), Montreal, QC, Canada, 8–13 October 2017.
68. Mangalvedhe, N.; Ratasuk, R.; Ghosh, A. NB-IoT deployment study for low power wide area cellular IoT. In Proceedings of the IEEE 27th Annual International Symposium on Personal, Indoor, and Mobile Radio Communications (PIMRC), Valencia, Spain, 4–8 September 2016.
69. Malik, H.; Pervaiz, H.; Mahtab Alam, M.; Le Moullec, Y.; Kuusik, A.; Ali Imran, M. Radio Resource Management Scheme in NB-IoT Systems. *IEEE Access* **2018**, *6*, 15051–15064. [CrossRef]
70. Yu, Y.; Tseng, S. Downlink Scheduling for Narrowband Internet of Things (NB-IoT) Systems. In Proceedings of the IEEE 87th Vehicular Technology Conference (VTC Spring), Porto, Portugal, 3–6 June 2018.
71. Liu, J.; Mu, Q.; Liu, L.; Chen, L. Investigation about the paging resource allocation in NB-IoT. In Proceedings of the 20th International Symposium on Wireless Personal Multimedia Communications (WPMC), Bali, Indonesia, 17–20 December 2017.
72. Kim, H.; Cho, S.C.; Lee, Y. Interference Analysis of Guardband NB-IoT System. In Proceedings of the International Conference on Information and Communication Technology Convergence (ICTC), Sydney, NSW, Australia, 4–7 June 2017.
73. Hsieh, B.; Chao, Y.; Cheng, R.; Nikaein, N. Design of a UE-specific uplink scheduler for narrowband Internet-of-Things (NB-IoT) systems. In Proceedings of the 3rd International Conference on Intelligent Green Building and Smart Grid (IGBSG), Yi-Lan, Taiwan, 22–25 April 2018.

74. Andres-Maldonado, P.; Ameigeiras, P.; Prados, J.; Ramos, J.; Navarro-Ortiz, J.M.; Lopez-Soler, J. Analytic Analysis of Narrowband IoT Coverage Enhancement Approaches. In Proceedings of the 2018 Global Internet of Things Summit (GIoTS), Bilbao, Spain, 4–7 June 2018; pp. 1–6.
75. Yu, C.; Yu, L.; Wu, Y.; He, Y.; Lu, Q. Uplink Scheduling and Link Adaptation for Narrowband Internet of Things Systems. *IEEE Access* **2017**, *5*, 1724–1734. [CrossRef]
76. Jiang, Z.; Han, B.; Chen, P.; Yang, F.; Bi, Q. On Novel Access and Scheduling Schemes for IoT Communications. *Mob. Inf. Syst.* **2016**, *2016*, 1–9. [CrossRef]
77. Oh, S.; Shin, J. An Efficient Small Data Transmission Scheme in the 3GPP NB-IoT System. *IEEE Commun. Lett.* **2017**, *21*, 660–663. [CrossRef]
78. Cluzel, S.; Franck, L.; Radzik, J.; Cazalens, S.; Dervin, M.; Baudoin, C.; Dragomirescu, D. 3GPP NB-IOT Coverage Extension Using LEO Satellites. In Proceedings of the IEEE 87th Vehicular Technology Conference (VTC Spring), Porto, Portugal, 3–6 June 2018.
79. Lauridsen, M.; Krigslund, R.; Rohr, M.; Madueno, G. An Empirical NB-IoT Power Consumption Model for Battery Lifetime Estimation. In Proceedings of the IEEE 87th Vehicular Technology Conference (VTC Spring), Porto, Portugal, 3–6 June 2018.
80. Lee, J.; Lee, J. Prediction-Based Energy Saving Mechanism in 3GPP NB-IoT Networks. *Sensors* **2017**, *17*, 2008. [CrossRef] [PubMed]
81. Bello, H.; Jian, X.; Wei, Y.; Chen, M. Energy-Delay Evaluation and Optimization for NB-IoT PSM with Periodic Uplink Reporting. *IEEE Access* **2019**, *7*, 3074–3081. [CrossRef]
82. Khan, S.; Malik, H.; Alam, M.; Le Moullec, Y. DORM: Narrowband IoT Development Platform and Indoor Deployment Coverage Analysis. In Proceedings of the 2nd International Workshop on Recent Advances in Cellular Technologies and 5G for IoT Environments (RACT-5G-IoT 2019), Leuven, Belgium, 29 April–2 May 2019.
83. Kim, T.; Kim, D.M.; Pratas, N.; Popovski, P.; Sung, D.K. An Enhanced Access Reservation Protocol With a Partial Preamble Transmission Mechanism in NB-IoT Systems. *IEEE Commun. Lett.* **2017**, *21*, 2270–2273. [CrossRef]
84. 3GPP. Architecture Enhancements to Facilitate Communications with Packet Data Networks and Applications. 2017. Available online: https://portal.3gpp.org/desktopmodules/Specifications/SpecificationDetails.aspx?specificationId=862 (accessed on 20 April 2019).
85. 3GPP. TS 23.401 V14.3.0. 2017. Available online: https://portal.3gpp.org/desktopmodules/Specifications/SpecificationDetails.aspx?specificationId=849 (accessed on 25 April 2019).
86. Andres-Maldonado, P.; Ameigeiras, P.; Prados-Garzon, J.; Navarro-Ortiz, J.; Lopez-Soler, J.M. Narrowband IoT Data Transmission Procedures for Massive Machine-Type Communications. *IEEE Netw.* **2017**, *31*, 8–15. [CrossRef]
87. Mohamed, M.O.; Abdelhamid, B.; El Ramly, S. Interference mitigation in heterogeneous networks using Fractional Frequency Reuse. In Proceedings of the International Conference on Wireless Networks and Mobile Communications (WINCOM), Fez, Morocco, 26–29 October 2016.
88. Monteiro, N.; Mihovska, A.; Rodrigues, A.; Prasad, N.; Prasad, R. Interference analysis in a LTE-A HetNet scenario: Coordination vs. uncoordination. In Proceedings of the Wireless VITAE 2013, Atlantic City, NJ, USA, 24–27 June 2013.
89. Palanisamy, P.; Nirmala, S. Downlink interference management in femtocell networks—A comprehensive study and survey. In Proceedings of the International Conference on Information Communication and Embedded Systems (ICICES), Chennai, India, 21–22 Feburary 2013.
90. Song, S.; Li, H.; Fan, Y.; Kong, W.; Zhang, W. Downlink Interference Rejection in Ultra Dense Network. In Proceedings of the 13th APCA International Conference on Control and Soft Computing (CONTROLO), Ponta Delgada, Portugal, 4–6 June 2018; pp. 361–364.
91. Chiumento, A.; Pollin, S.; Desset, C.; der Perre, L.V.; Lauwereins, R. Scalable HetNet interference management and the impact of limited channel state information. *EURASIP J. Wirel. Commun. Netw.* **2015**, *2015*, 74. [CrossRef]
92. Oo, T.; Tran, N.H.; Saad, W.; Niyato, D.; Han, Z.; Hong, C. Offloading in HetNet: A Coordination of Interference Mitigation, User Association, and Resource Allocation. *IEEE Trans. Mob. Comput.* **2017**, *16*, 2276–2291. [CrossRef]
93. Liou, R.H.; Lin, Y.B.; Tsai, S.C. An Investigation on LTE Mobility Management. *IEEE Trans. Mob. Comput.* **2013**, *12*, 166–176. [CrossRef]

94. Karandikar, A.; Akhtar, N.; Mehta, M. Mobility Management in LTE Networks. In *Mobility Management in LTE Heterogeneous Networks*; Springer: Berlin, Germany, 2017.
95. Francois, F.; Abdelrahman, O.H.; Gelenbe, E. Impact of Signaling Storms on Energy Consumption and Latency of LTE User Equipment. In Proceedings of the 2015 IEEE 17th International Conference on High Performance Computing and Communications, New York, NY, USA, 24–26 August 2015.
96. Koc, A.T.; Jha, S.C.; Vannithamby, R.; Torlak, M. Device Power Saving and Latency Optimization in LTE-A Networks Through DRX Configuration. *IEEE Trans. Wirel. Comm.* **2014**, *13*, 2614–2625.
97. Nikaein, N.; Krea, S. Latency for Real-Time Machine-to-Machine Communication in LTE-Based System Architecture. In Proceedings of the 17th European Wireless 2011—Sustainable Wireless Technologies, Vienna, Austria, 27–29 April 2011.
98. Laya, A.; Alonso, L.; Alonso-Zarate, J. Is the Random Access Channel of LTE and LTE-A Suitable for M2M Communications? A Survey of Alternatives. *IEEE Commun. Surv. Tutor.* **2013**, *16*, 4–16. [CrossRef]
99. Cheng, J.; Lee, C.; Lin, T. Prioritized Random Access with dynamic access barring for RAN overload in 3GPP LTE-A networks. In Proceedings of the 2011 IEEE GLOBECOM Workshops (GC Wkshps), Houston, TX, USA, 5–9 December 2011.

© 2019 by the authors. Licensee MDPI, Basel, Switzerland. This article is an open access article distributed under the terms and conditions of the Creative Commons Attribution (CC BY) license (http://creativecommons.org/licenses/by/4.0/).

Article

Aggregated Throughput Prediction for Collated Massive Machine-Type Communications in 5G Wireless Networks

Ahmed Adel Aly [1,*], Hussein M. ELAttar [2,*], Hesham ElBadawy [3,*] and Wael Abbas [1]

1. Department of Basic and Applied Sciences. Arab Academy for Science, Technology and Maritime Transport (AASTMT), Cairo P.O. Box 2033, Egypt; wael_abass@aast.edu
2. Department of Electronics and Communications Engineering. Arab Academy for Science, Technology and Maritime Transport (AASTMT), Cairo P.O. Box 2033, Egypt
3. Network Planning Department, National Telecommunication Institute (NTI), Cairo 11432, Egypt
* Correspondence: a.adel1992@aast.edu (A.A.A.); hattar@aast.edu (H.M.E.); heshamelbadawy@ieee.org (H.E.)

Received: 3 July 2019; Accepted: 17 August 2019; Published: 22 August 2019

Abstract: The demand for extensive data rates in dense-traffic wireless networks has expanded and needs proper controlling schemes. The fifth generation of mobile communications (5G) will accommodate these massive communications, such as massive Machine Type Communications (mMTC), which is considered to be one of its top services. To achieve optimal throughput, which is considered a mandatory quality of service (QoS) metric, the carrier sense multiple access (CSMA) transmission attempt rate needs optimization. As the gradient descent algorithms consume a long time to converge, an approximation technique that distributes a dense global network into local neighborhoods that are less complex than the global ones is presented in this paper. Newton's method of optimization was used to achieve fast convergence rates, thus, obtaining optimal throughput. The convergence rate depended only on the size of the local networks instead of global dense ones. Additionally, polynomial interpolation was used to estimate the average throughput of the network as a function of the number of nodes and target service rates. Three-dimensional planes of the average throughput were presented to give a profound description to network's performance. The fast convergence time of the proposed model and its lower complexity are more practical than the previous gradient descent algorithm.

Keywords: 5G; mMTC; IoT; CSMA; SINR; throughput; polynomial interpolation

---

1. Introduction

The evolution of the fifth generation of cellular mobile systems (5G) has become one of the most significant fields for commercial applications. The 5G system is promising to increase data rates by 10 times that of the traditional Long-Term Evolution (LTE) networks, to an average of 10 Gbps with a 1 ms round-trip latency. This high bandwidth is to accommodate an enormous number of connected devices per unit area under the Internet of Things (IoT) framework [1]. In fact, the 5G requirement covers a wide range of core services, specifically the massive Machine-Type Communications (mMTC) is one of the top three services. The other core services being the ultra-reliable low latency (URLLC) and the extreme mobile broadband (eMBB) communications [2]. The services in mMTC are defined by large numbers of linked devices that are generally transmit data traffic. It includes algorithms, mechanisms, and techniques that permit the exchange of information or data without explicit human involvement.

Recent research studies have shown that most of the existing machine-type communications suffer from limited coverage and access reservation. Currently, the procedure for reserving access is limited to a low number of devices and each device requires high data rates [3]. The main challenge in mMTC is the need for efficient connectivity for this massive number of devices that share packets of data. Additionally, mMTC suffers from losses in data packets due to heavy traffic and congestions. a proper way to overcome these data losses is to provide suitable quality of service (QoS) requirements such as high network throughput with low latency.

In [4], an overview of key radio resource management techniques for 5G dense small cells was studied. Preliminary system-level simulation results indicated that a mean throughput gain of around 63% and up to 84% in latency reduction can be achieved by utilizing resource management techniques. In [5], an efficient online scheme was proposed for predicting channel state information from historical data, in 5G wireless communication systems. The experiment results showed that the scheme not only obtained the predicted channel state information values very quickly but also achieved highly accurate predictions with up to 2.650%–3.457% average difference ratio between the prediction and measurements. In [6], a new Machine-to-Machine (M2M) communication paradigm based on cognitive radio technology was studied, namely the cognitive M2M communication. The cognitive M2M network architecture and cognitive machine model is presented and the coexistence of cognitive M2M devices in TV white spaces was discussed. Additionally, a spectrum exploration scheme motivated by energy-efficiency was introduced. Numerical results show important energy savings and efficiency in providing smart grid data transmission. In [7], device-to-device (D2D) energy-efficient resource allocation algorithm was introduced. To enhance QoS efficiency, a distributed interference mitigation mechanism consisting of a method for canceling interference and a method for optimizing transmission power constraint was discussed. Simulations analyze the achievable performance of the proposed algorithm and discuss implementation and complexity. Additionally, intelligent energy management based on the safe transfer of information between millions of sensors and actuators installed with little or no human involvement was developed in [8]. By investigating the inclusion of software-defined networking with machine-to-machine communication, this motivates the study of a coherent communication structure for intelligent energy management. The proposed software-defined machine-to-machine system was described, with a focus on its price reduction, resource allocation, and end-to-end service quality assurance. In [9], two-stage access control and resource allocation algorithm were developed. In the first phase, a contract-based incentive system was introduced to motivate some delay-tolerant machine-type communication equipment. a long-term cross-layer online resource allocation method was suggested in the second phase, which optimized rate control, energy allocation, and channel choice, without previous channel state information. Finally, under different simulation situations, the performance of the suggested algorithm was verified.

On the other hand, optimizations regarding carrier sense multiple access with collision avoidance (CSMA/CA) have met with great success in different applications. Research on CSMA/CA has a long tradition for years, on which a node senses the channel before transmitting on a shared transmission medium to avoid collision of data in wireless networks [10]. Recent papers [11–14] proposed various CSMA/CA scheduling algorithms that are able to optimize network QoS metrics, particularly the network throughput. In [11], a CSMA scheme was formulated in which throughput and power consumption of each node were optimized by controlling back-off and sleeping timers, while ensuring throughput optimality. In [12], link throughput was analyzed by taking back-off collisions into account; a model was formed to characterize the collision effect among the network's nodes. Results showed that their model was robust against different network topologies. In [13], the performance of CSMA network's throughput was studied under the signal to interference and noise ratio (SINR) model, where a packet was received as long as a certain SINR threshold was exceeded. In [14], they provided effective carrier sensing threshold adjustment algorithms for large wireless CSMA networks. Simulation for evaluating consistency and goodput guaranteed safe interference. They also introduced dynamic signal detection thresholds depending on neighboring transmission feedback. In [15], a distributed iterative algorithm was studied, which produced approximate solutions motivated by an approximation that allowed the expression of approximate solutions via a certain non-linear system with a polynomial size. Numerical results showed that the algorithm produced highly accurate solutions and converged much faster than previous studies. In [16], a distributed scheduling algorithm for the SINR model was studied and proved to be throughput optimal. Further, the algorithm was augmented by using a parallel update technique and the numerical results showed a good performance in terms of a supportable throughput and the convergence rate to steady-state. Moreover, a random access channel evaluation and load estimation of a large number of MTC devices were developed in [17]. a closed-form expression and an effective approach for achieving the Joint Probability Distribution Function (PDF) were extracted from the amount of effective and collided access requests within a random access opportunity. Numerical results justified their formulation's efficiency and demonstrated that the computational cost was smaller than that of other similar techniques.

There has been extensive consideration to the issue of connection scheduling for peak throughput performance with a focus on the maximum weight scheduling model established in [18,19]. Despite its optimization characteristic, a central controller is needed. In addition, solving for each schedule choice is a non-deterministic polynomial-time hard (NP-hard) problem. Various studies have tried to modify the algorithm of maximal weight to make it easier to deploy [20–22]. Such methods are greedy and might not, however, attain an optimized performance. a series of articles [23–25] optimized the throughput calculations for a group of connection scheduling schemes called adaptive CSMA, which can, thus, maintain any attainable rate. In particular, the transmission attempt rate was tuned by each connection to guarantee adequate average desired rates of service.

In [26], the transmission attempt rate was adjusted in the CSMA algorithm to support the required target service rates. This technique poses a problem in adjusting the transmission attempt rate parameter, as it is an NP-hard problem, which is difficult to handle. In addition, most of the research in this field aims at solving this problem using a stochastic gradient descent algorithm which is an iterative optimization method for differentiable objective functions [23]. The drawback of this method is that it consumes millions of iterations to converge. Unfortunately, this approach results in an impractical time of convergence depending on the size of the network. Few studies focus on using a proper SINR model for interference, as most of the studies just settled with using interference model based on conflict graph [23–25].

The work in this paper aimed to enhance the performance of global dense networks with a CSMA scheduling algorithm under a more practical and realistic SINR model, to adequately capture the complexity of wireless network interferences. In addition, by utilizing an approximation technique to overcome large network sizes, the optimization problem could be solved as the approximation technique distributes large networks into smaller ones. The size of the network and node distance with its neighbors is independent of the size of the whole large network. Thus, the solution to such optimization is achievable due to the scale of the small networks. This means that the global optimization function of the transmission attempt rate parameter in CSMA is distributed into local optimization functions for each node and its neighbors. The local optimization function is then solved using Newton's method of optimization instead of stochastic gradient descent, as it has a faster convergence rate [26]. The achievable service rate of each link is then calculated and its percentage error with the target service rate is formulated. In addition, the average throughput is finally calculated. The whole process is repeated under different network operational parameters as SINR thresholds, target service rates, and the number of nodes, to emphasize their effect on the average throughput. The main contribution of the presented work is giving a full description of the effect of changing network operational parameters on the average throughput and proving asymptotic relations using polynomial interpolation that describes the following.

- Throughput as a function of both target service rates and SINR threshold for a given number of nodes.
- Throughput as a function of both the number of nodes and the target service rates for a given SINR threshold.
- Maximum throughput for a different number of nodes at different SINR thresholds.

The proposed model proves its robustness against the increasing number of nodes due to its dependence on the local network size instead of the global one. The model also provides efficient asymptotic throughput relations to be used for estimating the performances of such wireless networks.

The rest of the paper is organized as follows. In Section 2, the system model is described, Section 2.1 describes the used CSMA scheduling algorithm and the research problem is explained. In Section 2.2, the global optimization function is introduced while the distributed local networks algorithm and its model are described in Section 2.3. In Section 2.4, the computational complexity is explained, Section 2.5 discusses the polynomial interpolation technique, and Section 2.6 analyzes the delay performance. In Section 3, the numerical analysis of the proposed model is obtained and described. Results are discussed in Sections 4 and 5 concludes the work and results.

The used parameters and variables in the following sections and their descriptions are summarized and given in Table 1.

Table 1. System parameters and their descriptions.

Parameter/Variable	Description
d_{ii}	Distance between transmitter and receiver of same node i.
d_{ij}	Distance between two nodes i and j.
N	Total number of nodes.
N_j	Number of nodes at node j neighborhood.
$x(t)$	Schedule of the network where $x_i(t) = 1$ means that link i is active and transmitting data and $x_i(t) = 0$ means that link i is not active.
R	Close-in-Radius distance where interference is neglected if distances between nodes exceeded it.
SINR	Signal to Interference and Noise Ratio.
p_i	Transmit power of link i.
α	Path loss exponent.
ω	The variance of the Gaussian thermal noise present at all receivers.
T	SINR threshold that has to be exceeded to ensure successful data reception.
I	List of all feasible schedules.
λ	Transmission attempt rate
s_i	The long-term service rate of node i which is the marginal probability that node i is active.
$1(x \in I)$	The indicator for the feasibility of the schedule.
z	Normalizing constant.
r	Transmission aggressiveness.
y_k	The feasible schedule such that $y_k = 0$ means that node k is inactive while $y_k = 1$ indicates that node k is active and meets the required SINR threshold.
Th	Average Normalized Throughput.
Th_max	Maximum Normalized Throughput.
$e(s^t)$	The approximate error between the achieved and the target service rates.
$P(x)$	the stationary distribution of the CSMA Markov chain

2. System Model

The model used in this article is based on a single-hop wireless network. Each node is formed from a pair of transmitter and receiver similar to the bipole model in [27], the distance between transmitter and receiver is d_{ii} for node i. N is defined as the total number of nodes in the network model. Let d_{ij} be the distance between two nodes i and j.

For scheduling data transmissions between nodes, $x(t)$ is defined as the schedule of the network; it can also be referred to as x. In other words, $x_i(t) = 1$ indicates that node i is active at time slot t and during data transmission. Two nodes are considered to be neighbors and interfere with each other if the distance between them is less than or equals to R (Close-in-Radius). Interference between nodes is neglected if they are not neighbors or, in other words, the distance between them is higher than R as shown in Figure 1.

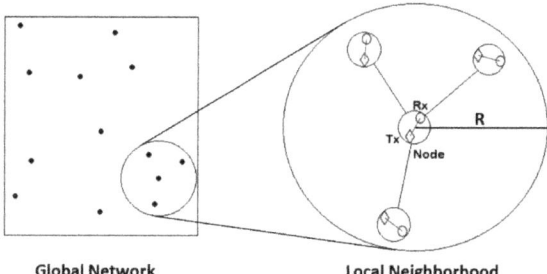

Figure 1. Local Neighborhood network as a part of the global network, the maximum distance between a node (a pair of transmitter and receiver) and its neighbors is the Close-in-Radius distance R

The *SINR* model at node i as a function of $x(t)$ is given by:

$$SINR = \frac{P_i d_{ii}^{-\alpha}}{\sum_{\{j \in N_j,\, j \neq I,\, x_j(t)=1\}} P_j d_{ji}^{-\alpha} + \omega}, \qquad (1)$$

where P_j is the transmitting power of node j, α is the path loss exponent of the standard path loss model $|d|^{-\alpha}$. ω is the variance of the Gaussian thermal noise present at all receivers. N_j is a set of node j and its neighbors.

The condition for $x_i(t) = 1$ (Node i is active and transmitting data) is SINR $\geq T$; T is the *SINR* presumed threshold constraint. If all active nodes in a schedule satisfy this condition, their schedule is called a feasible schedule. The list of all feasible schedules is defined as I, where receivers are able to receive data successfully.

2.1. CSMA Scheduling Algorithm

The main aim of the CSMA scheduling algorithm is to ensure proper data packets reception. The procedure objective is obtaining the transmission attempt rate λ, hence supporting the required target service rates of each node for acquiring the optimal throughput. The scheduling algorithm is based on the SINR readings of the previous time slot, to ensure successful transmission of data if a certain threshold T is exceeded. The process is described in the following CSMA SINR threshold scheduling Algorithm 1.

Algorithm 1. CSMA SINR threshold scheduling algorithm

-Each node is assigned with transmission attempt rate $\lambda > 0$.
-In each time slot, a randomly selected node i is allowed to update its schedule $x_i(t)$ based on the information in the previous time slot.
if $SINR_i(x(t-1)) < T$, **then**
$x_i(t) = 0$ and node i waits for another time slot to update its schedule again.
else if $SINR_i(x(t-1)) \geq T$, **then**
Node i exchanges messages with neighbors, to find if they can meet their *SINR* requirements if link i gets activated.
if any of its neighbors can't meet its requirement, **then**
$x_i(t) = 0$ and node i waits for another time slot to update its schedule again.
else if all neighbors can meet their *SINR* requirements if link i gets activated, **then**
$x_i(t) = 1$ with probability $\frac{\lambda_i}{1+\lambda_i}$ and $x_i(t) = 0$ with probability $\frac{1}{1+\lambda_i}$.
end if
end if

It can be shown in [25] that the adaptive CSMA algorithm induces a Markov chain on the state space of the schedules $\{0,1\}^N$. Further, the stationary distribution of the Markov chain, parametrized by the transmission attempt rate vector $\lambda = [\lambda_i]_{i=1}^N$, is given by:

$$p(x) = \frac{1}{z} \prod_{j:x_j=1} \lambda_j \, 1(x \in I), \forall\, x \in \{0,1\}^N, \tag{2}$$

where $1(x \in I)$ is an indicator of x being a feasible schedule, and z is the normalizing constant. Then, due to the ergodicity of the Markov chain, the long-term service rate of a node i denoted by s_i is equal to the marginal probability that node i is active, i.e., $p_i(x_i = 1)$. Thus, the service rates and the transmission attempt rates are related as follows:

$$s_i = p_i(1) = \sum_{x:\, x_i=1} \frac{1}{z} \prod_{j:x_j=1} \lambda_j, \forall\, i \in N, \tag{3}$$

where $p_i(1)$ denotes $p_i(x_i = 1)$. The adaptive CSMA algorithm can support any service provided that appropriate transmission attempt rates are used for the underlying distribution [28]. If the desired service rates are known, these transmission attempt rates can be obtained by solving the system of equations in Equation (3).

Assume that each node i has a capacity of 1. If node i transmits data all the time (without affecting other nodes), then its service rate is 1 (unit of data per unit time). Then, $s_i(r)$ is also the normalized service rate with respect to the node capacity.

The following concave function $G(r)$ can be maximized by choosing a suitable value of the transmission aggressiveness r, this maximization is equivalent to the minimization of the Kullback–Leibler divergence between the arrival rate and the service rate distribution functions as established in [29].

2.2. The Global Optimization Function

The global optimization function of the proposed network model is given by:

$$r = \underset{r \in R^N}{\mathrm{argmax}}\; G(r)G(r) = \sum_{k \in N} s_k r_k - \ln\left(\sum_{y \in I} \exp\left(\sum_{k \in N} y_k r_k \right) \right), \tag{4}$$

where r is the transmission aggressiveness vector of dimension N and it is a function of λ as $r = \ln(\lambda)$. let $y = [y_k]_{k \in N} \in \{0,1\}^N$ be the global feasible schedule such that $y_k = 0$ means that node k is inactive while $y_k = 1$ indicates that node k is active and meets the required SINR threshold. The desired service rates vector is denoted as $\{s_i\}_{i \in N}$.

For a proper understanding of the global optimization Equation (4), let $\frac{\partial G(r)}{\partial r_i} = 0$ to show that it solves Equation (3) and results in:

$$s_i = \frac{\sum_{y \in I:\, y_i=1} \exp\left(\sum_{k=1}^N y_k r_k \right)}{\sum_{y \in I} \exp\left(\sum_{k=1}^N y_k r_k \right)} \forall\, i \in N, \tag{5}$$

To solve Equation (4), the distributed stochastic gradient descent algorithm was used [23]. However, the gradient of (4) estimation was calculated in a distributed manner and took an impractical time of convergence in order to reach steady state. In order to rectify this problem, the proposed global optimization function was divided into separate and scalable approximated local optimization functions and, finally, these local functions were appropriately combined for estimating the solution to the global problem. The target service rates were assumed as predefined for each node.

2.3. The Local Optimization Function

The local optimization function was similarly structured as the global one with some parameters replaced by ones with a j index to represent its local attachment to node j and its neighbors. The local optimization function of node j was defined as:

$$r_j = \underset{r \in R^{N_j}}{argmax} \; F(r) \quad F(r) = \sum_{k \in N_j} s_k r_k - \ln\left(\sum_{y \in I_j} \exp\left(\sum_{k \in N_j} y_k r_k\right)\right), \quad (6)$$

where $r_j = [r_{jk}]_{k \in N_j}$ is the local transmission aggressiveness of node j, $y = [y_k]_{k \in N_j} \in \{0,1\}^{N_j}$ is the local feasible schedule at node j such that $y_j = 0$ means that node j is inactive and $y_j = 1$ indicates that node j is active and exceeds the SINR threshold. $s^{(j)} = \{s_k \mid k \in N_j\}$ is the local service rate vector of node j.

Due to the downscaling of the global network dimensions, the solution might be simplified by solving local optimization functions. The local solutions of the transmission attempt rate are then combined to produce a global transmission attempt rate that can be directly used in the CSMA algorithm. This process is chosen over the adaptation of transmission attempt rates using a stochastic gradient descent that requires extensive time to converge on the global function. Each node in the network executes (Algorithm 2) in parallel to get the average normalized throughput.

Algorithm 2. Proposed Algorithm to obtain Average Normalized Throughput

---

Input: $(s_k, k \in N_j)$; Output: Average Normalized Throughput $Th(s^t)$

(1) Neighbors of node j provide their target service rates.
(2) Local optimization function (6) is solved using Newton's method of optimization for node j with its surrounding neighbors and the maximum achievable service rates is locally obtained for all feasible schedules of the neighborhood.
(3) Node j and each neighbor from $k \in N_j$ provide their locally maximum achievable service rates.
(4) The average achievable service rate can be calculated for each node separately by averaging the service rates in each neighborhood contained in that node.
(5) Calculate the approximate error between the achieved and the target service rates:

$$e(s^t) = \frac{\sum_{i=1}^{N} |s_i^t - s_i^a|}{N}, \quad (7)$$

where $s^a = [s_i^a]_{i=1}^{N}$ are the service rates that can be achieved and $s^t = [s_i^t]_{i=1}^{N}$ is the given target service rate vector.

(6) Compute the average normalized throughput:

$$Th(s^t) = \frac{\sum_{i=1}^{N} s_i^t (1 - e(s^t))}{N}, \quad (8)$$

(7) Use Polynomial interpolation to form the average normalized throughput equations as a function of either—number of nodes and target service rates or SINR threshold and target service rates. Polynomial interpolation is also used to get the maximum normalized throughput as a function of the number of nodes.

The approximate global transmission attempt rate $\widetilde{\lambda}$ that can be directly applied to the CSMA SINR threshold scheduling algorithm (Algorithm 1) is given by:

$$\widetilde{\lambda} = \left(\frac{1-s_j}{s_j}\right)^{|N_j|-1} \prod_{k \in N_j} e^{r_{kj}}, \tag{9}$$

where s_j is the target service rate of link j, $|N_j|$ is the number of nodes in the local neighborhood of node j. The transmission aggressiveness is r_{kj}, $k \in N_j$ is the optimized parameter in Step (2) of node j and its presence in every local neighborhood.

Newton's method of optimization used in Step (2) is used to optimize the local optimization function (6) [30]. It can be computed without complexity due to its relatively smaller size as compared to the global function and is given by:

$$r^{(t)} = r^{(t-1)} - ([\nabla^2 F(r^{(t-1)})]_{ik})^{-1} \cdot [\nabla F(r^{(t-1)})]_k, \tag{10}$$

where $r^{(t)}$ is the transmission aggressiveness of the current iteration and $r^{(t-1)}$ is of the previous iteration. $[\nabla^2 F(r^{(t-1)})]_{ik}$ and $[\nabla F(r^{(t-1)})]_k$ are the Hessian matrix and the gradient vector of Newton's method, respectively. The gradient and Hessian of node j computations are done through the distribution:

$$\hat{b}_j(x^{(j)}) = \frac{1}{z_j} \exp\left(\sum_{k \in N_j} x_k r_k\right), \forall x^{(j)} \in I_j, \tag{11}$$

where z_j is the local normalization constant of node j and its neighbors. I_j is a list of all feasible schedules of node j and its neighbors. The gradient of (6) is given by:

$$[\nabla F(r)]_k = s_k - m_k(r), \ k \in N_j, \tag{12}$$

where $m_k(r) = p(x_k = 1)$ under $\hat{b}_j$ distribution for $k \in N_j$.

The Hessian of (6) is given by:

$$[\nabla^2 F(r)]_{ik} = \begin{cases} m_i(r)m_k(r) - m_{ik}(r), & i, k \in N_j, i \neq j \\ m_k(r)^2 - m_k(r), & i = k. \end{cases} \tag{13}$$

where $m_{ik}(r) = p(x_i = 1, x_k = 1)$ of the distribution $\hat{b}_j$.

2.4. Computational Complexity

It is possible to implement the Newton method in (Algorithm 2). This is because it is possible to analytically calculate the gradient and the Hessian of the local objective function $F(r)$ since the problem dimension is reduced from the global one.

The gradient and the Hessian calculations need information about the feasible local schedules at a link. This information is specifically needed to calculate the normalization constant z. These computations are feasible because of the complexity of $O(2^{|N_j|})$ associated with a computation scale of only a particular size of the local neighborhood that is independent of the network's global size, which could be considerably large.

2.5. Polynomial Interpolation

The polynomial cubic interpolation used in Step (7) is explained in [31–33] and has a common form of:

$$f(\theta) = a_0 + a_1\theta + a_2\theta^2 + \ldots + a_n\theta^n, \ a_n \neq 0, \tag{14}$$

where θ is the variable of the function f and n is the degree of the polynomial.

The solution of Equation (14) is computed using the Vandermonde matrix [33] formed in Equation (15), in order to calculate the coefficient a_n given both θ and f:

$$\begin{bmatrix} 1 & \theta_1 & \theta_1^2 & \cdots & \theta_1^m \\ 1 & \theta_2 & \theta_2^2 & \cdots & \theta_2^m \\ 1 & \theta_3 & \theta_3^2 & \cdots & \theta_3^m \\ \vdots & \vdots & \vdots & \ddots & \vdots \\ 1 & \theta_n & \theta_n^2 & \cdots & \theta_n^m \end{bmatrix} \begin{bmatrix} a_0 \\ a_1 \\ a_2 \\ \vdots \\ a_m \end{bmatrix} = \begin{bmatrix} f_0 \\ f_1 \\ f_2 \\ \vdots \\ f_n \end{bmatrix}, \quad (15)$$

Another form of the polynomial interpolation used in this work is the bicubic interpolation for a function of two variables, such as:

$$f(\theta, \varphi) = a_0 + a_1\theta + a_2\varphi + a_3\theta^2 + a_4\theta\varphi + a_5\varphi^2 + a_6\theta^3 + a_7\theta^2\varphi + a_8\theta\varphi^2 + a_9\varphi^3, \quad (16)$$

where θ and φ are the two variables of the function f.

The solution of Equation (16) is similar to Equation (14) and is computed using the Vandermonde matrix formed in Equation (17) to calculate the coefficient a_n given θ, φ, and f:

$$\begin{bmatrix} 1 & \theta_1 & \varphi_1 & \theta_1^2 & \theta_1\varphi_1 & \cdots & \varphi_1^m \\ 1 & \theta_2 & \varphi_2 & \theta_2^2 & \theta_2\varphi_2 & \cdots & \varphi_2^m \\ 1 & \theta_3 & \varphi_3 & \theta_3^2 & \theta_3\varphi_3 & \cdots & \varphi_3^m \\ \vdots & \vdots & \vdots & \vdots & \vdots & \ddots & \vdots \\ 1 & \theta_n & \varphi_n & \theta_n^2 & \theta_n\varphi_n & \cdots & \varphi_n^m \end{bmatrix} \begin{bmatrix} a_0 \\ a_1 \\ a_2 \\ \vdots \\ a_m \end{bmatrix} = \begin{bmatrix} f_0 \\ f_1 \\ f_2 \\ \vdots \\ f_n \end{bmatrix}, \quad (17)$$

Equations (15) and (17) are solved using the Gaussian elimination method [34] until a reduced echelon form is reached; hence, the values of the coefficients are computed. In Step (7), Polynomial interpolation is used to obtain the average normalized throughput as a function of either number of nodes and target service rates or SINR threshold and target service rates. Polynomial interpolation is also used to get the maximum normalized throughput as a function of the number of nodes.

2.6. Delay Performance

Considering the delay performance, it is known that the CSMA Markov chain's impractically slow mixing time leads to a bad delay performance [28,35]. Recent work such as [36,37] have enhanced delay performance by using several parallel CSMA Markov chain cases. These results [36] are demonstrated on the assumption that the ideal transmission attempt rate is pre-computed and is easily accessible to the algorithm. The proposed local algorithm can be used in combination with the methods in [36,37] to achieve a practical CSMA algorithm with an excellent throughput and a low delay, to effectively estimate these transmission attempt rates.

3. Results

In this section numerical analysis are used to estimate the performance of the proposed algorithm. Random topology graphs are considered with a different number of nodes to test the robustness of the model. Random networks are generated by placing nodes on a two-dimensional area of length of 12 unit distance. The system parameters are summarized in Table 2.

Random networks are generated with varying densities from 10 up to 100 nodes, and vertices are drawn when nodes are neighbors and interfere with each other. The total number of links is higher than the total number of nodes, but it does not reach the mesh topology where the number of links equals to $N(N-1)/2$. The interference graph of a 100-node random topology is shown in Figure 2.

Table 2. The system parameters used in numerical analysis.

System Parameter	Value
Dimensions of the interference graph (unit area)	12 × 12
Distance between transmitter and its corresponding receiver (unit distance)	0.5
Path loss exponent	3
Close in Radius (unit distance)	2.5
SINR threshold (dB)	9 up to 15
Target service rate (Unit of data per unit time)	0 up to 0.9
Transmit power (unit power)	1

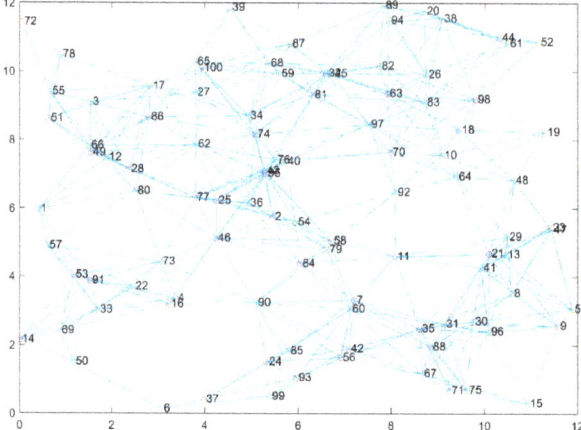

Figure 2. Interference graphs of 100-node random topology.

In order to prove the fast convergence rate of the proposed model, Figure 3 shows that the norm of the gradient in Newton's method of a random local network sample converges in 4 to 5 iterations.

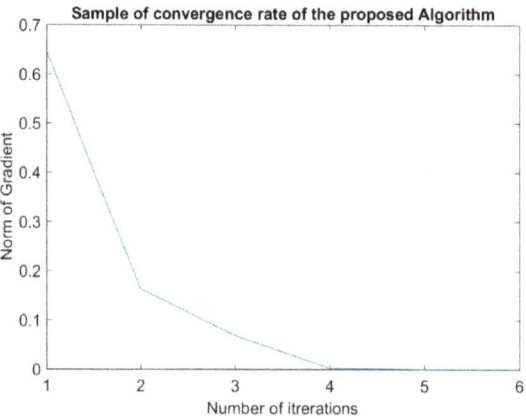

Figure 3. Sample of Convergence Rate of the proposed algorithm.

In Figure 4, the average normalized throughput as a function of the target service rate is shown for different network topologies of 10, 30, 50, and 100 randomly distributed nodes at 9, 12 and 15 dB SINR thresholds.

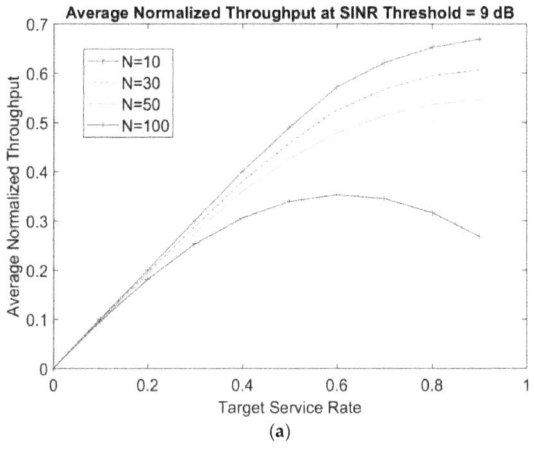

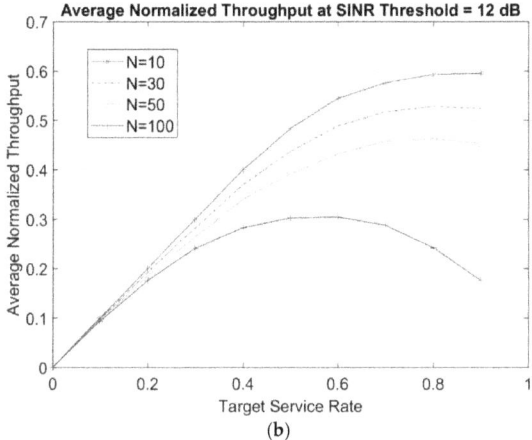

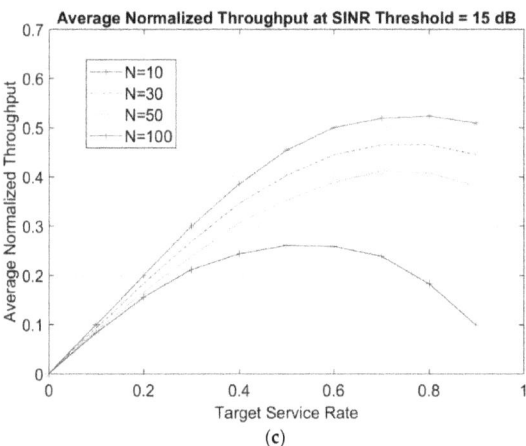

Figure 4. Average normalized throughput for a different number of nodes—(**a**) 9 dB SINR threshold, (**b**) 12 dB SINR threshold, and (**c**) 15 dB SINR threshold.

As shown in Figure 4, increasing the SINR threshold shows degradation in throughput, especially at high-target service rates. This is because the nodes will be unable to transmit until a higher SINR threshold is met. Meanwhile, it is observed that initially, as target service rate increases, the average normalized throughput increases until it reaches its maximum. This is where the network works under "stable operating conditions". After that point, the network enters "unstable operating conditions" in which collisions become more likely and the number of backlogged frames increases. This means that the arrival rate of new frames to the system will be larger than the capability of successful frames transmission, thus, leading to a decrease of the average normalized throughput.

In order to acquire the obtained results in Figure 4 as a system of asymptotic relations, polynomial bicubic interpolation algorithm is used. Therefore, the average normalized throughput can be estimated via the following proposed equation in the general form of:

$$Th(N,S) = a_0 + a_1 N + a_2 S + a_3 N^2 + a_4 NS + a_5 S^2 + a_6 N^3 + a_7 N^2 S + a_8 NS^2 + a_9 S^3, \quad (18)$$

where Th is the average normalized throughput as a function of the number of nodes N and target service rate is denoted as S. The constants $a_0, \ldots, a_9$ are the constant coefficients in Equation (18) and are given in Table 3.

Table 3. Coefficients of Equation (18) for different SINR thresholds.

SINR Threshold	a_0	a_1	a_2	a_3	a_4	a_5	a_6	a_7	a_8	a_9
$T = 9$	0.3476	-3×10^{-4}	0.559	3.5×10^{-6}	0.0029	1.096	10^{-9}	-3.4×10^{-5}	-0.00466	-1.016
$T = 10$	0.1716	-4.6×10^{-4}	-1.2147	3.4×10^{-6}	0.004496	5.442	10^{-9}	-3.26×10^{-5}	-0.0064	-3.865
$T = 11$	-0.314	3.65×10^{-4}	4.92	3.3×10^{-6}	-0.00486	-8.14	10^{-9}	-3×10^{-5}	0.00256	4.28
$T = 12$	-0.119	8.45×10^{-5}	2.49	3.18×10^{-6}	-0.00189	-2.97	10^{-9}	2.8×10^{-5}	7.5×10^{-4}	1.239
$T = 13$	-0.305	3.65×10^{-4}	4.896	4×10^{-6}	-0.00614	-8.66	10^{-9}	-3×10^{-5}	0.0038	4.81
$T = 14$	-0.3	1.9×10^{-4}	4.898	5.3×10^{-6}	-0.0059	-8.73	10^{-9}	-3.4×10^{-5}	0.004	4.847
$T = 15$	-0.0515	-4.25×10^{-4}	1.718	6.6×10^{-6}	-7.35×10^{-4}	-1.597	10^{-9}	-3.76×10^{-4}	-5.79×10^{-4}	0.447

The interpolation equations are used to describe the missing values between the calculated throughput results and also give more details in three-dimensional planes of the average normalized throughput as a function of both target service rates and the number of nodes. In Figure 5, the plane is represented at different SINR thresholds of 9 up to 15 dB. It can be observed that increasing the number of transmitting nodes in the network will cause more collisions, thus, leading to a decrease in the network throughput.

Another way to make use of the polynomial bicubic interpolation is to generate another form of the previous equation with different parameters, such as:

$$Th(T,S) = c_0 + c_1 T + c_2 S + c_3 T^2 + c_4 TS + c_5 S^2 + c_6 T^3 + c_7 T^2 S + c_8 TS^2 + c_9 S^3, \quad (19)$$

where Th is the average normalized throughput as a function of target service rate denoted as S and the $SINR$ threshold T. The constants $c_0, \ldots, c_9$ are the constant coefficients in Equation (19) and are given in Table 4.

Moreover, the previous step of interpolation is repeated to find an asymptotic relation of the average normalized throughput as a function of both target service rates and the SINR threshold. In Figure 6, the plane is shown with a number of nodes from 10 up to 100 nodes.

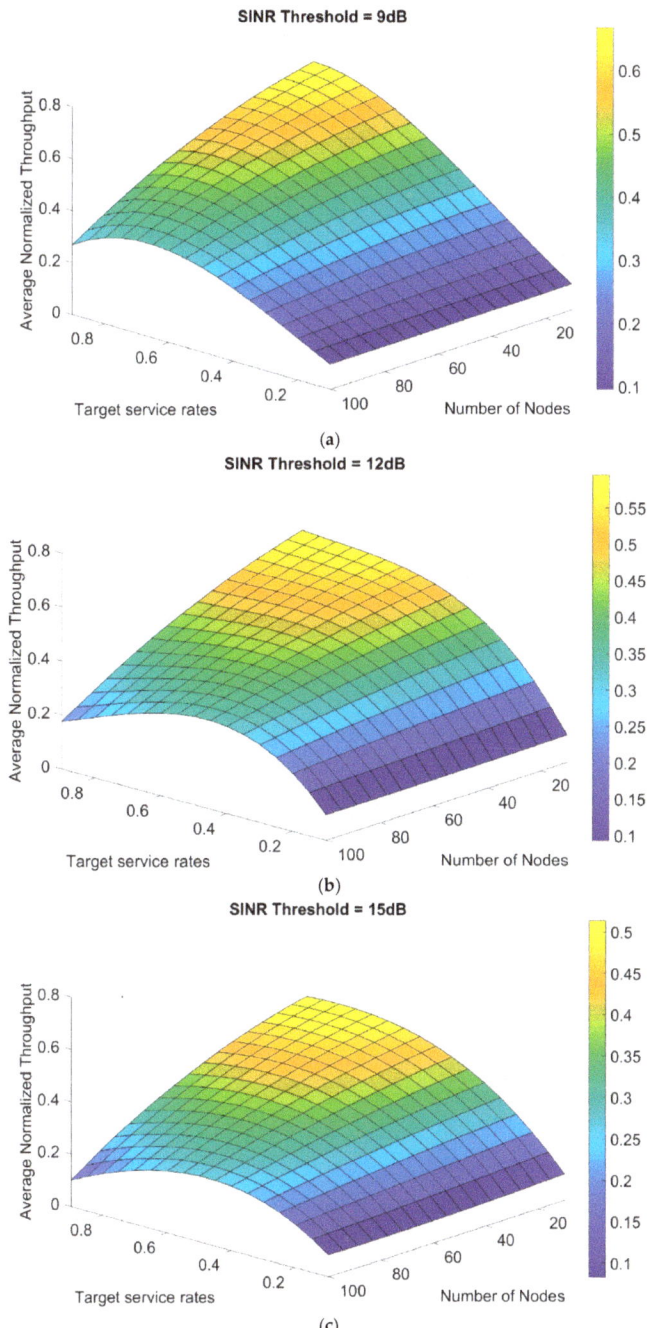

Figure 5. Plane of average normalized throughput generated from polynomial interpolation at a different number of nodes for—(**a**) 9 dB SINR threshold, (**b**) 12 dB SINR threshold, and (**c**) 15 dB SINR threshold.

Table 4. Coefficients of Equation (19) for different number of nodes.

No. of Nodes	c_0	c_1	c_2	c_3	c_4	c_5	c_6	c_7	c_8	c_9
$N = 10$	−0.028	−0.00138	1.32	8.95×10^{-5}	0.0166	−0.315	10^{-6}	8.95×10^{-4}	−0.028	−0.128
$N = 20$	−0.0367	0.0025	1.25	-1×10^{-4}	0.0015	−0.112	10^{-6}	-1.66×10^{-4}	−0.03	−0.219
$N = 30$	−0.069	0.0063	1.5	-2.89×10^{-4}	−0.0135	−0.566	10^{-7}	-5.66×10^{-4}	−0.0324	0.0755
$N = 40$	−0.077	0.01069	1.43	-4.8×10^{-4}	−0.0328	−0.313	10^{-7}	0.00129	−0.03	−0.0665
$N = 50$	−0.449	0.01425	0.8649	-6.74×10^{-4}	−0.0437	0.9377	10^{-8}	0.002	−0.0365	−0.752
$N = 60$	−0.0276	0.0143	0.64	-6.76×10^{-4}	−0.0428	1.339	10^{-8}	0.00197	−0.036	−1.025
$N = 70$	−0.0265	0.0143	0.623	-6.79×10^{-4}	−0.0419	1.29	10^{-8}	0.0019	−0.036	−1.04
$N = 80$	−0.134	0.0165	1.942	-6.8×10^{-4}	−0.0647	−1.38	10^{-8}	0.00185	−0.012	0.366
$N = 90$	−0.0599	0.0156	1.006	-6.8×10^{-4}	−0.052	0.516	10^{-9}	0.0018	−0.023	−0.77
$N = 100$	−0.061	0.0156	1.019	-6.86×10^{-4}	0.0517	0.409	10^{-9}	0.00175	−0.0229	−0.749

After obtaining the average normalized throughput from the numerical analysis, the maximum normalized throughput could be extracted. Polynomial cubic interpolation algorithm was used to concatenate the above figures and define a relation for acquiring the maximum normalized throughput in the general form of:

$$Th\_max(N) = k_0 + k_1 N + k_2 N^2, \qquad (20)$$

where Th_max is the maximum normalized throughput as a function of the number of nodes N. The constants k_0, k_1, and k_2 are the constant coefficients in Equation (20) and are given in Table 5.

Table 5. Coefficients of Equation (20) for different SINR thresholds.

SINR Threshold	k_0	k_1	k_2
$S = 9$	0.6989	−0.003	-4×10^{-7}
$S = 12$	0.6282	0.0033	-7.48×10^{-8}
$S = 15$	0.551	−0.00278	-1.2×10^{-7}

The interpolation equations graphs of the maximum normalized throughput are shown in Figure 7 as a function of the number of nodes at the different SINR thresholds from 9 up to 15 dB.

Another way to describe the maximum normalized throughput more generally, using the polynomial bicubic interpolation is to formulate a plane that is a function of, both, the number of nodes and the SINR threshold. Therefore, the maximum normalized throughput was estimated using the following equation as the general form:

$$Th\_max(N, T) = e_0 + e_1 N + e_2 T + e_3 N^2 + e_4 NT + e_5 T^2, \qquad (21)$$

where Th_max is the maximum normalized throughput as a function of the number of nodes N and the SINR threshold denoted as T. The constants $e_0, \ldots, e_5$ are the constant coefficients in Equation (21) and are given in Table 6.

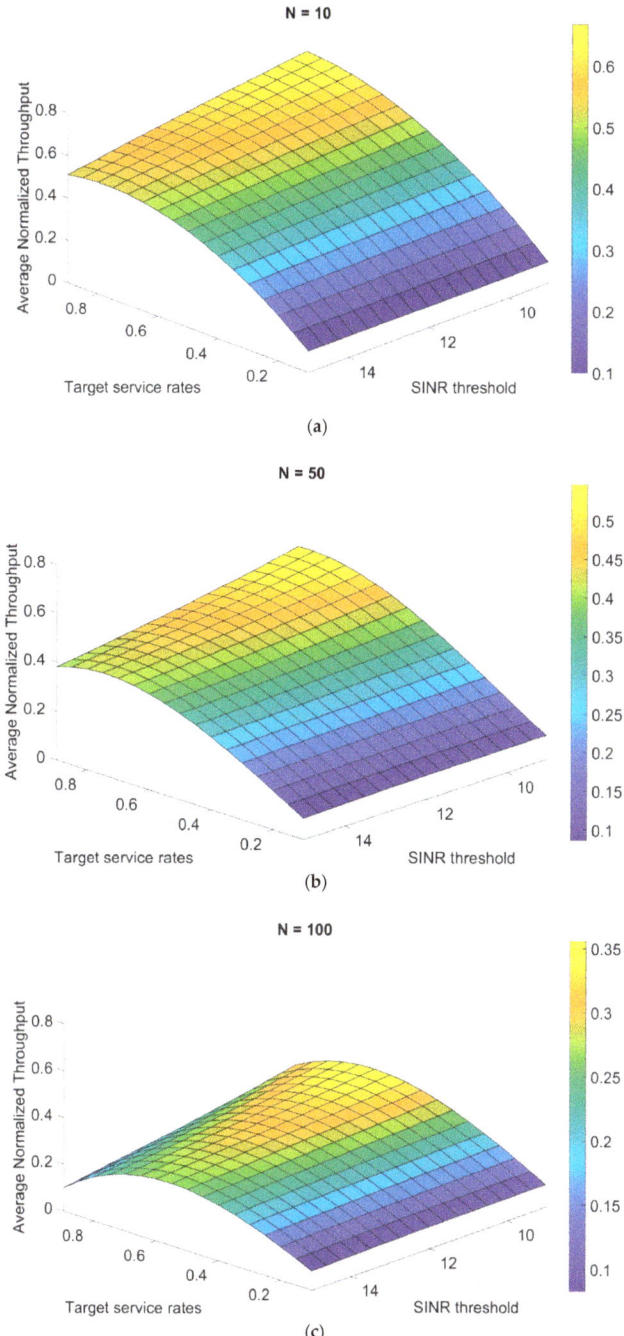

Figure 6. Plane of average normalized throughput generated from the polynomial interpolation for different SINR thresholds of—(**a**) 10-node topology, (**b**) 50-node topology, and (**c**) 100-node topology.

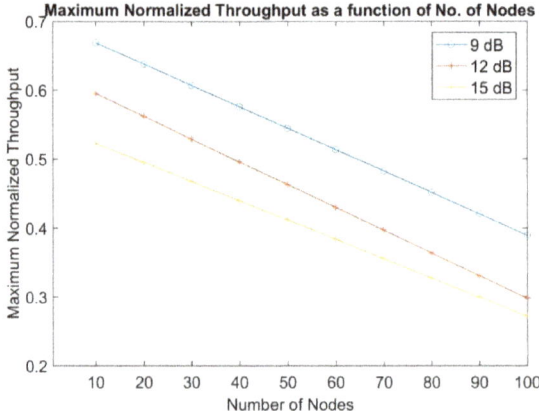

Figure 7. Maximum normalized throughput at different number of nodes for SINR threshold of 9 dB, 12 dB, and 15 dB.

Table 6. Coefficients of Equation (21).

Coefficient	e_0	e_1	e_2	e_3	e_4	e_5
Value	0.9213	−0.00345	−0.025	-8.47×10^{-6}	9.81×10^{-5}	10^{-5}

As a result to Equation (21), the maximum normalized throughput as a function of both the number of nodes and the SINR threshold can be generated approximately. Different from Equation (20), the new equation adds the dimension of the SINR threshold to give a more general description and eases the prediction to be based on two parameters instead of one. Equation (21) is shown in Figure 8 for different number of nodes and SINR thresholds.

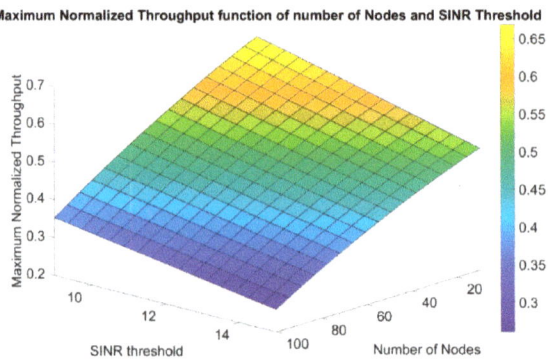

Figure 8. Maximum normalized throughput as a function of both the number of nodes and the SINR threshold.

A sample of error using Equation (7) between the proposed approximation technique and the stochastic gradient descent algorithm is shown in Figure 9. Where the local approximation algorithm that calculates the approximate transmission attempt rates uses their static values in the CSMA algorithm (i.e., they are not adapted during the algorithm). On the other hand, the stochastic gradient descent algorithm begins with some initial transmission attempt rates, and by observing the corresponding service rates, it adapts the transmission attempt rates [23].

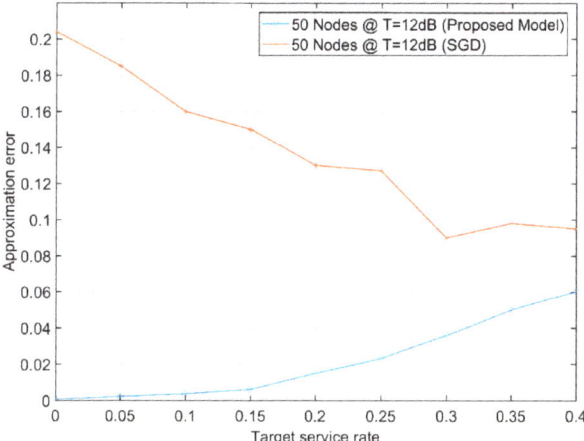

Figure 9. The approximate error of the achievable and the target service rates due to stochastic gradient descent and the proposed approximation algorithm with 50 nodes distributed with the same random topology used in this article.

For 5G networks, the error has to be minimum at all target service rates. Additionally, determining the transmission attempt rate is an NP-hard problem and the stochastic gradient descent is unpractical as it adapts those transmission attempt rates in several iterations, as shown in the Figure 9. This is why the proposed approach of pre-calculating those transmission attempt rates and using them directly in the network helps in reducing the error at most target service rates. The results would be comparable even at higher rates but there still exists an advantage in the early stages where the network works under "stable operating conditions". The error of the proposed model did not exceed 0.06, while the gradient descent started with an error exceeding 0.2 for a target service rate up to 0.4 unit data per unit time.

4. Discussion

The proposed model aims at providing possible ways of predicting random network performances under different circumstances. One thing to mention is that the proposed model is scalable, as it studied a 10- to 100-node random topology in an area of 12 × 12 square units of distance. This means that if a 400-node random topology is studied at 24 × 24 square units of distance, the performance will be similar to the proposed 100-node one. This is due to the dependence of the proposed model on the local neighborhood size, not the global one. Therefore, what matters is the performance degradation that would arise from increasing the number of nodes in the same limited area. Additionally, increasing the given area can allow for fitting a higher number of nodes, as the limited area considered in this work affects the interference to a great extent, due to the high-density neighborhoods interfering with each other. Additionally, increasing the number of nodes in the given area to a value higher than the studied ones might lead to a density (number of nodes/unit area) > 1, which is rarely found and should most of the time be ≤ 1. Additionally, the resulting model might be used for scalable networks. In other words, the resulted network could be used to estimate the performance of both small as well as large networks. The presented analysis is based on the normalized unit area and the normalized throughput, so it might provide a good performance whatever be the network size.

The proposed polynomial interpolation's throughput asymptotic relations here are based on the studied topologies and might lead to other performance prediction if different parameters were used. To sum up the used parameters here, the SINR threshold was set to 9, 12, and 15 dB, up to 0.9 unit target service rates, and up to 100 nodes randomly distributed in the unit area were considered.

Three-dimensional topologies were applicable and would not have differed much from the two-dimensional topology used in the proposed model, as the distance was the only matter. We previously mentioned the close-in-radius R distance as the distance where interference was neglected if two nodes were at $distance > R$ apart from each other. This radius could either be of a circle if a two-dimensional topology was considered or a radius of a sphere in case of the three-dimensional topology. Therefore, at the end what matters is distances at any directions and the three-dimensional topology could be equivalent to a dense network that is already studied in the proposed model. Additionally, the main scope of this paper was to calculate the effect of the increasing the number of nodes on the throughput in dense networks, which is suitable for 5G applications. The energy and the delay could be considered in details in future work.

5. Conclusions

A CSMA algorithm of a single-hop wireless network was considered under a realistic SINR model. An approximation algorithm of distributing the global network into downscaled local neighborhoods was used to calculate the transmission attempt rate to optimize the throughput of the global network. This was done by achieving target service rates of nodes while varying the number of nodes up to a random 100-node topology. The proposed model converged fast and proved its robustness against the increasing number of nodes as it depended only on the size of the local network instead of the global dense one. Three-dimensional planes of the average normalized throughput were obtained by polynomial interpolation that produced a complete description of the performance of the network. Maximum normalized throughput was obtained too as a function of the number of nodes using polynomial interpolation. The used approach of pre-calculating the transmission attempt rates and using them directly in the network helped in reducing the error at most values of the target service rates. Even at higher rates, results would be comparable with a gradient descent algorithm but there still would exist an advantage in the early stages where the network would work under "stable operating conditions". The proposed model is also scalable, so it might provide a network performance, whatever be its size. Additionally, the proposed model has a faster convergence time and is considered to be less complex and more practical than the previously used gradient descent algorithm.

Author Contributions: Conceptualization, A.A.A. and H.E.; Investigation, A.A.A. and H.M.E.; Methodology, A.A.A. and H.M.E.; Project administration, H.E.; Resources, A.A.A.; Software, A.A.A.; Supervision, H.E., H.M.E. and W.A.; Validation, A.A.A. and H.E.; Visualization, H.E.; Writing—original draft, A.A.A.; Writing—review and editing, A.A.A., H.M.E., H.E., and W.A.

Funding: This research received no external funding.

Conflicts of Interest: The authors declare no conflict of interest.

References

1. Agiwal, M.; Roy, A.; Saxena, N. Next generation 5G wireless networks: a comprehensive survey. *IEEE Commun. Surv. Tutor.* **2016**, *18*, 1617–1655. [CrossRef]
2. Marsch, P.; Bulakci, Ö.; Queseth, O.; Boldi, M. *5G System Design: Architectural and Functional Considerations and Long Term Research*; John Wiley & Sons: Hoboken, NJ, USA, 2018.
3. Bockelmann, C.; Pratas, N.K.; Wunder, G.; Saur, S.; Navarro, M.; Gregoratti, D.; Vivier, G.; De Carvalho, E.; Ji, Y.; Stefanović, Č. Towards massive connectivity support for scalable mMTC communications in 5G networks. *IEEE Access* **2018**, *6*, 28969–28992. [CrossRef]
4. Mahmood, N.H.; Lauridsen, M.; Berardinelli, G.; Catania, D.; Mogensen, P. Radio resource management techniques for eMBB and mMTC services in 5G dense small cell scenarios. In Proceedings of the IEEE 84th Vehicular Technology Conference, Montréal, QC, Canada, 18–21 September 2016; pp. 1–5.
5. Luo, C.; Ji, J.; Wang, Q.; Chen, X.; Li, P. Channel state information prediction for 5G wireless communications: a deep learning approach. *IEEE Trans. Network Sci. Eng.* **2018**. [CrossRef]
6. Zhang, Y.; Yu, R.; Nekovee, M.; Liu, Y.; Xie, S.; Gjessing, S. Cognitive machine-to-machine communications: Visions and potentials for the smart grid. *IEEE Netw.* **2012**, *26*, 6–13. [CrossRef]

7. Zhou, Z.; Dong, M.; Ota, K.; Wang, G.; Yang, L.T. Energy-efficient resource allocation for D2D communications underlaying cloud-RAN-based LTE-A networks. *IEEE Int. Things J.* **2015**, *3*, 428–438. [CrossRef]
8. Zhou, Z.; Gong, J.; He, Y.; Zhang, Y. Software defined machine-to-machine communication for smart energy management. *IEEE Commun. Mag.* **2017**, *55*, 52–60. [CrossRef]
9. Zhou, Z.; Guo, Y.; He, Y.; Zhao, X.; Bazzi, W.M. Access Control and Resource Allocation for M2M Communications in Industrial Automation. *IEEE Trans. Ind. Inform.* **2019**, *15*, 3093–3103. [CrossRef]
10. Wyglinski, A.M.; Pu, D. *Digital Communication Systems Engineering with Software-Defined Radio*; Artech House: Norwood, MA, USA, 2013.
11. Maatouk, A.; Assaad, M.; Ephrmmides, A. Energy efficient and throughput optimal CSMA scheme. *IEEE ACM Trans. Netw.* **2019**, *27*, 316–329. [CrossRef]
12. Kai, C.; Zhang, S.; Wang, L. Impacts of packet collisions on link throughput in CSMA wireless networks. *China Commun.* **2018**, *15*, 1–14. [CrossRef]
13. Sun, X. Maximum throughput of CSMA networks with capture. *IEEE Wirel. Commun. Lett.* **2016**, *6*, 86–89. [CrossRef]
14. Chau, C.; Ho, I.W.; Situ, Z.; Liew, S.C.; Zhang, J. Effective static and adaptive carrier sensing for dense wireless CSMA networks. *IEEE Trans. Mobile Comput.* **2016**, *16*, 355–366. [CrossRef]
15. Yun, S.Y.; Shin, J.; Yi, Y. CSMA using the Bethe approximation for utility maximization. In Proceedings of the 2013 IEEE International Symposium on Information Theory, Istanbul, Turkey, 7–12 July 2013; pp. 206–210.
16. Swamy, P.S.; Ganti, R.K.; Jagannathan, K. Spatial CSMA: a distributed scheduling algorithm for the SIR model with time-varying channels. In Proceedings of the IEEE Twenty First National Conference on Communications, Mumbai, India, 27 February–1 March 2015; pp. 1–6.
17. Tello-Oquendo, L.; Pla, V.; Leyva-Mayorga, I.; Martinez-Bauset, J.; Casares-Giner, V.; Guijarro, L. Efficient random access channel evaluation and load estimation in LTE-A with massive MTC. *IEEE Trans. Veh. Technol.* **2018**, *68*, 1998–2002. [CrossRef]
18. TassiulAs, L.; Ephremides, A. Stability properties of constrained queueing systems and scheduling policies for maximum throughput in multihop radio networks. In Proceedings of the 29th IEEE Conference on Decision and Control, Honolulu, HI, USA, 5–7 December 1990; pp. 2130–2132.
19. Tassiulas, L.; Ephremides, A. Dynamic server allocation to parallel queues with randomly varying connectivity. *IEEE Trans. Inform. Theory* **1993**, *39*, 466–478. [CrossRef]
20. Chaporkar, P.; Kar, K.; Luo, X.; Sarkar, S. Throughput and fairness guarantees through maximal scheduling in wireless networks. *IEEE Trans. Inform. Theory* **2008**, *54*, 572–594. [CrossRef]
21. Dimakis, A.; Walrand, J. Sufficient conditions for stability of longest-queue-first scheduling: Second-order properties using fluid limits. *Adv. Appl. Probab.* **2006**, *38*, 505–521. [CrossRef]
22. Wu, X.; Srikant, R.; Perkins, J.R. Queue-Length Stability of Maximal Greedy Schedules in Wireless Networks. Available online: http://ita.ucsd.edu/workshop/06/papers/262.pdf (accessed on 19 August 2019).
23. Jiang, L.; Walrand, J. a distributed CSMA algorithm for throughput and utility maximization in wireless networks. *IEEE/ACM Trans. Netw.* **2010**, *18*, 960–972. [CrossRef]
24. Rajagopalan, S.; Shah, D.; Shin, J. Network adiabatic theorem: An efficient randomized protocol for contention resolution. *ACM SIGMETRICS Perform. Eval. Rev.* **2009**, *37*, 133–144.
25. Ni, J.; Tan, B.; Srikant, R. Q-CSMA: Queue-length-based CSMA/CA algorithms for achieving maximum throughput and low delay in wireless networks. *IEEE/ACM Trans. Netw.* **2012**, *20*, 825–836.
26. Swamy, P.S.; Ganti, R.K.; Jagannathan, K. Adaptive CSMA under the SINR model: Efficient approximation algorithms for throughput and utility maximization. *IEEE/ACM Trans. Netw.* **2017**, *25*, 1968–1981. [CrossRef]
27. Baccelli, F.; Singh, C. Adaptive spatial Aloha, fairness and stochastic geometry. In Proceedings of the 2013 11th International Symposium and Workshops on Modeling and Optimization in Mobile, IEEE Ad Hoc and Wireless Networks, Ibaraki, Japan, 13–17 May 2013; pp. 7–14.
28. Jiang, L.; Leconte, M.; Ni, J.; Srikant, R.; Walrand, J. Fast mixing of parallel Glauber dynamics and low-delay CSMA scheduling. *IEEE Trans. Inform. Theory* **2012**, *58*, 6541–6555. [CrossRef]
29. Jiang, L.; Walrand, J. Scheduling and congestion control for wireless and processing networks. *Synth. Lect. Commun. Netw.* **2010**, *3*, 1–156. [CrossRef]
30. Boyd, S.; Vandenberghe, L. *Convex Optimization*; Cambridge University Press: Cambridge, UK, 2004.
31. Chapra, S.C.; Canale, R.P. *Numerical Methods for Engineers*, 7th ed.; McGraw-Hill Education: New York, NY, USA, 2015.

32. Rao, G.S. *Numerical Analysis*, 3rd ed.; New Age International (P) Ltd.: New Delhi, India, 2006.
33. Cheney, E.W.; Kincaid, D.R. *Numerical Mathematics and Computing*, 6th ed.; Cengage Learning: Mason, OH, USA, 2008.
34. Lay, D.C.; Lay, S.R.; McDonald, J. *Linear Algebra and Its Applications*, 5th ed.; Pearson: London, UK, 2016.
35. Shah, D.; David, N.C.; Tsitsiklis, J.N. Hardness of low delay network scheduling. *IEEE Trans. Inform. Theory* **2011**, *57*, 7810–7817. [CrossRef]
36. Lee, D.; Yun, D.; Shin, J.; Yi, Y.; Yun, S.Y. Provable per-link delay-optimal CSMA for general wireless network topology. In Proceedings of the IEEE Conference on Computer Communications (INFOCOM 2014), Toronto, ON, Canada, 27 April–2 May 2014; pp. 2535–2543.
37. Kwak, J.; Lee, C.H.; Eun, D.Y. Exploiting the past to reduce delay in CSMA scheduling: a high-order Markov chain approach. *ACM SIGMETRICS Perform. Eval. Rev.* **2013**, *41*, 353–354. [CrossRef]

© 2019 by the authors. Licensee MDPI, Basel, Switzerland. This article is an open access article distributed under the terms and conditions of the Creative Commons Attribution (CC BY) license (http://creativecommons.org/licenses/by/4.0/).

Article

Interference-Aware Subcarrier Allocation for Massive Machine-Type Communication in 5G-Enabled Internet of Things

Wenjun Hou [1], Song Li [1], Yanjing Sun [1,2,*], Jiasi Zhou [1], Xiao Yun [1] and Nannan Lu [1]

[1] School of Information and Control Engineering, China University of Mining and Technology, Xuzhou 221000, China; hauomenguanu@cumt.edu.cn (W.H.); lisong@cumt.edu.cn (S.L.); jiasi_zhou@cumt.edu.cn (J.Z.); yx.tong@163.com (X.Y.); lunannan@cumt.edu.cn (N.L.)
[2] school of Communication and Information Engineering, Xi'an University of Science and Technology, Xi'an 710054, China
* Correspondence: yjsun@cumt.edu.cn

Received: 3 August 2019; Accepted: 14 October 2019; Published: 18 October 2019

Abstract: Massive machine-type communication (mMTC) is investigated as one of three typical scenes of the 5th-generation (5G) network. In this paper, we propose a 5G-enabled internet of things (IoT) in which some enhanced mobile broadband devices transmit video stream to a centralized controller and some mMTC devices exchange short packet data with adjacent devices via D2D communication to promote inter-device cooperation. Since massive MTC devices have data transmission requirements in 5G-enabled IoT with limited spectrum resources, the subcarrier allocation problem is investigated to maximize the connectivity of mMTC devices subject to the quality of service (QoS) requirement of enhanced Mobile Broadband (eMBB) devices and mMTC devices. To solve the formulated mixed-integer non-linear programming (MINLP) problem, which is NP-hard, an interference-aware subcarrier allocation algorithm for mMTC communication (IASA) is developed to maximize the number of active mMTC devices. Finally, the performance of the proposed algorithm is evaluated by simulation. Numerical results demonstrate that the proposed algorithm outperforms the three traditional benchmark methods, which significantly improves the utilization of the uplink spectrum. This indicates that the proposed IASA algorithm provides a better solution for IoT application.

Keywords: 5G; internet of things; mMTC; eMBB

1. Introduction

In the future industrial internet of things (IIoT), a large number of devices including monitoring sensors and execution control units will be deployed to support factory automation and industry control system [1]. Massive periodic/non-periodic data will be transferred to a centralized control unit or adjacent devices via an industry wireless network, including video monitoring information, sensing data, operation instructions. However, due to the limited capacity and throughput of the current cellular system, it is insufficient in supporting future IoT applications with a tremendous number of devices and heterogeneous information traffic [2].

Massive machine-type communication (mMTC), as one of three typical application scenarios in the 5th-generation (5G) network, is investigated to support communication among a massive number of devices, which provides a feasible solution for future industrial IoT (IIoT) [3]. Due to limited spectrum resources in the cellular system, massive devices access the wireless network in a spectrum-sharing manner in which multiple devices are allocated in the same spectrum at the same time. Thus, the co-channel interference among devices restricts the number of devices connected to the cellular system. Effective interference management plays a vital role in mMTC to support the simultaneous access of more devices. The features and challenges of mMTC in IoT are as follows. First, devices in IoT need to exchange information with their neighbor devices frequently. In other words, the communication is performed between adjacent devices [4]. Second, the coding blocklengths for IoT are usually short, to reduce the transmission delay. The transmission rate cannot be estimated by the conventional Shannon's capacity, which assumes an infinite blocklength [5]. Third, a massive number of devices in IoT need to be supported. Thus, efficient resource allocation in mMTC needs to be investigated to address these challenges.

A wide range of works have contributed to the resource allocation problem in mMTC. In [6], the authors establish an interference model and a formulate resource allocation problem between users and machine-type communication (MTC) gateways in mMTC burst scenarios. In [7], the authors investigated the access management issues for MTC devices with heterogeneous quality of service (QoS) in the same cellular network. This work does not consider bandwidth utilization because transmission opportunities are reserved for a group of MTC devices at the same time. The authors in [8] propose two relay schemes and transmission protocols to specifically stimulate system capacity. In a multi-cell MTC system, Kwon et al. [9] establishes the interference model and analyzes the signal-to-interference-plus-noise-ratio (SINR) distributions and drives efficient resource allocation schemes. In [10], the authors propose a dynamic resource allocation algorithm based on the estimation of the number of MTC devices to handle massive and dynamic MTC devices while satisfying the random access delay requirement of MTC devices. To achieve effective resource utilization, a resource allocation metric based on statistical priority is proposed in [11]. In this way, effective resource utilization is achieved by letting MTC devices send a reduced set of their data. In [12], the authors consider a connectivity maximization problem for narrowband IoT with non-orthogonal multiple-access (NOMA). However, articles [6–12] assume that all devices communicate with the base station or centralized controller directly and do not consider the communications between adjacent devices.

D2D technology, as another promising technology in 5G, can establish communication between adjacent nodes, which can improve the spectrum efficiency and offload the load of Base Station (BS). D2D communication is introduced to mMTC system to stimulate spectrum efficiency and support more mMTC devices accessed with limited spectrum resources. However, designing better resource allocation algorithms to manage the inter-user interference between D2D users and cellulars is the key challenge for improving system performance. Resource allocation and interference problems of D2D communication have been investigated in many works [13–15]. The authors in [16] propose a cell sectorization scheme to alleviate the interference between cellular users and D2D users. In [17], the authors investigate interference coordination for downlink full-dimension multiple-output systems with underlying D2D communications.

Adopting D2D technology, the number of supported devices can be improved in mMTC scenarios [18]. By allowing unauthorized devices to reuse the frequency bands of authorized cellular users, bandwidth utilization can be improved [19–21]. The literature [22] proposes a heuristic subcarrier allocation method to set the user's signal-to-noise ratio (SNR) threshold to meet the QoS of the system. The authors in [23] propose a mobile traffic offloading scheme that combines small base stations with D2D offloading. The goal is to accommodate a large number of MTC connections by maximizing the throughput of the network system. The impact of radio frequency energy harvesting on the spectral efficiency of the D2D-assisted MTC system is analyzed first in [24]. In [25], the authors propose two solutions to manage the communication between D2D devices and the BS to lighten the overhead of MTC devices on the 5G network. However, resource allocation [19–25] mainly focuses on throughput maximization or interference minimization. In an IIoT enabled by D2D communications, massive devices demand access to the network via D2D mode. Thus, the connectivity maximization problem becomes a challenging issue to tackle. To support a system in which the number of users is higher than the number of subcarriers, a range of fair subcarrier allocation algorithms is proposed that always improves the reliability [26]. However, the author only considers the scenario of mobile users in the downlink and does not consider the influence of interference.

In this paper, we investigate a D2D-enabled internet of things in which some devices (enhanced Mobile Broadband (eMBB) devices) connect to the centralized controller while other devices (mMTC devices) communicate with their adjacent devices via D2D communication to promote inter-device cooperation in industry automation. Specifically, mMTC devices reuse the spectrum resource with eMBB devices. We establish the connectivity maximization problem of mMTC devices while guaranteeing the QoS of eMBB devices and mMTC devices. Furthermore, we propose an interference-aware subcarrier allocation algorithm to tackle the problem. The main contributions of this paper are as follows:

- We establish a problem of maximizing the number of accessed mMTC pairs subject to the constraints of QoS in a system with both eMBB and mMTC devices, which is proven to be a mixed-integer non-linear programming (MINLP) problem.
- We propose an interference-aware subcarrier allocation algorithm for mMTC (IASA) considering the interference range of each mMTC device.
- In order to evaluate the proposed algorithm, a simulation is conducted. The results demonstrate that the proposed algorithm outperforms two benchmark algorithms significantly in terms of the number of mMTC pairs accessed under the same constraints.

The remainder of the paper is organized as follows. The system model and assumptions is elaborated in the "System Model" section. The optimization problem and constraints are introduced in the "Problem Formulation" section. The proposed subcarrier allocation algorithm is presented in the "Interference-Aware Subcarrier Allocation for mMTC Communication Algorithm" section. Comprehensive simulation results are provided in the "Simulation result analysis" section. Finally, we conclude the paper in the "Conclusion" section.

2. System Model

In this paper, we investigate an industrial wireless network in which some devices, referred to as eMBB devices (such as monitoring cameras), transmit video information to a centralized controller, while devices referred to as mMTC devices (such as sensors and actuators), transmit short blocklength packets to their adjacent devices to promote inter-device cooperation and industrial automation. The system model is illustrated in Figure 1, where N eMBB devices and M mMTC devices are randomly distributed, represented by sets $\mathcal{N} = \{CU_1,...,CU_i,...,CU_N\}$ and $\mathcal{M} = \{MU_1,...,MU_j,...,MU_M\}$, respectively. The mMTC devices transmit information from the transmitters to receivers by the D2D communication mode, and the mMTC pairs reuse the eMBB devices' uplink resources in order to improve the spectrum efficiency. The mMTC pairs and eMBB devices are represented by MU_j and CU_i, respectively. Each mMTC pair is composed of one mMTC transmitter and one mMTC receiver represented by MU_j^t and MU_j^r, respectively. All available spectrum resources are divided into sub-carriers with the same bandwidth. Each eMBB device occupies mutually orthogonal sub-carriers. Therefore, there is no co-channel interference between the eMBB devices. We assume that the eMBB device CU_i occupies the subcarrier i. Considering the impact of devices on each other, each mMTC pair is allowed to access no more than one subcarrier, and each subcarrier can be accessed by multiple mMTC devices. All of the channels in the system are assumed to be quasi-static Rayleigh fading channels. The channel gain remains constant for each symbol transmission period but varies independently between different symbol periods. The parameters in the article are shown in Table 1.

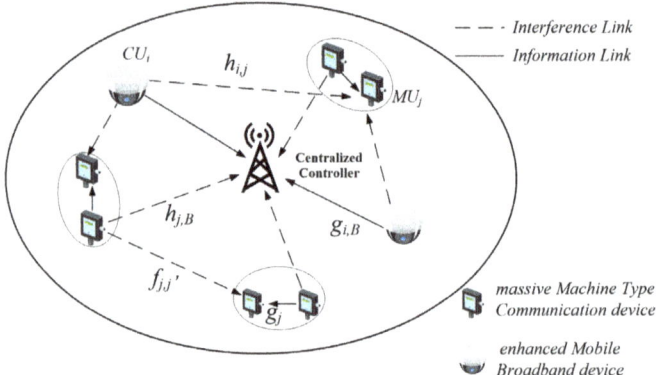

Figure 1. System model.

Table 1. System parameters.

Notation	Description
g_j	Gain between MU_j^t and MU_j^r
$g_{i,B}$	Gain between CU_i and the centralized controller
$h_{i,j}$	Interference gain between CU_i and MU_j^r
$h_{j,B}$	Interference gain between MU_j^t and the centralized controller
$f_{j',j}$	Interference gain between $MU_{j'}^t$ and MU_j^r
P_i^c	The transmit power of CU_i
P_j^d	The transmit power of MU_j^t

We define a binary subcarrier allocation matrix $A \in \{0,1\}^{M \times N}$, where $\alpha_{ij} = 1$ indicates that MU_j occupies subcarrier i, otherwise $\alpha_{ij} = 0$, $i \in \mathcal{N}, j \in \mathcal{M}$. Thus, the received signal of the base station on subcarrier i is

$$y_i = \sqrt{P_i^c} g_{i,B} x_i + \sum_{j \in \mathcal{M}} \alpha_{ij} \sqrt{P_j^d} h_{j,B} x_j + n_0, \qquad (1)$$

where x_i and x_j are the transmitted signals of the eMBB devices CU_i and the mMTC transmitter MU_j^t, respectively. n_0 represents normalized additive white Gaussian noise, $n_0 \sim \mathcal{CN}(0, \sigma_0^2)$. When the centralized controller receives the uplink signals of the eMBB device CU_i, the SINR at centralized controller can be calculated as

$$\gamma_{cu}^i = \frac{|g_{i,B}|^2 P_i^c}{I + \sigma_0^2 B}, \qquad (2)$$

where B is the subcarrier bandwidth and I represents the interference caused by the mMTC pairs which access the subcarrier i:

$$I = \sum_{j=1}^{M} \alpha_{ij} P_j^d |h_{j,B}|^2. \qquad (3)$$

Since the eMBB devices transmit long packet data, the achievable transmission rate of the eMBB device CU_i can be obtained by Shannon's theorem:

$$R_{cu}^i = B \log\left(1 + \gamma_{cu}^i\right). \qquad (4)$$

The signal received by the jth mMTC pair MU_j is

$$z_j = \sqrt{P_j^d} g_j x_j + \sum_{i \in \mathcal{N}} \alpha_{ij} \sqrt{P_i^c} h_{i,j} x_i + \sum_{\substack{j'=1 \\ j' \neq j}}^{M} \sum_{i \in \mathcal{N}} \alpha_{ij'} \sqrt{P_{j'}^d} f_{j',j} x_{j'} + n_0, \tag{5}$$

where the first term is the signal receiver MU_j^r received from mMTC transmitter MU_j^t. The second term is the interference from the eMBB device CU_i. The third term is the interference signal from the transmitter of the mMTC pair $MU_{j'}$ that occupies the same subcarrier with mMTC pair MU_j. According to (5), the SINR of the mMTC receiver MU_j^r can be derived as

$$\gamma_j = \frac{P_j^d |g_j|^2}{\sum_{i \in \mathcal{N}} \alpha_{ij} P_i^c |h_{i,j}|^2 + \sum_{\substack{j'=1 \\ j' \neq j}}^{M} \sum_{i \in \mathcal{N}} \alpha_{ij'} P_{j'}^d |f_{j',j}|^2 + \sigma_0^2}. \tag{6}$$

Since mMTC communication is mostly aimed at periodic monitoring data in IoT applications such as smart cities, the length of data packets transmitted is usually very short. According to information theory, the rate of short packets cannot achieve the Shannon limit. Therefore, the transmission rate of the mMTC devices is represented by the short packet rate [27], as shown in (7):

$$R_j = \log_2(1 + \gamma_j) - \sqrt{\frac{1}{m}} \log_2(e) Q^{-1}(\varepsilon) \sqrt{1 - \frac{1}{(1 + \gamma_j)^2}} \tag{7}$$

where m is the block length, ε is the transmission error probability, and $Q^{-1}(x)$ is the inverse of the Gaussian Q function.

3. Problem Formulation

The optimization goal of this paper is to maximize the total number of mMTC devices accessed under the QoS of each eMBB device. All of the mMTC devices access the network adhere to the following criteria:

(i) Each mMTC device is allowed to access no more than one subcarrier.
(ii) To ensure the transmission quality of the eMBB devices, the interference each subcarrier can suffer should be below a threshold.
(iii) Both eMBB devices and mMTC devices should satisfy their own transmission rate requirements.

When the jth mMTC pair is allowed to access the subcarrier i, the SINR of the receiver of MU_j is

$$\gamma_j = \frac{P_j^d |g_j|^2}{P_i^c |h_{i,j}|^2 + \sum_{\substack{j'=1 \\ j' \neq j}}^{M} \alpha_{ij'} P_{j'}^d |f_{j',j}|^2 + \sigma_0^2}. \tag{8}$$

Taking γ_j in (8) into (7), we can get the transmission rate R_j after mMTC pair MU_j is allowed to access the subcarrier i. In order to ensure the QoS of the mMTC pair MU_j, the achievable rate R_j of the mMTC pair MU_j should not be less than the minimum required rate R_{j_min}, i.e.,

$$R_j \geq R_{j\_min}. \tag{9}$$

When the mMTC pair MU_j is allowed to access the subcarrier i, the minimum transmission rate of the eMBB device CU_i is $R_{cu\_min}^i$. Under the QoS constraint of the eMBB devices and the mMTC pairs, we maximize the number of mMTC pairs accessed in the network. Mathematically, the optimization problem is formulated as

$$\max_{\alpha_{ij}} \sum_{i=1}^{N} \sum_{j=1}^{M} \alpha_{ij}, \tag{10a}$$

$$s.t \quad R_{cu}^i \geq R_{cu\_min}^i, \tag{10b}$$

$$R_j \geq R_{j\_min}, \tag{10c}$$

$$\alpha_{ij} \in \{0,1\}, \quad \forall i \in \mathcal{N} \ j \in \mathcal{M}, \tag{10d}$$

$$\sum_{i=1}^{N} \alpha_{ij} \leq 1. \tag{10e}$$

The constraints described in (10b) and (10c) indicate that the respective minimum transmission rate requirements of the eMBB devices and the mMTC pairs should be satisfied. Equation (10e) reveals that each mMTC pair is allowed to occupy at most one subcarrier. The optimization problem is a binary optimization problem, and the traversal complexity of the problem is 2^{MN}. In the following section, we propose a lower complexity algorithm named interference-aware subcarrier allocation for mMTC communication.

4. Interference-Aware Subcarrier Allocation for mMTC Communication Algorithm

According to the constraint condition (10b), the accumulated interference allowed by the eMBB device CU_i can be derived from the minimum transmission rate $R_{cu\_min}^i$

$$I_{cu\_max}^i = \frac{P_i^c |g_{i,B}|^2}{2^{R_{cu\_min}^i} - 1} - \sigma_0^2 B. \tag{11}$$

When the jth mMTC pair MU_j occupies the subcarrier i, the interference caused by MU_j to the base station is

$$I_{j,B} = P_j^d |h_{j,B}|^2. \tag{12}$$

To ensure the rate requirement of the eMBB device, the interference caused by the mMTC pair MU_j should not exceed the maximum interference allowed by the subcarrier i, i.e.,

$$I_{j,B} \leq I_{cu\_max}^i. \tag{13}$$

Therefore, a set of mMTC pairs allowed to occupy subcarrier i can be selected according to Equation (13). For all mMTC pairs that can be accessed, we first define the normalized interference caused by each pair as $I_{j,i} = I_{j,B}/I_{cu\_max}^i$. Define the interference matrix as

$$\Omega = \begin{bmatrix} I_1^1 & I_2^1 & \cdots & I_M^1 \\ I_1^2 & I_2^2 & \cdots & I_M^2 \\ \vdots & \vdots & \cdots & \vdots \\ I_1^N & I_2^N & \cdots & I_M^N \end{bmatrix}.$$

When the mMTC pair $MU_{j'}$, $j' \neq j$, attempts to access the subcarrier i, the following two conditions should be satisfied.

(i) The interference caused by $MU_{j'}$ and the total interference of mMTC pairs cannot exceed the maximum interference allowed by subcarrier i, i.e., $I_{cu\_max}^i$.
(ii) All of the accessed mMTC pairs should satisfy their own QoS.

When both conditions are satisfied, the mMTC pair is allowed to access the subcarrier. In order to facilitate the calculation, we convert the rate constraint of the eMBB device in (11) into an interference limit:

$$I_{cu}^i \leq I_{cu\_max}^i, \tag{14}$$

where

$$I_{cu}^i = \sum_{j \in M} I_{j,B} \quad \forall i \in \mathcal{N}. \tag{15}$$

Substituting (14) for (11), then the optimization problem is converted into

$$\max_{\alpha_{ij}} \sum_{i=1}^{N} \sum_{j=1}^{M} \alpha_{ij}, \tag{16a}$$

$$s.t \quad I_{cu}^i \leq I_{cu\_max}^i, \tag{16b}$$

$$R_j \geq R_{j\_min}, \tag{16c}$$

$$\alpha_{ij} \in \{0,1\}, \quad \forall i \in \mathcal{N} \quad j \in \mathcal{M}, \tag{16d}$$

$$\sum_{i=1}^{N} \alpha_{ij} \leq 1 \quad \forall j \in \mathcal{M}. \tag{16e}$$

The optimization variables of this problem are all binary variables, and the optimization problem (16a) is an NP-hard problem that cannot be solved by the convex optimization method. The traditional way to solve an NP-hard problem is to carry out an exhaustive search, which involves unaccepted computational complexity. Thus, it is difficult to obtain the optimal result by direct solution. This paper proposes a heuristic algorithm with lower complexity to tackle the problem, referred to as interference-aware subcarrier allocation for mMTC.

Firstly, the mMTC pair with minimal interference to the base station is selected in the two-dimensional matrix Ω, and the access conditions (16b) and (16c) are updated according to the QoS of the eMBB devices and other mMTC pairs. Once an mMTC pair occupies a subcarrier, the mMTC pair is prohibited from accessing other subcarriers. When the sum of the interference ratio accumulated on subcarrier i is not less than 1, other mMTC pairs will be no longer allowed to access the subcarrier i.

For the constraint condition (16c), we can estimate whether the QoS requirements are still satisfied after each mMTC pair occupies the subcarrier according to (7) and (9). Due to the interference between mMTC pairs accessing the same channel, the interference range of the mMTC pair is defined in this paper to suppress interference between mMTC pairs. The interference range of an mMTC pair MU_j is defined as the range in which the mMTC pairs suffer from the interference of MU_j. Thus, the mMTC pairs in the interference range of MU_j cannot access the same subcarrier with MU_j to avoid interference. In other words, when mMTC pair MU_j accesses subcarrier i, the other pairs within the interference range of MU_j cannot access subcarrier i to reduce the interference between the mMTC pairs and to ensure the QoS of each mMTC pair.

The IASA algorithm is summarized in Algorithm 1. Lines 1–5 of the algorithm calculate the maximum interference that all subcarriers can support (line 3) and the proportion of the interference from each mMTC pair (line 5). Then a two-dimensional matrix is formed. Lines 6–12 of the algorithm sort the data in the two-dimensional matrix in ascending order one-dimensionally. Firstly, we find the mMTC pair and subcarrier corresponding to the minimum interference ratio. Then we estimate whether the accumulated interference caused by the mMTC pair exceeds the maximum interference allowed by the subcarrier so as to determine whether the mMTC pair can access the sub-carrier (line 7). After the mMTC pair MU_j accesses subcarrier i, the mMTC pair within the interference range of MU_j is prohibited from accessing subcarrier i (line 9).

Algorithm 1 Interference-Aware Spectrum Allocation for mMTC Communication Algorithm.

1: Initialize: B, P_i^c, P_j^d, $R_{cu\_min}^i$, R_j, $\alpha_{ij} = 0$ $\forall i \in \mathcal{N}, j \in \mathcal{M}$.
2: Calculate $I_{cu\_max}^i$, $i \in \mathcal{N}$ in accordance with (11).
3: Calculate the interference $I_{j,B}$ caused by all mMTC pairs accessing subcarriers according to (10c).
4: Calculate the proportion of all mMTC pairs to the maximum interference that each subcarrier can withstand $I_{cu\_max}^i$ and obtain the interference matrix Ω.
5: Select the smallest element (i^*, j^*) in matrix Ω. For CU_i^*, MU_j^*, judge whether (9) and (14) is established.
6: If $I_{cu}^i \le I_{cu\_max}^i$, the mMTC pair is allowed to access subcarriers, $\alpha_{ij} = 1$.
   If $I_{cu}^i > I_{cu\_max}^i$, the mMTC pair is not allowed to access the subcarrier, $\alpha_{ij} = 0$.
7: $\alpha_{i'j} = 0, \forall i' \in \mathcal{N}, i' \ne i$.
8: Filter out the mMTC pair set J within the pair MU_j interference range. Assign all the α_{ij} $j \in J$ corresponding to subcarrier i to 0.
9: Assign the ratio $\Omega(:,j)$ of the mMTC pair to the corresponding subcarrier in this cycle to $s(s \ge 1)$.
10: Repeat 6–9 until all of the mMTC pairs have been assigned.
11: Output matrix A.

The IASA algorithm is a centralized algorithm that can be implemented by a centralized controller. First, the centralized controller collects the information from mMTC devices who want to transmit packets with their neighborhood devices, including the channel state information and the required transmission rate. Then the centralized controller obtains the spectrum allocation results according to the IASA algorithm and broadcasts the allocation results to each mMTC pair. Then each mMTC transmitter completes the packets transmission on its allocated subcarrier.

5. Simulation Result Analysis

In this section, we present numerical results to verify the performance of the proposed IASA algorithm. We compare the number of mMTC pairs accessing the network successfully according to the proposed subcarrier allocation algorithm, the random access algorithm, and the sequential access algorithm. The two benchmark algorithms are described as follows.

Random access algorithm (RAA): Firstly, an interference ratio is randomly selected in the two-dimensional interference matrix. Then, we find out the corresponding mMTC devices and the access to subcarrier i. According to (9) and (14), it can be judged whether the QoS of the eMBB devices and the mMTC devices is satisfied, that is, whether the mMTC devices can access the subcarrier i. In the end, we repeat the above selection and access process until all subcarriers achieve the limit of interference they can sustain.

Sequential access algorithm (SAA): All of the mMTC devices sequentially judge whether the QoS of the eMBB devices and the mMTC devices are satisfied. When the sum of the interference ratio of subcarrier i is more than 1, other mMTC devices are prohibited from accessing the subcarrier i.

Greedy algorithm (GA): In the greedy algorithm, each subcarrier gives priority to its own access number. Specifically, the mMTC pair with minimum interference to a certain subcarrier is firstly accessed, when the QoS of the eMBB device and mMTC pair can be satisfied. When the cumulative interference of subcarrier i exceeds 1, other mMTC pairs are forbidden from accessing the subcarrier.

It is assumed that the eMBB and mMTC devices are evenly distributed in a circular region where the radius is 200 m. All of the devices are served by the same base station that controls the allocation of subcarriers. This paper considers a flat Rayleigh fading channel. The distance-dependent path loss PL(D) is [12]

$$PL(D) = 120.9 + 37.6 \log(D/1000) + L + AG, \tag{17}$$

where D is the communication distance and AG is the antenna gain. The value of AG is 0.4 dB. L is the indoor penetration loss. we assume that 80% of mMTC equipment is indoor equipment, where L takes 20 dB; 20% is outdoor mMTC equipment, where L takes 0 dB.

Figure 2 is a location distribution diagram of eMBB devices and mMTC devices. We consider an area with a radius of 200 m, in which the eMBB devices and mMTC devices are evenly and randomly distributed in the area, and the base station is set at the origin.

Figure 3 shows the number of mMTC pairs that successfully access subcarriers for different eMBB devices with $P_i^c = 10$ dB, $P_j^d = 7$ dB, $R_{cu\_min}^i = 10$ bps, $R_j = 5$ bps. The total number of mMTC pairs is 150, and the interference range of the mMTC pair is 80 m. Among the four algorithms, the number of mMTC pairs accessed increases as the number of subcarriers increases. The reason is that when the number of subcarriers increases, the mMTC pairs will be more likely to access the subcarriers. Some mMTC pairs with large interference also have the opportunity to access the subcarriers. Compared to the three contrastive algorithms, the IASA algorithm can realize more mMTC devices accessed in the system. The reason is that under the premise of guaranteeing the QoS of the eMBB devices and the mMTC pairs, the mMTC pair with the least interference to the subcarriers is selected first according to the IASA algorithm. The distance limitation is established to reduce the interference between the adjacent mMTC pairs. Meanwhile, the complexity of the algorithm is reduced. For GA, each subcarrier gives priority to the access number optimization of itself rather than the access performance of the whole system, so the access number of the system cannot be maximized. However, since GA considers the mMTC priority access with less interference in each subcarrier, its performance is better than RAA and SAA. For RAA and SAA, there may be an mMTC pair with large interference accessing the subcarrier at any time, which occupies a large proportion of the interference space that the subcarrier can sustain. Under these circumstances, some mMTC pairs with small interference cannot access the subcarrier because the space for interference is finite. Therefore, the IASA algorithm enables the system to accommodate more mMTC pairs.

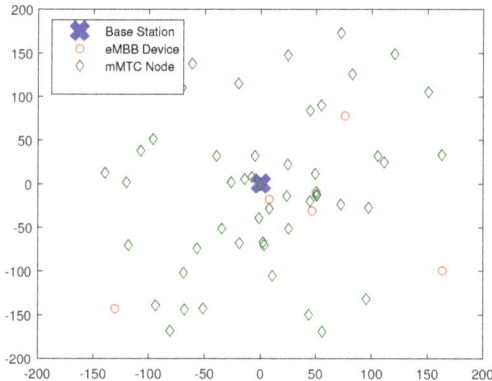

Figure 2. Distribution of eMBB devices and mMTC devices in cellular systems.

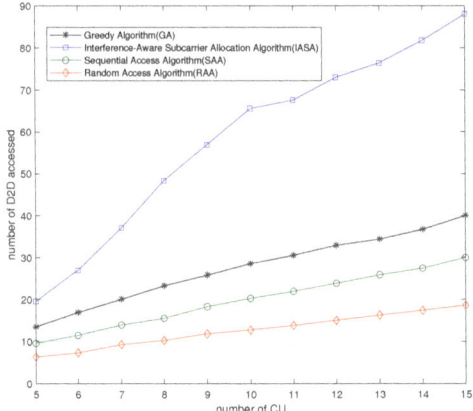

Figure 3. Number of mMTC pairs accessed versus different eMBB devices using different algorithms.

Figure 4 depicts the number of mMTC pairs that access subcarriers for different eMBB devices with $P_i^c = 10$ dB, $P_j^d = 7$ dB, $R_j = 5$ bps. The interference range of the mMTC pairs is 80 m. The number of mMTC pairs that access the subcarriers decreases gradually when the minimum transmission rate of the eMBB devices gradually increases. The reason is that when the minimum transmission rate of the eMBB devices increases, the maximum interference that each subcarrier can sustain is reduced. In the case where the interference caused by the mMTC pair is unchanged, the number of mMTC pairs that can access the subcarriers is reduced. In the low-rate phase, the RAA exhibits much lower access performance than the IASA algorithm. And the SAA exhibits a comparable access performance with IASA algorithm. However, as the rate of eMBB devices increase, the performance of SAA have dropped significantly compared to the IASA algorithm.

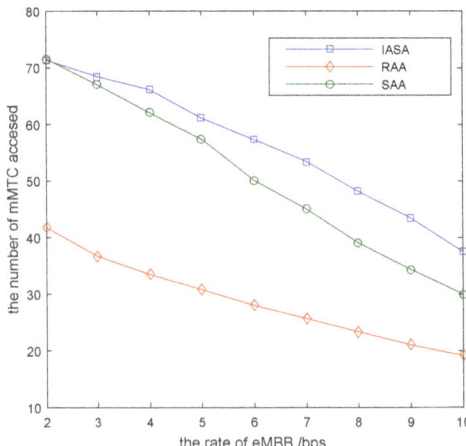

Figure 4. Number of mMTC pairs accessed versus different rate requirements of eMBB devices.

Figure 5 presents the number of mMTC pairs accessed when the power of the mMTC pairs transmitter change with $P_i^c = 10$ dB, $R_{cu\_min} = 10$ bps, $R_j = 5$ bps. The interference range of the mMTC pairs is 80 m. The number of accessed mMTC devices gradually decreases as the power of the mMTC transmitter increases. This is due to the fact that the interference to subcarriers increases while the power of the mMTC transmitting increases. Therefore, the number of mMTC devices accessed is relatively reduced. Compared to the variables such as the number of eMBB devices and the minimum transmission rate of eMBB devices, the power of the mMTC transmitter has a relatively small impact on the number of mMTC pairs accessed in the system.

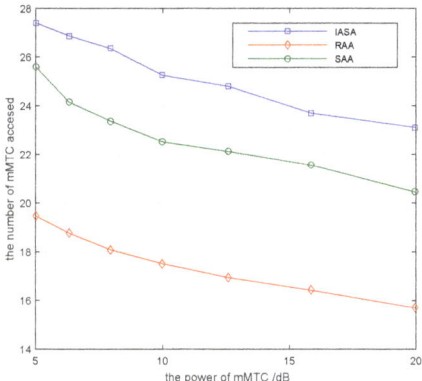

Figure 5. Number of mMTC pairs accessed versus different mMTC device transmit powers.

Figure 6 investigates the number of mMTC pairs accessed when there are different numbers of mMTC pairs in the system. The parameter settings are the same as in Figure 3. The number of mMTC devices accessed increases as the number of mMTC pairs in the system increases. Before the number of subcarriers is saturated, the more mMTC devices in the system, the greater the opportunity to access devices that satisfy the QoS requirements of the eMBB devices. Thus, the total number of mMTC pairs accessed will increase. However, when the power and the minimum transmission rate of the eMBB devices are fixed, the number of mMTC devices accessed by all subcarriers is constant, so the number of mMTC pairs accessed will gradually become saturated. Among the three algorithms shown in Figure 6, the RAA achieves the access saturation state first.

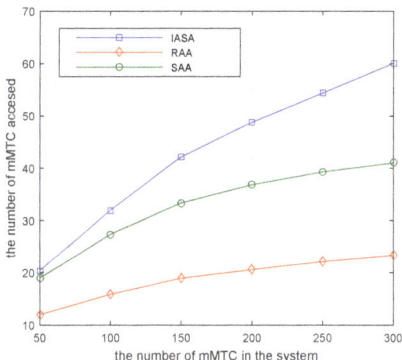

Figure 6. Number of mMTC pairs accessed versus different number of mMTC devices.

6. Conclusions

In this paper, an interference-aware subcarrier allocation algorithm for mMTC is proposed for the subcarrier allocation problem of D2D communication in mMTC scenarios. Initially, we establish a model maximizing the number of mMTC pairs accessed. When carrying out subcarrier allocation, we calculate the maximum interference that each subcarrier can sustain. Then we determine the mMTC pair with the least interference to the subcarrier and estimate the access property according to whether the QoS of the mMTC pair is satisfied. When the accumulated interference caused by the mMTC pair is greater than the maximum interference limit that the subcarrier can sustain, the subcarrier will be not be accessible to other mMTC pairs anymore. Simulation results demonstrated the effectiveness of the proposed algorithm.

Author Contributions: W.H. designed the algorithm, performed the theoretical analysis, and wrote the manuscript. S.L. and Y.S. implemented the simulation and contributed to the manuscript preparation. J.Z. analyzed the data. X.Y. and N.L. contributed to polishing the revised manuscript and provided suggestions on simulation evaluation.

Acknowledgments: This work was supported by the Science and Technology Project of Xuzhou (KC18105 KC18068), the National Natural Science Foundation of China (61771417 51734009 51804304), the National Key Research and Development Program (2016YFC0801403), and the Fundamental Research and Development Foundation of Jiangsu Province.

Conflicts of Interest: The authors declare no potential conflicts of interest with respect to the research, authorship, and/or publication of this article.

References

1. Meng, Z.; Wu, Z.; Gray, J. A Collaboration-Oriented M2M Messaging Mechanism for the Collaborative Automation between Machines in Future Industrial Networks. *Sensors* **2017**, *17*, 2694. [CrossRef] [PubMed]
2. Park, J.H. Advances in Future Internet and the Industrial Internet of Things. *Symmetry* **2019**, *11*, 244. [CrossRef]
3. Al-Sakran, A.; Qutqut, M.H.; Almasalha, F.; Hassanein, H.S.; Hijjawi, M. An Overview of the Internet of Things Closed Source Operating Systems. In Proceedings of the 2018 14th International Wireless Communications Mobile Computing Conference (IWCMC), Limassol, Cyprus, 25–29 June 2018; pp. 291–297. [CrossRef]
4. Laya, A.; Alonso, L.; Alonso-Zarate, J. Is the random access channel of LTE and LTE-A suitable for M2M communications? A survey of alternatives. *IEEE Commun. Surv. Tutor.* **2013**, *16*, 4–16. [CrossRef]
5. Kawabata, H.; Ishibashi, K.; Vuppala, S.; de Abreu, G.T. Robust relay selection for large-scale energy-harvesting IoT networks. *IEEE Internet Things J.* **2016**, *4*, 384–392. [CrossRef]
6. Hu, X.; Sun, J. Interference Analysis and Resource Allocation of Burst Scenario in Massive Machine-Type Communications. In Proceedings of the 2018 IEEE 18th International Conference on Communication Technology (ICCT), Chongqing, China, 8–11 October 2018; pp. 822–826. [CrossRef]
7. Lien, S.; Chen, K. Massive Access Management for QoS Guarantees in 3GPP Machine-to-Machine Communications. *IEEE Commun. Lett.* **2011**, *15*, 311–313. [CrossRef]
8. Tefek, U.; Lim, T.J. Relaying and Radio Resource Partitioning for Machine-Type Communications in Cellular Networks. *IEEE Trans. Wirel. Commun.* **2017**, *16*, 1344–1356. [CrossRef]
9. Kwon, T.; Choi, J. Multi-Group Random Access Resource Allocation for M2M Devices in Multicell Systems. *IEEE Commun. Lett.* **2012**, *16*, 834–837. [CrossRef]
10. Oh, C.; Hwang, D.; Lee, T. Joint Access Control and Resource Allocation for Concurrent and Massive Access of M2M Devices. *IEEE Trans. Wirel. Commun.* **2015**, *14*, 4182–4192. [CrossRef]
11. Mostafa, A.E.; Gadallah, Y. Uniqueness-Based Resource Allocation for M2M Communications in Narrowband IoT Networks. In Proceedings of the 2017 IEEE 86th Vehicular Technology Conference (VTC-Fall), Toronto, ON, Canada, 24–27 September 2017; pp. 1–5. [CrossRef]
12. Mostafa, A.E.; Zhou, Y.; Wong, V.W. Connectivity maximization for narrowband IoT systems with NOMA. In Proceedings of the 2017 IEEE International Conference on Communications (ICC), Paris, France, 21–25 May 2017; pp. 1–6.
13. Han, S.; Xu, X.; Zhao, L.; Tao, X. Joint time and power allocation for uplink cooperative non-orthogonal multiple access based massive machine-type communication Network. *Int. J. Distrib. Sens. Netw.* **2018**, *14*, 1550147718778215. [CrossRef]
14. Li, J.; Zhang, X.; Feng, Y.; Li, K.C. A Resource Allocation Mechanism Based on Weighted Efficiency Interference-Aware for D2D Underlaid Communication. *Sensors* **2019**, *19*, 3194. [CrossRef] [PubMed]

15. Li, Z.; Gui, J.; Xiong, N.; Zeng, Z. Energy-Efficient Resource Sharing Scheme With Out-Band D2D Relay-Aided Communications in C-RAN-Based Underlay Cellular Networks. *IEEE Access* **2019**, *7*, 19125–19142. [CrossRef]
16. Ningombam, D.D.; Lee, C.G.; Shin, S. Interference Mitigation for Multicast D2D Communications Underlay Cellular Networks. In Proceedings of the 2019 International Conference on Artificial Intelligence in Information and Communication (ICAIIC), Okinawa, Japan, 11–13 February 2019; pp. 1–4. [CrossRef]
17. Li, X.; Qin, N.; Sun, T. Interference coordination for FD-MIMO cellular network with D2D communications underlaying. *China Commun.* **2018**, *15*, 75–88.
18. Asadi, A.; Wang, Q.; Mancuso, V. A survey on device-to-device communication in cellular networks. *IEEE Commun. Surv. Tutor.* **2014**, *16*, 1801–1819. [CrossRef]
19. Kai, Y.; Wang, J.; Zhu, H.; Wang, J. Resource Allocation and Performance Analysis of Cellular-Assisted OFDMA Device-to-Device Communications. *IEEE Trans. Wirel. Commun.* **2019**, *18*, 416–431. [CrossRef]
20. Lu, B.; Lin, S.; Shi, J.; Wang, Y. Resource Allocation for D2D Communications Underlaying Cellular Networks Over Nakagami- *m* Fading Channel. *IEEE Access* **2019**, *7*, 21816–21825. [CrossRef]
21. Chour, H.; Jorswieck, E.A.; Bader, F.; Nasser, Y.; Bazzi, O. Global Optimal Resource Allocation for Efficient FD-D2D Enabled Cellular Network. *IEEE Access* **2019**, *7*, 59690–59707. [CrossRef]
22. Huang, Y.F.; Tan, T.H.; Liu, S.P.; Liu, T.Y.; Chen, C.M. Performance of subcarrier allocation of D2D multicasting for wireless communication systems. In Proceedings of the 2018 Tenth International Conference on Advanced Computational Intelligence (ICACI), Xiamen, China, 29–31 March 2018; pp. 193–196.
23. Cao, W.; Feng, G.; Qin, S.; Liang, Z. D2D Communication Assisted Traffic Offloading for Massive Connections in HetNets. In Proceedings of the 2016 IEEE Global Communications Conference (GLOBECOM), Washington, DC, USA, 4–8 December 2016; pp. 1–6. [CrossRef]
24. Atat, R.; Liu, L.; Mastronarde, N.; Yi, Y. Energy Harvesting-Based D2D-Assisted Machine-Type Communications. *IEEE Trans. Commun.* **2017**, *65*, 1289–1302. [CrossRef]
25. Bagaa, M.; Ksentini, A.; Taleb, T.; Jantti, R.; Chelli, A.; Balasingham, I. An efficient D2D-based strategies for machine type communications in 5G mobile systems. In Proceedings of the 2016 IEEE Wireless Communications and Networking Conference, Doha, Qatar, 3–6 April 2016; pp. 1–6. [CrossRef]
26. Shi, J.; Yang, L.L. Novel subcarrier-allocation schemes for downlink MC DS-CDMA systems. *IEEE Trans. Wirel. Commun.* **2014**, *13*, 5716–5728. [CrossRef]
27. Xu, S.; Chang, T.; Lin, S.; Shen, C.; Zhu, G. Energy-Efficient Packet Scheduling With Finite Blocklength Codes: Convexity Analysis and Efficient Algorithms. *IEEE Trans. Wirel. Commun.* **2016**, *15*, 5527–5540. [CrossRef]

© 2019 by the authors. Licensee MDPI, Basel, Switzerland. This article is an open access article distributed under the terms and conditions of the Creative Commons Attribution (CC BY) license (http://creativecommons.org/licenses/by/4.0/).

Article

K-Means Spreading Factor Allocation for Large-Scale LoRa Networks

Muhammad Asad Ullah [1], Junnaid Iqbal [1], Arliones Hoeller [1,2,3], Richard Demo Souza [2] and Hirley Alves [1,*]

1. Centre for Wireless Communications, University of Oulu, 90014 Oulu, Finland; muhammad.asadullah@oulu.fi (M.A.U.); junnaid.iqbal@oulu.fi (J.I.)
2. Department of Electrical and Electronics Engineering, Federal University of Santa Catarina, Florianópolis 88040-900, Brazil; richard.demo@ufsc.br
3. Department of Telecommunications Engineering, Federal Institute for Education, Science, and Technology of Santa Catarina, São José 88103-310, Brazil; arliones.hoeller@ifsc.edu.br
* Correspondence: Hirley.Alves@oulu.fi

Received: 30 September 2019; Accepted: 28 October 2019; Published: 30 October 2019

Abstract: Low-power wide-area networks (LPWANs) are emerging rapidly as a fundamental Internet of Things (IoT) technology because of their low-power consumption, long-range connectivity, and ability to support massive numbers of users. With its high growth rate, Long-Range (LoRa) is becoming the most adopted LPWAN technology. This research work contributes to the problem of LoRa spreading factor (SF) allocation by proposing an algorithm on the basis of K-means clustering. We assess the network performance considering the outage probabilities of a large-scale unconfirmed-mode class-A LoRa Wide Area Network (LoRaWAN) model, without retransmissions. The proposed algorithm allows for different user distribution over SFs, thus rendering SF allocation flexible. Such distribution translates into network parameters that are application dependent. Simulation results consider different network scenarios and realistic parameters to illustrate how the distance from the gateway and the number of nodes in each SF affects transmission reliability. Theoretical and simulation results show that our SF allocation approach improves the network's average coverage probability up to 5 percentage points when compared to the baseline model. Moreover, our results show a fairer network operation where the performance difference between the best- and worst-case nodes is significantly reduced. This happens because our method seeks to equalize the usage of each SF. We show that the worst-case performance in one deployment scenario can be enhanced by 1.53 times.

Keywords: stochastic geometry; resource allocation; Internet of Things

1. Introduction

The Internet of Things (IoT) is the integration of modern electronic devices, smart sensors, internet protocols, and wireless communications technologies. IoT applications are rapidly gaining popularity in many domains such as industrial operations, smart parking, augmented maps, healthcare, smart cars, and smart homes [1–5]. According to a Gartner Inc. report, there will be around 26 billion IoT devices deployed worldwide by 2020 [6]. In the Statista report, it is predicted that there will be over 75 billion IoT devices worldwide by 2025 [7].

In the modern era, the spectacular growth and transformation of wireless connectivity are driven by the IoT paradigm, with technologies having attributes of large-scale network infrastructure with low-cost sensors connected to the Internet. In this context, low-power wide-area networks (LPWANs) are quite popular in terms of prototypes, standards, and on the commercial level because of their significance with respect to power efficiency along with long range [8,9]. Within this context, LoRA, SigFox, NB-IoT, Weightless, RPMA and DASH7 [10,11] are the most distinguished technologies.

This paper focuses on LoRa, which provides good performance in terms of reliability and energy consumption. The network architecture contains end-devices, gateways, and a network server (NS), forming a star topology. It operates at unlicensed frequency ISM (Industrial, Scientific, Medical) bands of 863–870 MHz and 915 MHz in Europe and the U.S., respectively [12,13]. In Europe, the duty cycle limitations range from 0.1% to 10%, following European Telecommunications Standards Institute (ETSI) standards. In addition, LoRa works on variable and adaptive data rates by using different spreading factors. This is achieved by the NS controlling the spreading factors (SFs) and bandwidth (BW) of the end-devices. Higher SFs allow larger coverage areas; however, as a drawback, they reduce the data rate and increase the time-on-air (ToA) of LoRa packets [14].

Notably, the gateway has the ability to receive data from multiple nodes at the same time because of the orthogonality of sub-bands and the quasi-orthogonality of different SFs. The LoRa MAC layer, known as LoRaWAN [15], is a type of ALOHA protocol controlled by the NS. LoRaWAN defines three classes of devices depending upon the application. Class A devices may wait for acknowledgments (ACK) only in their receiving windows during downlink transmission and consume the least power. Class B devices are able to open extra receiving windows at scheduled times, thus reducing downlink latency. Class C nodes consume the most energy because they leave the receiver enabled all the time, allowing for the lowest latency time [16].

For instance, extensive measurement campaigns show that the communication range of LoRa reaches up to 30 km over the water and more than 15 km on the ground [11]. LoRa is suitable for a wide range of telemetry applications (e.g., sensing and monitoring), which can be used in several industry verticals, such as smart grids and cities, and smart agriculture up to industrial IoT applications [17,18]. During the past few years, many studies have contributed by proposing new algorithms, systems models, analyses, and by designing new approaches for performance enhancement of LoRa networks. However, only a few considered resource allocation.

The major contribution of this work is the modeling of an approach for SF allocation for a large scale LoRa network based on K-means clustering and the analysis of connection, capture, and coverage probabilities. Instead of using constant steps of distance from the gateway to define SF areas [19,20], the proposed algorithm assigns a maximum range of individual SF regions, which allows for distinct user distribution. Then, we evaluate the performance of the proposed algorithm over the uplink of a large-scale LoRa network with a single gateway based on the model introduced in [19].

The remainder of this article is structured as follows. Section 2 discusses related work and a short overview of LPWAN technologies. Section 3 introduces the system model, and Section 3.1 details the outage probabilities of the baseline model, used to examine the performance of proposed SF allocation approach. The proposed algorithm is presented in Section 4. Simulation results are discussed in Section 5. Finally, Section 6 concludes the paper and proposes future work.

2. Related Work

Overviews of LoRa and LPWAN technologies are provided in [21,22]. Usually, LoRa operates with a bandwidth of 125 kHz, but it also allows for bandwidths of 250 kHz and 500 kHz. The wider bands promote resistance to fading, channel noise, Doppler effects, and long-term relative frequency [23]. Chirp spread spectrum (CSS) modulation, which enables high receiver sensitivity, makes LoRa more robust against the interference when compared to Sigfox, which employs ultra-narrowband (UNB) communication [24]. As a tradeoff, the use of wider bands for the transmission of narrowband signals makes less efficient use of the spectrum. A realistic SigFox communication model is implemented and tested in [25]; it evaluates the performance of a high-density large-scale wireless sensor network (WSN). From the obtained results, one can observe that the performance of the SigFox network significantly degrades by increasing the number of sensors, and some solutions are presented to improve the performance.

Unlike Sigfox, LoRa can be deployed locally, i.e., without the need for a cellular infrastructure, and has higher bit rates. By contrast, NB-IoT is an expensive technology having the pros of low latency and high quality of service (QoS) [26]. In [27], the authors compare different LPWAN technologies (Bluetooth, ZigBee, SigFox, and LoRa) and discuss LoRa with respect to code rate (CR), bandwidth (BW), and SF but without considering the influence of Rayleigh fading and path loss attenuation. Theoretical and simulation results show that SF, BW, and CR influence the ToA of a packet. Larger SFs and CRs result in higher ToA of LoRa packets. Conversely, ToA reduces with larger bandwidths.

The work in [28] proposes two different algorithms named EXPLoRa-SF and EXPLoRa-AT and shows in simulation results that these algorithms perform considerably better than the LoRaWAN adaptive rate strategy (ADR). EXPLoRa-AT delivers higher bit rates in the event of higher traffic loads, while EXPLoRa-SF allocates SFs at the different subgroups of end-devices depending on the received signal strength indicator (RSSI). The results demonstrate that the data extraction rate (DER) drops dramatically for higher SFs and larger numbers of end-devices. The authors, however, assume a short range and dense network in their analysis.

The EXPLoRa approach is further extended in [29], K-means is applied to identify the non-circular crowded region, and all the nodes inside that area are assumed to have same SF. On the other hand, in the proposed work the geometry of network is circular, with six annuluses representing the range of individual SFs. We have analyzed the scalability and the performance of the uplink LoRa model considering Rayleigh fading, connection H_1, capture Q_1, and coverage probabilities H_1Q_1 in the presence of interfering signals using the same SF. The considerations of H_1 and H_1Q_1 are missing in [28,29]. Moreover, in our model, we consider a dense and wide network (radius of several kilometers) and analyze the performance by considering the maximum distance of individual SF boundaries from the gateway.

Another scientific study used K-means for the classification of end-devices into three groups based on traffic characteristics with different priorities. The grouping of end-devices was computed in terms of priority-based transmission instead of SF allocation [30].

In [31,32], SF distribution is mainly based on the power level of the signals that the gateway receives from the end-devices and gateway sensitivity, without considering the location of end-devices. As a drawback, SF allocation was disturbed because of high-density buildings, and 53.2% of the end-devices were forced to use SF12. Furthermore, in [28–31], only network-level simulators such as ns-3 and LoRaSim are used, which abstracts some characteristics of the physical layer that are incorporated in our analysis. Conversely, our study evaluated the performance of the proposed SF allocation algorithm considering the analytical model, realistic parameters, and averaging over 10^5 random deployment of the Poisson point process (PPP) by Monte Carlo computer simulations, which match with the theoretical results.

The tree-based spreading factor clustering algorithm (TSCA) for SF allocation in multihop LoRa networks is introduced in [33]. This approach offloads the data traffic in many sub-networks, which are linked to a sink node assigning a specific SF according to network clustering, thus enabling parallel frame transmission with multiple SFs. The authors show that TSCA increases the network performance in a network with rectangular geometry.

A single gateway uplink model considering path loss attenuation and Rayleigh fading is designed in [19], utilizing stochastic geometry to model network interference and then disconnection and collision probabilities. Such a model is further extended in [20], in which the authors propose a scheme that considers message replication and gateways with multiple receive antennas/decoders to attain time and spatial diversity. They demonstrate that the number of users and traffic density directly affects the performance of the LoRa network and that sending multiple message copies is beneficial for low-density networks. Both of these studies adopt equal radius SF allocation approaches. Unlike [19,20], our work considers K-means-based fair SF allocation of nodes in LoRa networks.

Recently, several studies have addressed the problems associated with automatic repeat request (ARQ) and contributed to downlink reliability in LoRaWAN applications. The sequential transmission of downlink frames, saturation of duty cycle, and half-duplex nature of LoRa gateway radios are marked as the major shortcomings for the downlink transmission [34,35]. Furthermore, these works also highlight the significance of gateway selection algorithm to prevent traffic losses due to sequential transmission of downlink frames and duty cycle limitations.

One experimental study evaluates the performance of a LoRa network at a 125 kHz bandwidth and SF7 for a sailing monitoring model, and the measurements show a 60.49% packet loss at the maximum distance of 3284 m [36]. Another LoRaWAN-based indoor environment monitoring system composed of 331 sensor nodes is deployed at the University of Oulu, where the gateway is installed at a distance of ~180 m and 24 m above the ground [37]. The measurements performed at SF7 show a maximum 11.33% packet error rate (PER), which can be due to co-spreading factor interference because all 331 end-devices use the same SF. As illustrated in [19], nodes using the same SF face co-spreading factor interference. The motivation behind our work is to propose a suitable SF allocation algorithm for a large-scale LoRa network to efficiently utilize the different data rates. To enhance SF allocation, we propose a novel algorithm, based on the machine learning technique called K-means clustering, for effectively allocating the SFs.

3. System Model

We consider $\tilde{N}$ uniformly distributed smart devices inside an uplink class-A LoRaWAN network without retransmissions, utilizing a single channel within radio range of R km and a circular area of $V = \pi R^2$ around a single gateway. Figure 1 illustrates a deployment with $\tilde{N} = 500$ and $R = 3$ km. The gateway is at the origin, and nodes are distributed uniformly in $V = 28.26$ km^2. Note that such model captures the characteristics of telemetry applications such as those in smart cities and smart buildings. For instance, the University of Oulu Smart Campus has a LoRaWAN network constantly monitoring several sensors such as temperature, luminosity, and CO_2 [37].

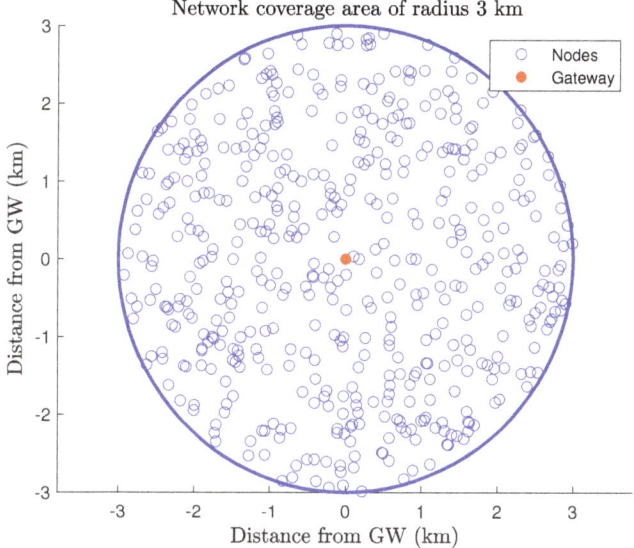

Figure 1. Uniform distribution of $\tilde{N} = 500$ nodes in a circular network area of radius $R = 3$ km, with the gateway (GW) at the origin.

The LoRa modulation bit rate is defined as [14]

$$R_b = \frac{4}{4+CR} \frac{BW}{2^{SF}}, \qquad (1)$$

where $\frac{4}{4+CR}$ is the effective coding rate, ranging from $\frac{4}{5}$ to $\frac{4}{8}$, while CR denotes the LoRa coding rate configuration, varying from 1 to 4. In our work, we assume $CR = 1$, and the LoRa uplink channel aggregated bit rate is expressed as $bitrate_U = \sum_{i=7}^{12} Rb_i = 12.17$ kbps. For instance, Table 1 shows the characteristics of 9 byte LoRa packets with explicit header and CRC modes enabled and $BW = 125$ kHz.

Table 1. Characteristics of the LoRa uplink model containing packets of 9 bytes at BW = 125 kHz.

SF (i)	Bit Rate kbps (Rb_i)	Receiver Sensitivity dBm	SNR dB (q_{SF})	Range km
7	5.47	−123	−6	$l_0 - l_1$
8	3.13	−126	−9	$l_1 - l_2$
9	1.76	−129	−12	$l_2 - l_3$
10	0.98	−132	−15	$l_3 - l_4$
11	0.54	−134.5	−17.5	$l_4 - l_5$
12	0.29	−137	−20	$>l_5$

3.1. Uplink Outage Probability

The uplink transmission of nodes is based on the ALOHA protocol, and the probability of collision in ALOHA networks is high when many stations are connected [38]. In LoRa, simultaneous signals of different SFs are quasi-orthogonal because the inter-SF rejection gain varies from 16 to 36 dB [39]. Therefore, for the sake of simplicity, our work does not inspect inter-SF interference and focuses on co-SF interference only, which is stronger.

In this paper, the uplink model includes the influence of Rayleigh fading and path loss attenuation as the baseline model [19] for performance analysis, where $g(d_k) = \frac{\lambda}{(4\pi d_k)^\eta}$ is the path loss attenuation function, $\eta \geq 2$ is the path loss exponent, λ is the wavelength, and h_k is the fading in the link between the k-th node and the gateway. Let us consider the transmitted signal of a single LoRa node $s_1(t)$ to examine the impact of co-SF interference originated due to simultaneous transmission of nodes with same SF. The mathematical expression of the received signal at the gateway can be expressed as

$$r_1(t) = g(d_1) h_1 * s_1(t) + \sum_{k=2}^{N} \chi_k^{SF}(t) g(d_k) h_k * s_k(t) + n(t), \qquad (2)$$

where $n(t)$ is additive white Gaussian noise with zero mean and variance $\mathcal{N} = -174 + NF + 10\log_{10}(BW)$ dBm, NF is the noise figure of the receiver, and -174 dBm/Hz is the thermal noise spectral density constant.

We consider that an outage of the received signal in an uplink channel can take place in the two scenarios [19]. First, if the signal-to-noise ratio (SNR) of the received packet is less than the SF specific threshold q_{SF}, then the node is considered disconnected. Second, if the signal-to-interference ratio (SIR) between the target-received packet and any other concurrent signals of the same SF and frequency channel is less than 6 dB, then it is considered as a collision.

3.1.1. Outage Condition I

The distance of the end-device to the gateway in a wireless transmission domain is crucial. The instantaneous SNR can be expressed as SNR = $\frac{\mathcal{P}_1 |h_1|^2 g(d_i)}{\mathcal{N}}$, where $\mathcal{P}_1$ is the transmit power of end-device 1 in mW and $|h_1|^2$ is the squared envelop of the channel coefficient. Communication is only possible when the SNR of the received signal at the gateway is less than the reception threshold q_{SF}. Thus, the first outage condition, the connection probability, is defined as [19]

$$H_1 = \exp\left(\frac{\mathcal{N}q_{SF}}{\mathcal{P}_1 g(d_1)}\right), \quad (3)$$

where d_1 (in meters) is the distance of the desired end-device from the gateway.

3.1.2. Outage Condition II

A collision in LoRa end-device transmission takes place if the SIR of the desired signal with respect to interference from the same SF and frequency channel is less than 6 dB, i.e., if the desired signal is at least four times stronger than the interference. We model this outage condition based on [19], where interference is approached by considering the strongest interfering device. According to [19], the highest interference comes from the end-device k^*.

The probability that no collision occurs or that the strongest interfering signal is at least 6 dB below the desired one, termed the capture probability, is

$$Q_1 = \mathbb{P}\left[\frac{|h_1|^2 g(d_1)}{|h_{k^*}|^2 g(d_{k^*})} \geq 4 \,\middle|\, d_1\right] = \mathbb{E}_{|h_1|^2}\left[\mathbb{P}\left[X_{k^*} < \frac{|h_1|^2 g(d_1)}{4} \,\middle|\, |h_1|^2, d_1\right]\right]. \quad (4)$$

The probability above depends on the distribution of $X_{k^*} = |h_{k^*}|^2 g(d_{k^*})$. The cumulative distribution function (CDF) of X_{k^*} is derived in [19] and is denoted as $F_{X_{k^*}}$. Thus,

$$Q_1 = \mathbb{E}_{|h_1|^2}\left[F_{X_{k^*}}\left(\frac{|h_1|^2 g(d_1)}{4}\right)\right] = \int_0^\infty e^{-z} F_{X_{k^*}}\left(\frac{z g(d_1)}{4}\right) dz. \quad (5)$$

Moreover, in [19] the authors present an approximation for (5) that is only accurate at the edges of each annulus. This paper considers only the exact probability in (5).

3.1.3. Coverage Probability

The probability that defines whether a selected end-device is in coverage and can successfully communicate with the gateway is termed the coverage probability. It is the product of H_1 and Q_1. The average coverage probability $\wp_c$ can be achieved by deconditioning the location of the individual node by averaging over the network coverage area $V = \pi R^2$, i.e., [19]

$$\wp_c = \frac{2}{R^2} \int_0^R H_1(d_1) Q_1(d_1) d_1 \, dd_1. \quad (6)$$

The average coverage probability of a individual SF annulus is also inspected. It indicates the probability of an end-device at distance d_1 in the annulus i by considering the connection and capture probabilities and is defined as [20]

$$\wp_{c,i} = \frac{2}{(l_{i+1} - l_i)^2} \int_{l_i}^{l_{i+1}} H_1(d_1) Q_1(d_1) (d_1 - l_i) \, dd_1, \quad (7)$$

where l_{i+1} is the radius of the outer circle and l_i is the radius of the inner circle of the ith annulus.

4. Proposed SF Allocation Algorithm

In this paper, we propose an SF allocation algorithm, i.e., an algorithm to define the range of each SF annulus. Our solution uses the K-means machine learning algorithm [40], used in the process of vector quantization in data mining by clustering. It is a non-deterministic, numerical, and iterative approach. The main objective of the K-means algorithm is to find the minimum cost function, defined as the distance between each point in the data set and its nearest centroid. The distance between the cluster centers and data elements typically assumes the Euclidean distance. K-means clustering method can efficiently achieve robust clustering results when dealing with large data sets. The K-means algorithm first arbitrarily chooses K points from the data set, which indicate the initial centroids. The remaining points are then clustered to the closest centroid, and the coordinates of centroids are recalculated, iteratively, until the cost function converges.

Consequently, it is important to choose the appropriate number of centroids during the initialization procedure because the area of each annulus $\pi(l_{i+1}^2 - l_i^2)$ increases towards the higher SFs in a strategy based on equal distance steps per SF, which results in the growth of node density due to uniform distribution. That is why it is essential to select the sequences for K-means iterations that can provide larger values of K clusters for higher SFs. In order to avoid an extensive number of nodes in an individual SF, there should be a fair difference between the inner (l_i) and outer (l_{i+1}) radii of annulus. In the proposed work, the annulus area is directly dependent on the difference between the K clusters for two consecutive iterations.

In our approach, we use five iterations of K-means. We start by computing the boundaries of the outermost SF ring, SF12, and then proceed to define the inner boundaries for lower SFs. For each iteration, K clusters are selected to develop the centroids of end-devices in the LoRa network covered by a single gateway. Four different mathematical sequences listed in Table 2—a Fibonacci series, square numbers, Wythoff array, and arithmetic series—are used to assign the values of K for each iteration.

Table 2. K Cluster Values for K-Means Iterations.

Iteration	1st	2nd	3rd	4th	5th
Series	SF12	SF11	SF10	SF9	SF8
Fibonacci series	34	21	13	8	5
Square number	49	36	25	16	9
Arithmetic series	34	28	22	16	10
Wythoff array	37	32	24	16	11

In our work, K-means operates iteratively. Each iteration defines the set of nodes at the outer SF ring. In each iteration, the algorithm seeks the set of K centroids C that minimizes the average of the distances between any node and its closest centroid, i.e.,

$$C = \arg \min_{C_k \in C} \frac{1}{|ED_k|} \sum_{X_i \in ED_k} dist(C_k, X_i)^2, \tag{8}$$

where ED_k is the set of devices at the k-th iteration, X_i is a device in ED_k, and C_k is the closest centroid of X_i. The function $dist(x_1, x_2)$ computes the Euclidean distance between x_1 and x_2. This procedure returns the collection of K centroids of network nodes, whereas $C = \{C_1, \ldots C_K\}$. After computing the centroids, the algorithm determines the boundary of C, so that $[C_x, C_y] = boundary(C)$, which determines the 2-D vector of border points around the Cartesian coordinates of the centroids. Then, it separates the nodes that are inside of the centroid boundary, forming the set $I = [I_x, I_y]$, where I_x and I_y are vectors storing the coordinates of the inner nodes in each of the Cartesian dimensions. In the next step, the maximum absolute value of each dimension of I is calculated to set the radius as $l_i = \frac{max|I_x| + max|I_y|}{2}$, which defines the limit of the SF ring i. The procedure repeats to determine the boundaries of the remaining SF rings ($l_5, \ldots l_1$).

The steps involved in the proposed SF allocation technique are described in Algorithm 1. It repeats the process five times to allocate nodes for SF12–SF8. At the end, the remaining nodes use SF7. Initially, the SF12 outer limit is set to the network radius. In each iteration, the number of clusters is assigned to K depending on the chosen mathematical series (as mentioned in Table 2). For the first iteration, the algorithm considers all of the $\tilde{N}$ nodes inside the set ED. In line 4, it computes the K-means of ED, which returns the centroids $C = \{C_1, \ldots C_K\}$ by (8). Since nodes in the set I are inside the boundary of centroids, line 7 computes the inner limit of SF ring l_i. This process is repeated iteratively until l_1 is calculated for the allocation of SF8 and the remaining nodes are assigned to SF7 (line 11). Note that the set of nodes ED is updated at the end of each iteration by removing the nodes that were already allocated to an SF (line 9).

Algorithm 1 K-Means-based SF Allocation

Input: $ED := \tilde{N}$ uniformly deployed nodes

Output: $L := \{l_0, l_1, \ldots, l_6\}$

1: $l_6 := R$
2: **for** i in $\{5, \ldots, 1\}$ **do** ▷ For each SF ring, starting from the outermost ring
3: $\quad K := $ GetKfromSeries(i) ▷ Set number of centroids for this iteration
4: $\quad C := $ Kmeans(ED, K) ▷ Compute the centroids
5: $\quad B := $ boundary(C) ▷ Compute the boundary of C
6: $\quad I := \{x \in ED \mid x \in \mathbf{conv}B\}$ ▷ Select nodes that are inside the boundary B
7: $\quad l_i := \frac{\max(|I_x|) + \max(|I_y|)}{2}$ ▷ Compute the new SF ring limit
8: $\quad SF_{i+7} := \{x \in ED \mid x \notin \text{Ball}[(0,0), l_i]\}$ ▷ Allocate SF_{i+7} to nodes outside the circle of radius l_i
9: $\quad ED := \{x \in ED \mid x \in \text{Ball}[(0,0), l_i]\}$ ▷ Remove nodes outside the circle of radius l_i
10: $l_0 := 0$
11: $SF_7 := ED$ ▷ Allocate SF_7 to remaining nodes
12: **return** L

All of the iterations of the proposed algorithm for an example network are demonstrated in Figure 2. The radius of the network circular area is $R = 3$ km, and therefore, the outer limit of SF12 is $l_6 = R = 3$ km. The first iteration of the algorithm defines the inner boundary of SF12, l_5, as shown in Figure 2. After excluding the devices inside the SF12 ring from ED_k, the algorithm runs a new iteration and defines l_4, i.e., the inner boundary of the SF11 ring. The iterations continue until l_1 is defined and the complete network geometry is obtained.

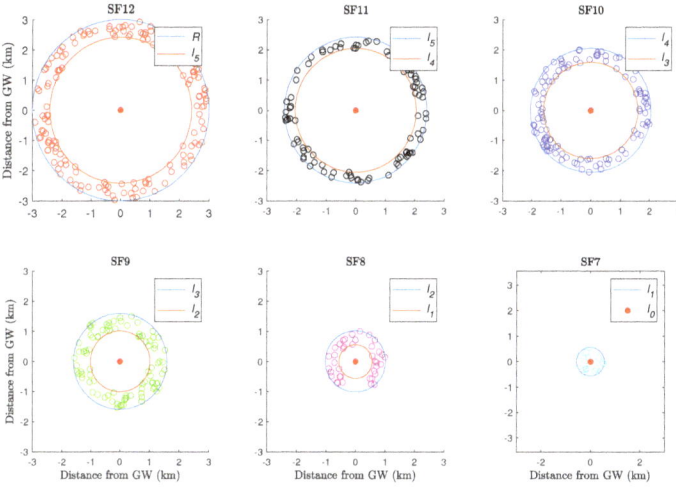

Figure 2. K-means iterations for SF allocation based on Fibonacci series and 500 nodes, where l_i and l_{i+1} are the inner and outer radii of ith annulus, respectively.

Figure 3 depicts the clusters of nodes, centroids, and gateway at the origin for the last K-means iteration. The nodes outside l_1 are allocated to SF8, and the remaining nodes are assigned to SF7. The SF distribution of $\bar{N} = 500$ nodes based on the proposed approach considering the Fibonacci series for clustering is shown in Figure 4. The number of nodes in SF rings depends on the chosen series because of the distinct number of clusters for each sequence.

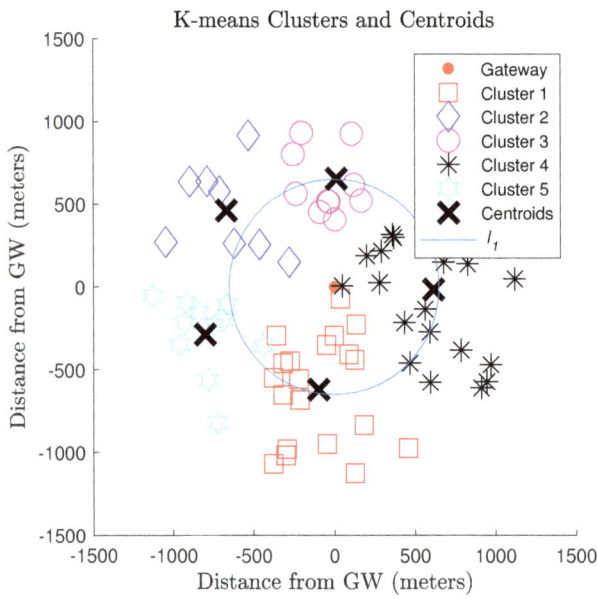

Figure 3. 5th iteration with Fibonacci series. Nodes outside l_1 use SF8, those inside use SF7.

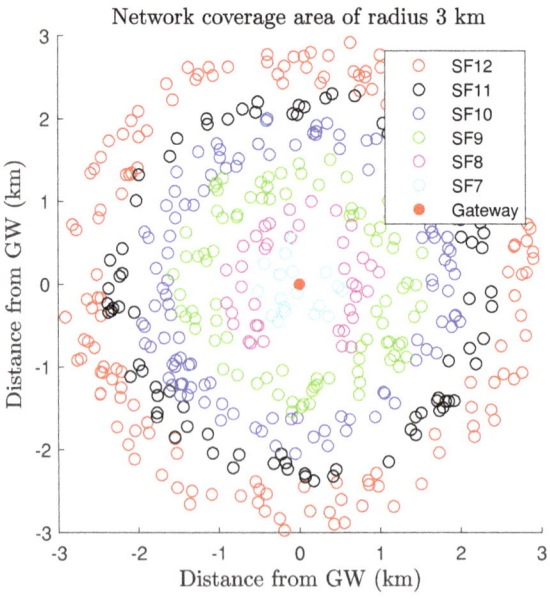

Figure 4. SF allocation of 500 nodes with Fibonacci series.

The area of each annulus also varies according to the mathematical series, which also affects the network performance. An important aspect to take into consideration is the selection of the number of clusters (K); if the difference between the clusters of two consecutive iterations is too big, that will result in a large number of end-devices in that specific region and, as a consequence, the probability of collisions and of co-SF interference will be high. In the same way, SF7 will have a larger coverage area and more nodes if K is high for the last iteration.

5. Numerical Results and Discussion

In this section, we evaluate the scalability and performance of the proposed methodology by means of computer simulations. The results are based on a $p_0 = 1\%$ duty cycle, $BW = 125$ kHz, $\eta = 2.75$ path loss exponent, 868 MHz European frequency band, and network radius of $R = 3$ km. Table 3 summarizes the parameters considered for the results.

Table 3. System Parameters.

Parameter	Symbol	Value
Nodes	$\bar{N}$	300–700
Spreading Factor	SF	7–12
Bandwidth	BW	125 kHz
Carrier frequency	f	868 MHz
Noise figure	NF	6 dBm
Transmit power	$\mathcal{P}_1$,	14 dBm
Duty cycle	p_0	1%
Path loss exponent	η	2.75

5.1. SF Allocation and Scalability Analysis

As discussed in Section 4, we used mathematical series to assign the numbers of clusters for K-means iterations. The Fibonacci series has the shortest ranges for SF7, 715 m for $\bar{N}$ = 500. The distance between the SFs and the distribution of nodes can be changed by modifying the number of clusters (K). On the other hand, for the same number of nodes, the Square series has a longer range for SF7, which is 1201 m, and it contains more end-devices. This type of configuration is due to the higher value of K clusters for iterations (see Table 2). While in the case of Fibonacci and equal-distance-based SF allocation in [19,20], the network has fewer nodes in SF7, and the number of nodes in each SF region increases considerably towards the higher SFs, as shown in Table 4. The average SF ranges for the Fibonacci series, square series, Wythoff array and arithmetic series are shown in Table 5.

Table 4. Comparison of the proposed approach with the reference method, based on the number of nodes.

Series	$\bar{N}$	SF7	SF8	SF9	SF10	SF11	SF12
	300	18	19	48	66	75	77
Fibonacci series	500	29	35	79	107	124	129
	700	40	51	110	148	173	180
	300	44	24	43	52	63	77
Arithmetic series	500	69	42	70	87	107	129
	700	94	61	98	121	149	180
	300	53	33	55	44	58	61
Square numbers	500	81	57	87	75	99	105
	700	111	81	121	105	137	148
	300	49	24	45	52	60	73
Wythoff array	500	77	42	75	87	100	123
	700	103	62	103	121	140	173
	300	9	25	42	59	76	92
Reference model	500	14	42	70	98	126	152
	700	20	59	97	137	176	214

115

Table 5. Comparison of proposed approach with the reference method based on the distance of individual SF outer boundaries from the gateway (values in meters).

Series	$\tilde{N}$	SF7	SF8	SF9	SF10	SF11	SF12
Fibonacci series	300	735	1049	1586	2112	2588	3000
	500	715	1060	1591	2112	2586	3000
	700	707	1071	1601	2112	2587	3000
Arithmetic series	300	1144	1412	1806	2194	2590	3000
	500	1110	1403	1795	2183	2584	3000
	700	1099	1409	1801	2188	2588	3000
Square numbers	300	1248	1591	2037	2336	2680	3000
	500	1201	1568	2004	2316	2670	3000
	700	1190	1568	2002	2313	2667	3000
Wythoff array	300	1209	1469	1872	2246	2613	3000
	500	1168	1453	1857	2237	2607	3000
	700	1150	1450	1851	2231	2604	3000
Reference model	300	500	1000	1500	2000	2500	3000
	500	500	1000	1500	2000	2500	3000
	700	500	1000	1500	2000	2500	3000

The boxplot is a standard process to quantify the variability of data on the basis of five parameters, i.e., the minimum, first quartile (25%), median, third quartile (75%), and maximum. Th distance of the SFs boundary from the gateway for $\tilde{N} = 500$ is demonstrated in Figure 5 for each of the considered series.

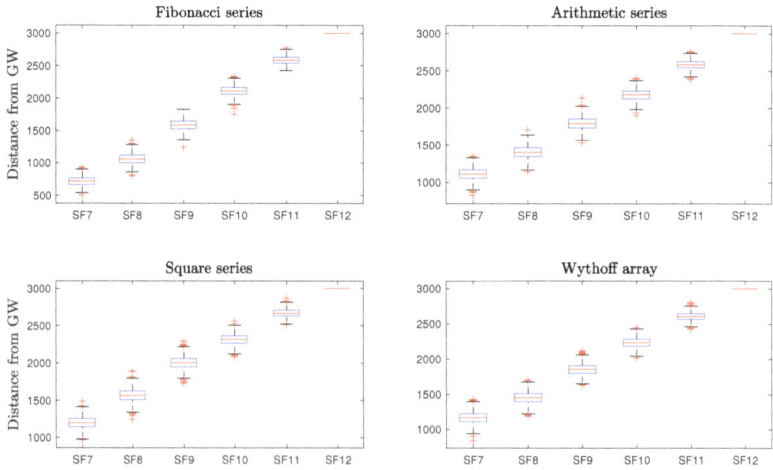

Figure 5. Boxplots representing the distance of SFs from the gateway. The red "+" signs indicate outliers. SF12 has a constant distance of 3 km for all of the series.

The median of every SF is identical to distances provided in Table 5. As depicted in the graphical results, SF7 and SF8 have a large disparity in range and number of nodes for different scenarios, while SF11 and SF12 have nearly close coverage areas for all scenarios. Furthermore, we can also clearly observe that variation in K clusters selection has a direct effect on SF allocation based on the proposed methodology. The number of nodes in each SF are illustrated in Table 4. Square-series-based networks demonstrate five times more nodes in SF7 as compared to the reference model.

The large values of K clusters for the last iterations result in longer radii that directly increase the region of SF7 and keep SF8 further away from the gateway. These different scenarios can be used according to different situations and requirements of LoRa applications. An approach based on the Fibonacci series is beneficial for applications where fewer nodes in lower SFs are required, while the Wythoff array, square series, and arithmetic series have wider regions for SF7, and thus can be used in setups where more nodes are required in SF7 inside the radio range of nearly 1200 m to provide highest data rate (R_b = 5.47 kbps, see Table 1). Several studies examined the performance of LoRa networks and show that the success probability of data packets decreases for higher SFs. In our work, we consider the effect of changing the coverage range and varying the number of devices for individual SFs.

5.2. Performance Analysis

After the application of the proposed SF allocation algorithm, we investigated the performance of the resulting LoRa network. The theoretical results were verified by Monte Carlo simulations. In the figures, each marker represents the average over 10^5 random deployments of the Poisson point process (PPP) for a single gateway LoRa uplink model, considering an end-device at d_1 meters from the gateway. In Figure 6, the solid lines demonstrate the theoretical results, while marker points of the same color illustrate the simulation outputs. The simulated results align with the theoretical ones. Within the context of the previously discussed mathematical sequences, we considered and examine the impact of different SF allocation scenarios on connection probability H_1, capture probability Q_1, and coverage probability H_1Q_1 against the distance from the gateway.

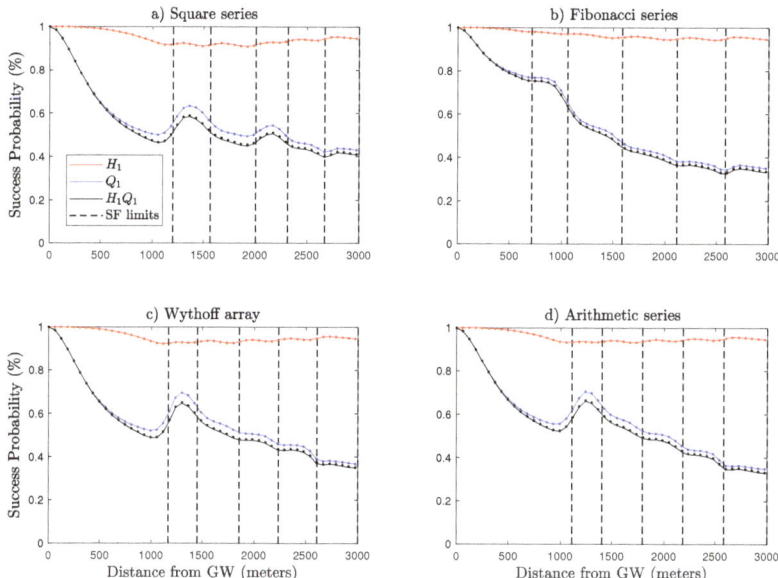

Figure 6. LoRa uplink performance with the proposed SF allocation algorithm for different series and $\tilde{N}$ = 500 nodes. (**a**) Square series with $K = \{49, 36, 25, 16, 9\}$. (**b**) Fibonacci series with $K = \{34, 21, 13, 8, 5\}$. (**c**) Wythoff with $K = \{37, 32, 24, 16, 11\}$. (**d**) Arithmetic series with $K = \{34, 28, 22, 16, 10\}$.

As expected, the distance of the end-device from the gateway has considerable influence on connection probability H_1. In the case of the Fibonacci series, the model has a better success probability for lower SFs as compared to the square series, arithmetic series, and Wythoff array. This fact is due to the difference in a range of individual SF boundaries for said SF allocation schemes and the distance-dependent SNR threshold q_{SF}. Furthermore, in the scenario of the square series, SF7 has large coverage areas of 1201 m (see Table 5), which affects path loss attenuation and the instantaneous SNR. The SNR threshold q_{SF}, however, remains the same at −6 dB (see Table 1). As a consequence, the outage condition in (3) slightly degrades the network connection probability. Although a performance boost is illustrated during transitions of end-devices into the next SF because of the lower value of q_{SF}, the performance of the previous SF has a direct consequence on the next SF.

Moving towards the capture probability Q_1, unlike H_1, it considers co-SF interference. Q_1 declines gradually with increasing SF, as illustrated in Figure 6. This trend is because of two major factors including ToA and the number of nodes in each annulus. ToA grows exponentially with SF, thus for the higher SFs, the wireless channel remains occupied for a long time slot, which increases the risk of collisions between simultaneously transmitted LoRa packets. In the same way, the number of end-devices in an individual annulus increases for higher SFs due to the uniform distribution of nodes in the circular coverage area, as demonstrated in Table 4. As a result, the network experiences co-SF interference that degrades the quality of transmission. For the cases with the square series, arithmetic series, and Wythoff array (Figure 6a,c,d), the model has a larger coverage area and more nodes in SF7, resulting in higher co-spreading factor interference, which is the major reason behind the lower performance of the network for these specific scenarios. Although we are sacrificing network quality for lower SFs with fewer nodes and high success probabilities, as presented in Figure 7, we improved the performance of the network for the higher SFs and regions with more nodes, where the network performance was weak in the baseline model from [19], which considers fixed distance steps from the gateway to define the SF allocation. Moreover, SF allocation based on the square series (Figure 6a) has better network performance, which happens because of improved gain in Q_1 for higher SFs as compared to the Fibonacci series, arithmetic series, and Wythoff array.

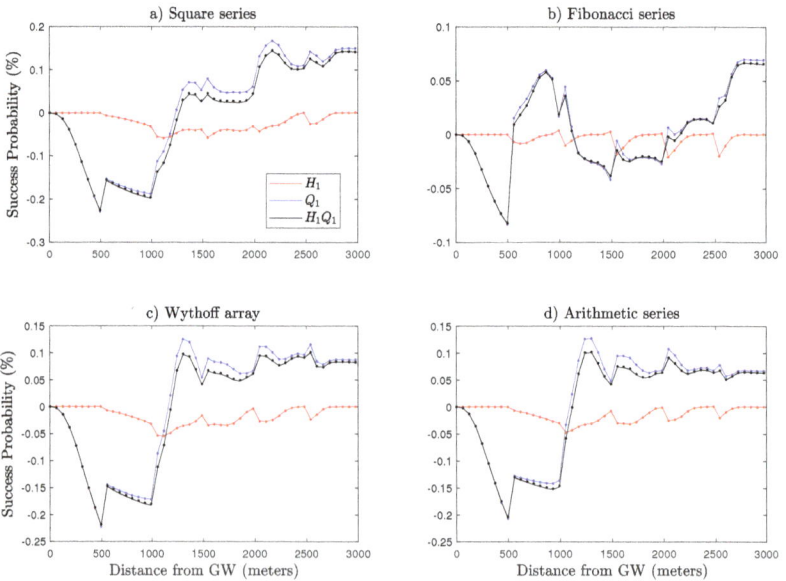

Figure 7. Comparison of performance gain of different series with respect to the baseline model for $\bar{N} = 500$, network size of $R = 3$ km with $p_0 = 1\%$, and path loss exponent $\eta = 2.75$.

In Figure 8, we present the performance of the square series based on the proposed SF allocation algorithm in comparison with the baseline model. First, we observe that the capture probability Q_1 and coverage probability $H_1 Q_1$ of the baseline model outperform the proposed algorithm in the region of radius $R > 1000$ m from the gateway. Here, it is worth mentioning that there are fewer nodes in this region as compared to the remaining area of the network. The proposed algorithm surpasses the baseline model with a gain in capture probability Q_1 (up to 53% for SF12 in Figure 8b). In addition, the nodes closer to the gateway always have better behavior in contrast with nodes far away, so that the network can tolerate a lower success probability with a boost in performance of farther end-devices. As expected, there was more change in success probability in the higher SFs region as compared to baseline model because of the fair distribution of SF by the proposed algorithm. Figure 8b shows the difference between the baseline model and the square series in terms of outage probabilities through the course of distance. The zero level on the y-axis (success probability) shows no difference, while there is a positive/negative gain either on the upper or lower side of that level. There is up to 16.73% growth in Q_1 demonstrated by the end-devices present in higher SFs.

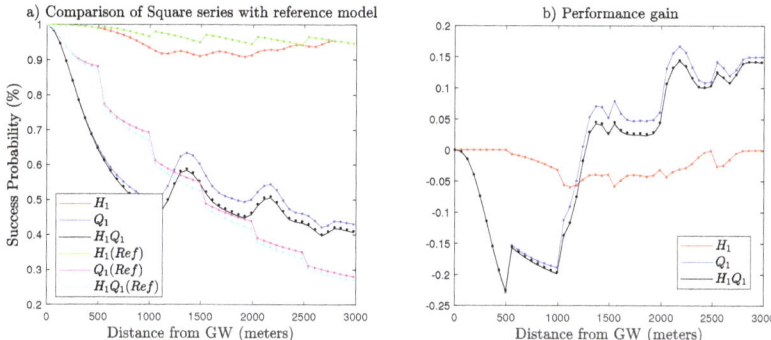

Figure 8. (a) Performance comparison of square series with the reference model for $\bar{N} = 500$. (b) Performance gain (success probability of proposed model − success probability of reference model). The proposed SF allocation approach sacrifices Q_1, $H_1 Q_1$ for the lower SFs but achieves greater performance for the higher SFs.

Moreover, we also consider the evaluation of the average coverage probability of the networks for different numbers of nodes ($\bar{N}$) ranging from 300 to 700, and results demonstrate that $\wp_c$ drops exponentially towards higher $\bar{N}$. Figure 9 depicts the average coverage probability of different numbers of end-devices. The SF allocation schemes deployed using the square series demonstrated a better performance gain than all other scenarios, including the reference model. It was followed by the Wythoff array, arithmetic series, and Fibonacci series, in that order. The proposed SF allocation scheme overcomes the performance of the baseline model by an overall growth of around 5% in its $\wp_c$. For instance, taking the square series into account, at $\bar{N} = 500$ there is a boost in the average coverage probability (6) from $\wp_c$ = 41.9% to 46.81% compared with the reference model. On the other hand, SF allocation schemes deployed using the Fibonacci series showed the least-improved network coverage probability compared to all other user distribution series.

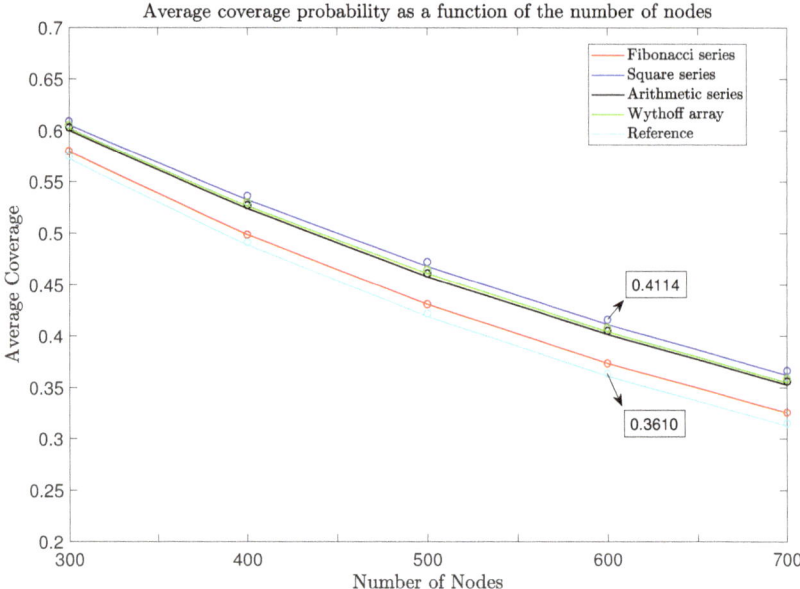

Figure 9. Comparison of the coverage probabilities of the LoRa uplink as a function of the number of nodes ranging from $\bar{N}$ = 300 to 700 nodes for the network size of R = 3 km with p_0 = 1% and path loss exponent η = 2.75.

We also investigated the performance of the proposed model taking into account the variation of different parameters on Fibonacci series. As seen in Figure 10a,b, H_1 is agnostic to the number of nodes and duty cycle. It is assessed at 0.1% and 1%, which is within the duty cycle range specified by ETSI for LoRa applications [34,35]. Furthermore, node density and the duty cycle demonstrate a negative impact on Q_1 because of co-SF interference caused by increasing medium usage from $\bar{N}$ = 501 nodes to $\bar{N}$ = 1005 nodes. Likewise, in Figure 10c, the path loss exponent η illustrates network connection degradation at 2.65 from 2.5 because H_1 depends on the distance, while the the capture probability (Q_1) is not dependent on the path loss exponent. In the case of one frequency channel, the transmit power $\mathcal{P}_k$ of nodes can be up to 20 dBm (100 mW) [11,41]. In order to evaluate the effect of different transmit powers, in Figure 10d we raised the transmit power from 14 dBm to 19 dBm. The results demonstrate that the transmit power of 19 dBm causes a better connection probability as compared to 14 dBm. Nevertheless, variations of the path loss exponent and transmit power do not affect Q_1 considerably because it is much more dependent on the number of nodes.

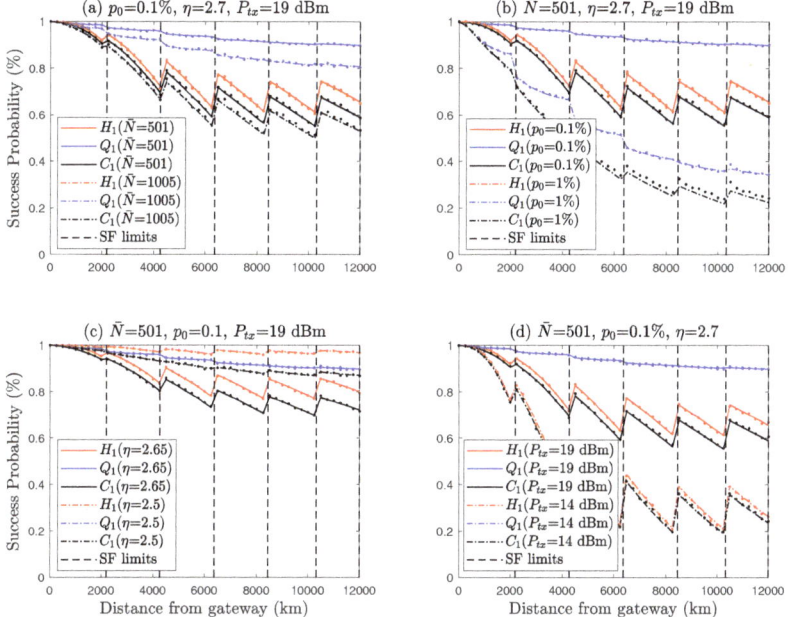

Figure 10. Performance LoRa uplink considering Fibonacci series K-means clustering, and the impact of different parameters on success probabilities. (**a**) Density of users increased from $\bar{N} = 501$ to $\bar{N} = 1005$. (**b**) Duty cycle increased from $p_0 = 0.1\%$ to $p_0 = 1\%$. (**c**) Path loss exponent from $\eta = 2.5$ to $\eta = 2.65$. (**d**) Transmit power from ($P_{tx} = 14$) dBm to $P_{tx} = 19$ dBm.

5.3. Discussion

The architecture of LoRaWAN consists of end-devices, a gateway, network server (NetServer), and application server [15]. The NetServer is mainly responsible for the overall management of the network. The dynamic configuration of the SF by the NetServer is already possible in LoRaWAN during the network join procedure or through specific MAC commands. In fact, these features are used in LoRaWAN when the adaptive data rate (ADR) mechanism is active. For our approach to be implemented in practice, the NetServer could run our algorithm periodically or when the number of connected devices changes significantly, and then issue the required MAC commands to reconfigure the devices that need to change their SF. Therefore, as the current LoRaWAN specification is already able to dynamically allocate SFs, our proposal only changes the way the proper allocation is calculated at the NetServer, and is therefore feasible in practice.

6. Conclusions

This paper has presented a novel SF allocation technique for a large-scale LoRa network using the K-means clustering machine learning algorithm. The authors also analyzed the impact of the distance of end-devices from the gateway and the number of nodes in each SF on network performance. In this work, four different scenarios are considered, which have different distances for the SF boundaries and variations in the number of nodes in an individual SF. Such fair distribution results in a better average coverage probability in the higher SFs, while dealing with the maximum number of nodes. Numerical findings show that our SF allocation algorithm outperforms the reference model not only in terms of success probability but also in regards to fair resource distribution. The evaluated theoretical and simulation results are useful for an in-depth understanding of large and dense LoRa networks. Our resource allocation method can handle dense and large circular coverage areas for LoRa sensors using distinct numbers of clusters instead of equal-radius-based SF allocation [19,20], while the techniques in [28,29,33] are designed for short-range networks. The studies [19,20,28,29,33] only highlighted the performance of networks for fixed parameters. In contrast to them, our work inspects different scenarios by obeying the restrictions of ETSI standards.

Author Contributions: Concept and methodology, H.A.; software, M.A.U. and J.I.; validation, A.H. and H.A.; Resources, A.H. and R.D.S.; writing original draft, M.A.U. and J.I.; writing revision, H.A., A.H., and R.D.S.; supervision, H.A.; Project Administration, H.A.

Funding: This research has been supported in Finland by the Academy of Finland: 6Genesis Flagship (Grant no318927) and EE-IoT (no319008); as well as the BusinessFinland MOSSAF project. This work has been supported in Brazil by CNPq, PrInt CAPES-UFSC "Automation 4.0", and INESC P&D Brasil (Project F-LOCO, Energisa, ANEEL PD-00405-1804/2018).

Conflicts of Interest: The authors declare no conflict of interest.

References

1. Chen, S.; Xu, H.; Liu, D.; Hu, B.; Wang, H. A Vision of IoT: Applications, Challenges, and Opportunities With China Perspective. *IEEE Internet Things J.* **2014**, *1*, 349–359. [CrossRef]
2. Sisinni, E.; Saifullah, A.; Han, S.; Jennehag, U.; Gidlund, M. Industrial Internet of Things: Challenges, Opportunities, and Directions. *IEEE Trans. Ind. Inf.* **2018**, *14*, 4724–4734. [CrossRef]
3. Sotres, P.; Lanza, J.; Sánchez, L.; Santana, J.R.; López, C.; Muñoz, L. Breaking Vendors and City Locks through a Semantic-enabled Global Interoperable Internet-of-Things System: A Smart Parking Case. *Sensors* **2019**, *19*, 229. [CrossRef] [PubMed]
4. Rodrigues, J.J.P.C.; De Rezende Segundo, D.B.; Junqueira, H.A.; Sabino, M.H.; Prince, R.M.; Al-Muhtadi, J.; De Albuquerque, V.H.C. Enabling Technologies for the Internet of Health Things. *IEEE Access* **2018**, *6*, 13129–13141. [CrossRef]
5. Asadullah, M.; Ullah, K. Smart home automation system using Bluetooth technology. In Proceedings of the 2017 International Conference on Innovations in Electrical Engineering and Computational Technologies (ICIEECT), Karachi, Pakistan, 5–7 April 2017; pp. 1–6.
6. Saha, H.N.; Mandal, A.; Sinha, A. Recent trends in the Internet-of-Things. In Proceedings of the 2017 IEEE 7th Annual Computing and Communication Workshop and Conference (CCWC), Las Vegas, NV, USA, 9–11 January 2017; pp. 1–4.
7. Internet of Things (IoT) Connected Devices Installed Base Worldwide from 2015 to 2025 (in Billions). Available online: https://www.statista.com (accessed on 16 October 2019).
8. Yasmin, R.; Petäjäjärvi, J.; Mikhaylov, K.; Pouttu, A. On the integration of LoRaWAN with the 5G test network. In Proceedings of the 2017 IEEE 28th Annual International Symposium on Personal, Indoor, and Mobile Radio Communications (PIMRC), Montreal, QC, Canada, 8–13 October 2017; pp. 1–6.
9. Mikhaylov, K.; Petrov, V.; Gupta, R.; Lema, M.A.; Galinina, O.; Andreev, S.; Koucheryavy, Y.; Valkama, M.; Pouttu, A.; Dohler, M. Energy Efficiency of Multi-Radio Massive Machine-Type Communication (MR-MMTC): Applications, Challenges, and Solutions. *IEEE Commun. Mag.* **2019**, *57*, 100–106. [CrossRef]
10. Vangelista, L.; Zanella, A.; Zorzi, M. Long-range IoT technologies: The dawn of LoRa. In Proceedings of the Future Access Enablers for Ubiquitous and Intelligent Infrastructures, Cham, Switzerland, 23–25 September 2015; pp. 51–58.
11. Petajajarvi, J.; Mikhaylov, K.; Roivainen, A.; Hanninen, T.; Pettissalo, M. On the coverage of LPWANs: Range evaluation and channel attenuation model for LoRa technology. In Proceedings of the 2015 14th International Conference on ITS Telecommunications (ITST), Copenhagen, Denmark, 2–4 December 2015; pp. 55–59.
12. Lauridsen, M.; Vejlgaard, B.; Kovacs, I.Z.; Nguyen, H.; Mogensen, P. Interference Measurements in the European 868 MHz ISM Band with Focus on LoRa and SigFox. In Proceedings of the 2017 IEEE Wireless Communications and Networking Conference (WCNC), San Francisco, CA, USA, 19–22 March 2017; pp. 1–6.
13. Cattani, M.; Boano, C.A.; Romer, K. An experimental evaluation of the reliability of lora long-range low-power wireless communication. *J. Sens. Actuator Netw.* **2017**, *6*, 7. [CrossRef]
14. AN1200.22 LoRa^{TM} Modulation Basics, Rev. 2. Semtech, 2015. Available online: http://www.semtech.com/uploads/documents/an1200.22.pdf (accessed on 31 July 2019).

15. LoRa Alliance. Available online: http://www.lora-alliance.org (accessed on 30 September 2019).
16. Phung, K.H.; Tran, H.; Nguyen, Q.; Huong, T.T.; Nguyen, T.L.; Nguyen, H. Analysis and assessment of LoRaWAN. In Proceedings of the 2018 2nd International Conference on Recent Advances in Signal Processing, Telecommunications & Computing (SigTelCom), Ho Chi Minh City, Vietnam, 29–31 January 2018; pp. 241–246.
17. LoRa Technology Is Connecting Our Smart Planet. Available online: https://www.semtech.com/lora/lora-applications (accessed on 31 July 2019).
18. de Castro Tomé, M.; Nardelli, P.H.J.; Alves, H. Long-Range Low-Power Wireless Networks and Sampling Strategies in Electricity Metering. *IEEE Trans. Ind. Electron.* **2019**, *66*, 1629–1637. [CrossRef]
19. Georgiou, O.; Raza, U. Low-Power Wide-Area Network Analysis: Can LoRa Scale? *IEEE Wirel. Commun. Lett.* **2017**, *6*, 162–165. [CrossRef]
20. Hoeller, A.; Souza, R.D.; Alcaraz López, O.L.; Alves, H.; de Noronha Neto, M.; Brante, G. Analysis and Performance Optimization of LoRa Networks With Time and Antenna Diversity. *IEEE Access* **2018**, *6*, 32820–32829. [CrossRef]
21. Augustin, A.; Yi, J.; Clausen, T.; Townsley, W. A Study of LoRa: Long-Range & Low-Power Networks for the Internet-of-Things. *Sensors* **2016**, *16*, 1466.
22. Haxhibeqiri, J.; De Poorter, E.; Moerman, I.; Hoebeke, J. A Survey of LoRaWAN for IoT: From Technology to Application. *Sensors* **2018**, *18*, 3995. [CrossRef] [PubMed]
23. Mikhaylov, K.; Petaejaejaervi, J.; Haenninen, T. Analysis of Capacity and Scalability of the LoRa Low-Power Wide-Area Network Technology. In Proceedings of the 22nd European Wireless Conference, Oulu, Finland, 18–20 May 2016; pp. 1–6.
24. Haxhibeqiri, J.; Karaagac, A.; Van den Abeele, F.; Joseph, W.; Moerman, I.; Hoebeke, J. LoRa indoor coverage and performance in an industrial environment: Case study. In Proceedings of the 2017 22nd IEEE International Conference on Emerging Technologies and Factory Automation (ETFA), Limassol, Cyprus, 12–15 September 2017; pp. 1–8.
25. Lavric, A.; Petrariu, A.I.; Popa, V. Long Range SigFox Communication Protocol Scalability Analysis Under Large-Scale, High-Density Conditions. *IEEE Access* **2019**, *7*, 35816–35825. [CrossRef]
26. Mekki, K.; Bajic, E.; Chaxel, F.; Meyer, F. A comparative study of LPWAN technologies for large-scale IoT deployment. *ICT Express* **2019**, *5*, 1–7. [CrossRef]
27. Noreen, U.; Bounceur, A.; Clavier, L. A study of LoRa low-power and wide-area network technology. In Proceedings of the 2017 International Conference on Advanced Technologies for Signal and Image Processing (ATSIP), Fez, Morocco, 22–24 May 2017; pp. 1–6.
28. Cuomo, F.; Campo, M.; Caponi, A.; Bianchi, G.; Rossini, G.; Pisani, P. EXPLoRa: Extending the performance of LoRa by suitable spreading factor allocations. In Proceedings of the 2017 IEEE 13th International Conference on Wireless and Mobile Computing, Networking and Communications (WiMob), Rome, Italy, 9–11 October 2017; pp. 1–8.
29. Cuomo, F.; Gámez, J.C.C.; Maurizio, A.; Scipione, L.; Campo, M.; Caponi, A.; Bianchi, G.; Rossini, G.; Pisani, P. Towards traffic-oriented spreading factor allocations in LoRaWAN systems. In Proceedings of the 2018 17th Annual Mediterranean Ad Hoc Networking Workshop (Med-Hoc-Net), Capri, Italy, 20–22 June 2018; pp. 1–8.
30. Kim, D.Y.; Kim, S. Data Transmission Using K-Means Clustering in Low Power Wide Area Networks with Mobile Edge Cloud. *Wirel. Pers. Commun.* **2018**, *105*, 567–581. [CrossRef]
31. Farhad, A.; Kim, D.; Pyun, J. Scalability of LoRaWAN in an Urban Environment: A Simulation Study. In Proceedings of the 2019 Eleventh International Conference on Ubiquitous and Future Networks (ICUFN), Zagreb, Croatia, 2–5 July 2019; pp. 677–681.
32. Magrin, D.; Centenaro, M.; Vangelista, L. Performance evaluation of LoRa networks in a smart city scenario. In Proceedings of the 2017 IEEE International Conference on Communications (ICC), Paris, France, 21–25 May 2017; pp. 1–7.
33. Zhu, G.; Liao, C.-H.; Sakdejayont, T.; Lai, I.-W.; Narusue, Y.; Morikawa, H. Improving the Capacity of a Mesh LoRa Network by Spreading-Factor-Based Network Clustering. *IEEE Access* **2019**, *7*, 21584–21596. [CrossRef]
34. Vincenzo, V.D.; Heusse, M.; Tourancheau, B. Improving Downlink Scalability in LoRaWAN. In Proceedings of the ICC 2019—2019 IEEE International Conference on Communications (ICC), Shanghai, China, 20–24 May 2019; pp. 1–7.

35. Hasegawa, Y.; Suzuki, K. A Multi-User ACK-Aggregation Method for Large-Scale Reliable LoRaWAN Service. In Proceedings of the ICC 2019—2019 IEEE International Conference on Communications (ICC), Shanghai, China, 20–24 May 2019; pp. 1–7.
36. Li, L.; Ren, J.; Zhu, Q. On the application of LoRa LPWAN technology in sailing monitoring system. In Proceedings of the 13th Annual Conference on Wireless On-demand Network Systems and Services (WONS), Jackson, WY, USA, 21–24 February 2017; pp. 77–80.
37. Yasmin,R.; Petäjäjärvi, J.; Mikhaylov, K.; Pouttu, A. Large and Dense LoRaWAN Deployment to Monitor Real Estate Conditions and Utilization Rate. In Proceedings of the 2018 IEEE 29th Annual International Symposium on Personal, Indoor and Mobile Radio Communications (PIMRC), Bologna, Italy, 9–12 September 2018.
38. Abramson, N. The Aloha System—Another alternative for computer communications. In Proceedings of the FallJoint Computer Conference (AFIPS '70 (Fall)), New York, NY, USA, 17–19 November 1970; pp. 281–285.
39. Goursaud, C.; Gorce, J.M. Dedicated networks for IoT: PHY/MAC state of the art and challenges. *EAI Endorsed Trans. Internet Things* **2015**, *1*. [CrossRef]
40. Na, S.; Xumin, L.; Yong, G. Research on k-means Clustering Algorithm: An Improved k-means Clustering Algorithm. In Proceedings of the 2010 Third International Symposium on Intelligent Information Technology and Security Informatics, Jinggangshan, China, 2–4 April 2010; pp. 63–67.
41. Bouguera, T.; Diouris, J.F.; Chaillout, J.J.; Jaouadi, R.; Andrieux, G. Energy Consumption Model for Sensor Nodes Based on LoRa and LoRaWAN. *Sensors* **2018**, *18*, 2104. [CrossRef] [PubMed]

© 2019 by the authors. Licensee MDPI, Basel, Switzerland. This article is an open access article distributed under the terms and conditions of the Creative Commons Attribution (CC BY) license (http://creativecommons.org/licenses/by/4.0/).

Article

User Association and Power Control for Energy Efficiency Maximization in M2M-Enabled Uplink Heterogeneous Networks with NOMA

Shuang Zhang [1,2] and Guixia Kang [1,2,*]

[1] Key Laboratory of Universal Wireless Communications, Ministry of Education, Beijing University of Posts and Telecommunications, Beijing 100876, China; zhangshuang2015@bupt.edu.cn
[2] Wuxi BUPT Sensory Technology and Industry Institute CO. LTD, Wuxi 214000, China
* Correspondence: gxkang@bupt.edu.cn

Received: 8 October 2019; Accepted: 25 November 2019; Published: 2 December 2019

Abstract: To support a vast number of devices with less energy consumption, we propose a new user association and power control scheme for machine to machine enabled heterogeneous networks with non-orthogonal multiple access (NOMA), where a mobile user (MU) acting as a machine-type communication gateway can decode and forward both the information of machine-type communication devices and its own data to the base station (BS) directly. MU association and power control are jointly considered in the formulated as optimization problem for energy efficiency (EE) maximization under the constraints of minimum data rate requirements of MUs. A many-to-one MU association matching algorithm is firstly proposed based on the theory of matching game. By taking swap matching operations among MUs, BSs, and sub-channels, the original problem can be solved by dealing with the EE maximization for each sub-channel. Then, two power control algorithms are proposed, where the tools of sequential optimization, fractional programming, and exhaustive search have been employed. Simulation results are provided to demonstrate the optimality properties of our algorithms under different parameter settings.

Keywords: M2M; heterogeneous networks; non-orthogonal multiple access; energy efficiency; MU association; power control

1. Introduction

The increase of smartphones, laptops, and other mobile devices as well as data-hungry applications, need huge demands for ubiquitous coverage and very high data rates in cellular networks. However, homogeneous networks cannot satisfy these requirements [1]. Then, two-fold efforts have been spent to meet the stringent requirements. On one hand, researchers have proposed heterogeneous networks (HetNets) where different types of base stations (BSs), e.g., macro BSs (MBSs) and small BSs (SBSs) are deployed in a multi-tier hierarchical structure. In this structure, all BSs have seamless coverage and reuse frequencies to achieve higher data rate [2,3]. On the other hand, the so-called non-orthogonal multiple access (NOMA) has been investigated as a potential technique to further improve the throughput of network [4–7]. Different from conventional orthogonal multiple access (OMA), NOMA serves multiple users at the same time/frequency/codes resource by allocating different powers for them, and the superposition coded signal can be decoded at receivers by successive interference cancellation (SIC). Therefore, the combination of HetNets and NOMA will exhibit great potential to satisfy the 1000-times increase of mobile broadband data for the upcoming fifth generation (5G) communication systems and beyond [3].

However, the severe inter-tier and intra-tier interference make the NOMA-enabled HetNets challenging to achieve. Resource management plays an important role to alleviate these interference [8].

For downlink communication, specifically, some work focuses on the sum rate maximization and shows higher spectral efficiency (SE) can be achieved by NOMA when considering the intercell interference [9–12]. Besides SE, energy efficiency (EE) is also a key performance metric investigated for resource allocation in NOMA-enabled HetNets [8,13,14]. Moreover, EE is more important in uplink than in downlink NOMA-enabled HetNets since the devices in uplink communications are often battery-limited. It is a fact that the battery capacity has been improved at a very slow pace over the past decades [15], and hence this increase cannot scale with the high energy consumption caused by the increasing traffic demands. Meanwhile, EE has emerged as a new prominent performance metric for wireless communication networks designs due to the economic, operational, and environmental concerns [16,17]. Therefore, it is a stringent work to improve EE for uplink transmission.

Machine-to-machine (M2M) communications, also known as machine-type communications (MTC), enable pervasive connections to support IoT. M2M communications are one of the potential applications of NOMA-enabled HetNets [18], since NOMA-enabled HetNets provide a practical infrastructure to offer massive access opportunities for such a huge number of devices, especially for the cases in which each device only needs to send a small amount of data periodically in uplink. One of the challenges for HetNets with M2M communications is the access control, which can manage the engagement of massive MTC devices (MTCDs) to the core network. Among the existing access solutions, deploying MTC gateways (MTCGs) is an effective approach to connect M2M communication and cellular communication [19–21]. When mobile user (MU) has more power and storage space than MTCDs (e.g., smart sensors), the MU can be configured as the MTCG, as proposed in [22].

Since 5G will be HetNets including various network models (e.g., cellular networks, wireless networks (WSNs), and low power wireless area networks) to support high data rate and massive devices [3], our work combing M2M communication and cellular network has a large significance for this heterogeneous scenario. The short distance communication in our system model can be realized with WSNs, which provide a new way to help the sink nodes in WSNs communicate to the core network. For example, the MTCDs can be the sensors in an environmental monitoring WSN, and they can transmit the collected data to the core network through a mobile device in cellular networks with NOMA. Therefore, our work also has a practical significance for sensors work.

Recently, there have been some studies addressing the aforementioned challenges of applying NOMA in HetNets for EE maximization. In [23], a distributed user association algorithm based on inter-cell interference plus noise ratios of BS and a centralized user association based on the popular size of BS were both proposed. After user association was determined, a power control algorithm was proposed based on Lagrangian dual method, then a one-dimensional search algorithm was used to search Lagrangian multiplier, which added algorithm complexity. Two specific examples were provided to demonstrate the effectiveness of unified NOMA-enabled heterogeneous ultra-dense networks with user association and power control in [18]. An alternated energy efficient resource allocation algorithm based on fixed power allocation was first proposed in [13]. Then, two iterative energy-efficient resource allocation algorithms were proposed to update for better EE based on Lagrangian dual method. Joint base station association and power control optimization algorithms were proposed based on coalition formation games and interior-point method in [24], but sub-channel allocation and fractional equation for EE maximization had not been considered. Moreover, the user association algorithms in the aforementioned work were all considered with fixed power allocation firstly, whereafter iterative algorithms were used to obtain the final optimal value.

There are also some studies on the usage of M2M communications in NOMA systems. For example, energy-efficient resource allocation with hybrid division multiple-access NOMA for cellular-enabled M2M communications was researched in [25,26]. With MTCDs cluster formation known beforehand, standard convex optimization and Lagrange duality methods were employed respectively for power control in [25,26]. User clustering in NOMA-aided cellular M2M communication systems was researched in [27,28] with millimeter-wave and narrow-band IoT separately. A joint power and sub-channel allocation for secrecy capacity algorithm was proposed in [29] to obtain the suboptimal solution of the optimization problem. However, the aforementioned work deploys M2M user in single-cell networks. The trend of more and more intensive network deployment motivates us to deploy M2M-enabled NOMA in the scenarios with multi-tier HetNets and new resource allocation needs to be considered with the non-convexity caused by inter-cell interference in HetNets.

In this paper, we focus on the uplink EE maximization via user association and power control for M2M-enabled HetNets using NOMA. In this scenario, one macro base station (MBS) is located in the cell center. Each small cell has one small base station (SBS) located in the cell center. MUs are distributed randomly in the cell. An MU acting as an MTCG can decode and forward both the information of MTCDs and its own data to the BS. The EE (bits/Joule) maximization problem is formulated and solved to obtain the optimal MU association and power allocation. The main contributions of this paper are summarized as follows:

- We propose a new framework of M2M-enabled HetNets with NOMA. In this framework, control data separation architecture, i.e., control information and data message are separated, which can reduce the signal overhead [30]. NOMA is adopted by the MTCDs to transmit the information to MUs which is regarded as the relay. MUs decode the overlaid information and simultaneously transmit received data to the BSs based on the NOMA principle.
- In order to solve the EE maximization optimization problem, a BS and a sub-channel are included in a couple, since a MU can only associate one BS at one sub-channel. Then, a many-to-one MU association algorithm is proposed based on matching game [31]. Through swap operation among each couple, the EE maximization problem can be tackled by solving the power control problem at each sub-channel. Compared with the previous studies on the algorithms (user association and power allocation) [13,18,23,24], our algorithms are jointly optimized and fixed power allocation is not required for initialization.
- Two power control algorithms are proposed based on sequential optimization [32,33]. The fractional programming [34] and sequential optimization are combined to develop a novel sequential fractional power control algorithm (SFPCA), from which the original problem is transformed to be convex and requires less computational complexity. The other algorithm combines the exhaustive search method with sequential optimization, which can verify the correctness of SFPCA.

The rest of this paper is organized as follows. The system model and problem formulation are focused in Section 2. The MU association matching algorithm is proposed in Section 3. The power control problem is solved in Section 4. Numerical results are provided in Section 5, and concluding remarks are given in Section 6.

Notations: Lowercase and uppercase boldface letters denote vectors and matrices, respectively. We use uppercase decorated letters to denote sets. For an arbitrary set $\mathcal{M}$, we always have the corresponding uppercase M to the denote the cardinality of $\mathcal{M}$, i.e., $|\mathcal{M}| = M$, $[\cdot]^T$ denotes the transpose operator.

2. System Model and Problem Formation

2.1. System Model

As shown in Figure 1, we consider uplink HetNets with M2M communications, where all MUs are anchored to the control base station (CBS). The CBS performs the MU association algorithm to select the best serving BS for MUs and establishes a high BS–MU connection through backhaul links. Each MTCD selects the nearest MU as an MTCG. Since NOMA is adopted between MTCDs that select the same MU as their MTCGs, SIC is performed at the MU to gather the interference and channel gain will be obtained by channel estimation at the MU. The HetNets consist of a set $\mathcal{F} = \{0, \cdots, F\}$ of BSs and a set $\mathcal{K} = \{1, \cdots, K\}$ of MUs. Each MU is regarded as an MTCG, which can acts as a relay for some MTCDs. Denote $\mathcal{U}_k$ as the specific set of MTCDs served by MU k (MU$_k$). The index 0 denotes the MBS and other indexes stand for the SBSs in set $\mathcal{F}$. Without special explanation, we always have $f \in \mathcal{F}$,

$F = |\mathcal{F}|$. The system bandwidth shared by all BSs is divided into N orthogonal sub-channels, and each one is assigned with bandwidth B. For convenience, hereinafter we always have $n \in \mathcal{N} = \{1, 2, \cdots N\}$ to denote the sub-channel. MUs are served by BSs according to the BSs' coverage.

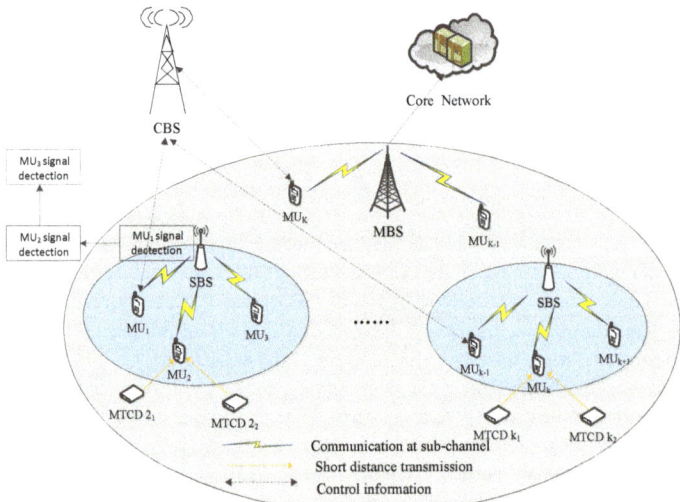

Figure 1. System Model.

2.2. NOMA Strategy

Multiple MTCDs can simultaneously transmit signals to the MU using NOMA. Since MUs and MTCDs use different transmission modes, we ignore the interference between MUs and MTCDs. The interference between MTCDs in different BSs is also not considered. According to the NOMA principle, the received signal of MU_k is

$$Y_k = \sum_{j \in \mathcal{U}_k} h_{j_k k} \sqrt{q_{j_k}} s_{j_k} + n_k, \quad (1)$$

where h_{j_k} is the channel between MTCD j_k and MU_k; q_{j_k} and s_{j_k} denote the transmit power and message of MTCD j_k; and n_k represents the additive zero-mean Gaussian noise with variance σ^2. $\mathcal{U}_k$ represents the set of MTCDs which are served by MU_k. Without loss of generality, the channels are sorted by $|h_{1_k k}|^2 / \sigma_k^2 > |h_{2_k k}|^2 / \sigma_k^2 > \cdots |h_{U_k k}|^2 / \sigma_k^2 > 0$. Applying SIC in NOMA [33], the achievable data throughput for MTCD j_k at MU_k is given by

$$R_{j_k} = \log_2 \left(1 + \frac{H_{j_k k} q_{j_k}}{1 + I_{j_k}}\right), \quad (2)$$

where $H_{j_k k} = |h_{j_k k}|^2 / \sigma_k^2$, $I_{j_k} = \sum_{i \in \{U_k | H_{i_k k} < H_{j_k k}\}} q_{i_k k} H_{i_k k}$, and we define $I_{j_k} = 0$ for $j_k = U_k$. After MUs successfully decode the messages from MTCDs, all MUs simultaneously transmit data to the BS based on the NOMA principle. Denote $h_{kfn} = g_{kfn} \sqrt{d_{kf}^{-\alpha}}$ as the channel gain between MU_k and BS f at sub-channel n (SC_n). g_{kfn} denotes the corresponding Rayleigh fading channel gain; α is the path loss factor; and d_{kf} is the distance between MU_k and BS f. In order to split the superimposed signals on SC_n in BS f, SIC is carried out at BS f. Based on the uplink NOMA protocol [35], the signal of MU with the highest channel gain will be first decoded at BS f and experiences interference from other MUs having relatively weaker channel gains on SC_n. Therefore, the channel gains of MUs over SC_n in BS f

are sorted as $|h_{1fn}|^2/\sigma_{fn}^2 > |h_{2fn}|^2/\sigma_{fn}^2 > \cdots > |h_{kfn}|^2/\sigma_{fn}^2 > \cdots |h_{S_{fn}fn}|^2/\sigma_{fn}^2$, where $S_{fn} = |\mathcal{S}_{fn}|$. Then the transmit data rate of MU_k associated with BS f over SC_n can be expressed as

$$R_{kfn} = B\log_2\left(1 + \frac{p_{kfn}H_{kfn}}{1 + I_{kfn} + \phi_{kfn}}\right), \tag{3}$$

where $H_{kfn} = |h_{kfn}|^2/\sigma_{fn}^2$. I_{kfn} is the interference that MU_k receives from other MUs whose channel gains are smaller than that on SC_n of BS f, which can be given by

$$I_{kfn} = \sum_{i \in \{\mathcal{S}_{fn}|H_{ifn} < H_{kfn}\}} p_{ifn}H_{ifn}. \tag{4}$$

$\phi_{kfn} = \sum_{f' \in \{\mathcal{F}\setminus f\}} \sum_{i \in \mathcal{S}_{f'n}} p_{if'n}H_{ifn}$ is the interference from MUs associated to other BSs on SC_n. Then the data rate of MUs at SC_n is

$$R_n = B \sum_{f \in \mathcal{F}} \sum_{k \in \mathcal{S}_{fn}} \log_2\left(1 + \frac{p_{kfn}H_{kfn}}{1 + I_{kfn} + \phi_{kfn}}\right). \tag{5}$$

2.3. Problem Formation

In this paper, we focus on the EE maximization problem for all MUs considering the minimum data rate requirements of them. The MU association contains two parts: BS selection and sub-channel allocation. For a given MU association, the out-of-cell interference only come from the MUs associated with different BSs at the same sub-channel due to the orthogonality among the sub-channels. Then, each MU may not concern the whole EE, but the sub-channel EE it chooses. Therefore, the optimization problem is converted into solving the EE maximization of each sub-channel by appropriate MU association including BS selection and sub-channel allocation and power control.

From a physical standpoint, the efficiency with which a system uses a given resource is the ratio between the benefit obtained by using the resource and the corresponding incurred cost [17]. Applying this general definition to the uplink communication at SC_n, then EE of SC_n can be written as

$$EE_n = \frac{R_n}{\sum_{f \in \mathcal{F}} \sum_{k \in \mathcal{S}_{fn}} p_{kfn} + P_c}, \tag{6}$$

where P_c is the additional circuit power consumption over each sub-channel. Then the considered EE optimization problem can be formulated as

$$\max_{\mathbf{P}} \sum_{n \in \mathcal{N}} EE_n \tag{7a}$$

$$\text{s.t.} \quad R_{kfn} \geq \sum_{j \in \mathcal{U}_{j_k}} R_{j_k} + R_{req}, \forall k \in \mathcal{K}, n \in \mathcal{N}, \tag{7b}$$

$$p_{kfn} > 0, p_{kfn} \leq P_{max}, \forall k \in \mathcal{K}, f \in \mathcal{F}, n \in \mathcal{N}, \tag{7c}$$

where $\mathbf{P}$ is the transmit power vector with elements p_{kfn}; P_{max} is the maximum transmit power of each MU; and R_{req} is the minimum data rate requirement of a MU. Since each MU is regarded as an MTCG for MTCDs, they should ensure the data rate that MTCDs can be uploaded to the SBS, therefore we have constraint (7b) as the data rate requirement of MU_k associated to BS f at SC_n [22]. Constraint (7c) is used to guarantee the feasible value ranges of $\mathbf{P}$.

3. MU Association

The MTCDs associated to the corresponding MU are known beforehand. Since solving the optimization problem is equal to obtain the optimal EE of each sub-channel, the MU association will become the matching problem among BSs, sub-channels and MUs to achieve sub-channel EE maximization. Thus, we propose a MU association algorithm using matching game [30] in the following parts.

3.1. Matching Problem Formulation

To develop a low-complexity MU association algorithm, we first regard a sub-channel and a BS as a couple, denoted as (n, f). Then, the optimization problem is transformed to match the MUs to the couples and allocate power appropriately, such that the EE can be maximized. Finally, the matching problem is a many-to-one problem between MUs and couples based on matching game, which is described as follows.

Definition 1. *Given two disjoint sets, $\mathcal{K} = \{1, \cdots, K\}$ denotes the set for MUs, and $\mathcal{M} = \{(1,1)\,(1,2)\cdots,(2,1)\cdots,(n,f),\cdots(N,F)\}$ represents the couples. A many-to-one matching Ψ is a mapping from the set $\mathcal{K} \cup \mathcal{M}$ into the set of all subsets of $\mathcal{K} \cup \mathcal{M}$ for $f \in \mathcal{F}, k \in \mathcal{K}, n \in \mathcal{N}$ satisfying*

i) $\Psi(k) \in \mathcal{M}$;
ii) $\Psi(n, f) \subset \mathcal{K}$;
iii) $|\Psi(k)| = 1 |, \Psi(SC_n, f)| = S_{fn}$;
iv) $(n, f) = \Psi(k) \Leftrightarrow k \in \Psi(n, f)$.

Condition i indicates that each MU matches with a sub-channel-BS couple. On the other hand, each couple matches a subset of MUs, which is illustrated in condition ii. Condition iii states a MU can only associate one BS and choose one sub-channel while each couple matches S_{fn} MUs.

The aim of each couple is to maximize its own EE. To this end, we exploit the swap operation into our matching algorithm. A swap operation means two MUs matching with different couples exchange their matchings based on different cases, while other MUs remain their matchings. The EE of the exchanged couples will be recomputed by the power control algorithm. Note that how to allocate power to obtain the optimal EE for a given sub-channel will be presented in the next section, and we assume it is known in advance. A swap operation will be approved and the matching will be exchanged only when all EE of the sub-channels belonging to the exchanged couples increase if the swap is performed. The swap operation will be continued until no swap is further preferred. More details are described in Algorithm 1.

3.2. Matching Algorithm

Algorithm 1 contains a initialization phase and a swap matching phase. Considering the user fairness, the number of MUs accommodated by one sub-channel in a given BS is at most $\left\lceil \frac{K}{FN} \right\rceil$. In the initialization step, the basic idea is to associate the MU to the couple providing the largest channel gain. This will lead to either a higher data rate for the MU, or a lower transmit power. Since the value of sub-channel gain between MU and the uncovered BSs is invalid and the maximized sub-channel gain is always chosen, there is no need to know whether the MUs are in the coverage of the exchanged BSs. However, in the swap matching phase, this judgment should be considered at first to avoid the invalid swap. Then, exchange will happen in the three cases. Iterations will continue until no swap operation can be approved in a new round.

Algorithm 1 The MU association matching algorithm.

Initialization phase: $L = \left\lceil \frac{K}{NF} \right\rceil, \hat{\mathcal{K}} = \mathcal{K}$
1: **for** $l = 1 : L$ **do**
2:    $\hat{\mathcal{M}} = \mathcal{M}, Count = 1$
3:    **while** $(Count \leq M)$ **do**
4:       $H_{k^*f^*n^*} = argmax\left\{H_{kfn}\right\}_{\forall k \in \hat{\mathcal{K}},(f,n) \in \hat{\mathcal{M}}}$. Assign k^* to the couple (f^*, n^*), $\hat{\mathcal{K}} = \hat{\mathcal{K}} \backslash k^*$, $\hat{\mathcal{M}} = \hat{\mathcal{M}} \backslash (f^*, n^*)$, and set $Count = Count + 1$
5:    **end while**
6: **end for**
   Swap matching phase: $Indicator = 1$
7: **while** $(Indicator)$ **do**
8:    **for** $u = 1 : K$ **do**
9:      **for** $k = 1 : K$ **do**
10:        **if** $\Psi(k) = \Psi(u)$ **then**
11:           continue;
12:        **else if** MU_k and MU_u are both in the coverage of the BSs of each other **then**
13:           **switch** $(\Psi(k), \Psi(u))$
14:           **case** MU_k and MU_u belonging to the same BS and different sub-channels:
15:               Calculate and compare the EE of the two sub-channels before and after the swap using the power control algorithm. If the EE of the two-subchannels both improve, exchange the sub-channel, form the new couple, and set $Indicator = 1$.
16:           **case** MU_k and MU_u belonging to the different BSs and different sub-channels:
17:               Calculate and compare the EE of the two sub-channels before and after the swap using the power control algorithm. If the EE of the two sub-channels both improve, exchange the couple, form the new couple, and set $Indicator = 1$.
18:           **case** MU_k and MU_u belonging to the different BSs and same sub-channels:
19:               Calculate the EE of the sub-channel before and after the swap using the power control algorithm. If the EE of the sub-channel has been improved, exchange the BS, form the new couple, and set $Indicator = 1$.
20:           **end switch**
21:        **end if**
22:      **end for**
23:    **end for**
24: **end while**

3.3. Convergence and Complexity

Theorem 1. *The proposed MU association and power control algorithm converges after a finite number of swap operations.*

Proof of Theorem 1. For each swap operation, the matching changes from Ψ_{ex} to Ψ_{now}. We have $EE_{n,ex}$ and $EE_{n,now}$ to denote the corresponding EE of Ψ_{ex} and Ψ_{now} on SC_n. Based on the aim of swap operation, we have $EE_{n,now} > EE_{n,ex}$, that is, the EE of each sub-channel increases after each swap matching. Since each sub-channel is orthogonal to each other, the system EE will increase owing to the improved EE of each sub-channels. Moreover, the system EE has an upper bound due to the limited transmit power of each MU. Therefore, the MU association algorithm and power allocation converge after a finite number of swaps. □

4. Power Control

In this section, we will investigate the optimal power control design appearing in Algorithm 1 to obtain the maximum EE of SC_n. Before we present the optimization problem for EE_n maximization, we first deal with R_n, which can be rewritten as

$$R_n(\mathbf{p}_{fn}) = B \sum_{f \in \mathcal{F}} \log_2 \left(1 + \frac{\sum_{k \in \mathcal{S}_{fn}} p_{kfn} H_{kfn}}{1 + \phi_{kfn}}\right), \quad (8)$$

and we can also obtain

$$R_{tot,k} = \sum_{j \in \mathcal{U}_k} R_{j_k} = \log_2 \left(1 + \sum_{j \in \mathcal{U}_k} H_{j_k k} q_{j_k}\right). \quad (9)$$

Due to the multi-interference in the sum-rate function in (8), EE_n in (6) is non-convex and cannot be directly solved by the generalized fractional programming approach. Then, we first transform the numerator into the difference of two non-negative functions and the EE_n maximization can be rewritten as

$$\max_{\mathbf{p}_{fn}} \tilde{\eta}_n = \frac{F^+(\mathbf{p}_{fn}) - F^-(\mathbf{p}_{fn})}{\sum_{f \in \mathcal{F}} \sum_{k \in \mathcal{S}_{fn}} p_{kfn} + P_c} \quad (10a)$$

$$\text{s.t.} \quad (7c), (11), \quad (10b)$$

with

$$C^+_{kfn}(\mathbf{p}_{fn}) - C^-_{kfn}(\mathbf{p}_{fn}) \geq 0, \quad \forall k \in \mathcal{K}, \forall f \in \mathcal{F}, \forall n \in \mathcal{N}, \quad (11)$$

where $\mathbf{p}_{fn} = [p_{1fn}, p_{2fn}, \cdots, p_{kfn}, \cdots, p_{S_{fn}F_n}]^T$ denotes the transmit power vector for MUs on SC_n. Moreover, we have

$$\begin{aligned}
F^+(\mathbf{p}_{fn}) &= B \sum_{f \in \mathcal{F}} \log_2 \left(1 + \sum_{k \in \mathcal{S}_{fn}} p_{kfn} H_{kfn} + \phi_{kfn}\right), \\
F^-(\mathbf{p}_{fn}) &= B \sum_{f \in \mathcal{F}} \log_2 \left(1 + \phi_{kfn}\right), \\
C^+_{kfn}(\mathbf{p}_{fn}) &= B\log_2 \left(1 + p_{kfn} H_{kfn} + I_{kfn} + \phi_{kfn}\right) \\
&\quad - R_{tot,k} - R_{req}, \\
C^-_{kfn}(\mathbf{p}_{fn}) &= B\log_2 \left(1 + I_{kfn} + \phi_{kfn}\right).
\end{aligned} \quad (12)$$

Note that $F^+, F^-, C^+,$ and C^- are concave functions regarding to $\mathbf{p}_{fn}$, then the numerator of (10a) and the constraint functions in (10b) are expressed as the difference of concave functions, which are not concave in general. Motivated by [31,32], where sequential optimization is used to solve the similar problem as (10), we adopt this method and combine it with fractional programming and exhaustive search to propose two power control algorithms. Before introducing the two algorithms, we first present the details of the sequential optimization theory in the next sub-section.

4.1. Sequential Optimization Theory

Sequential optimization is a powerful tool that can tackle a difficult optimization problem by solving a sequence of approximate problems in simple forms with affordable complexity. Specifically, we give a formal maximization problem $\tilde{F}$ with a compact feasible set as [31,32], shown as

$$\max_{\mathbf{x}} f_0(\mathbf{x}) \tag{13a}$$

$$\text{s.t.} \quad f_i(\mathbf{x}) \geq 0, \forall i \in \{1, \cdots, I\}, \tag{13b}$$

where $f_0(\mathbf{x})$ is the differentiable objective with constraints $f_i(\mathbf{x}) \geq 0$. Let $G^{(v)}$ be the problem solved in the v-th iteration by the sequential method to tackle problem $\tilde{F}$, which can be written as

$$\max_{\mathbf{x}} g_0^{(v)}(\mathbf{x}) \tag{14a}$$

$$\text{s.t.} \quad g_i^{(v)}(\mathbf{x}) \geq 0, \forall i \in \{1, \cdots, I\}, \tag{14b}$$

where $g_0^{(v)}(\mathbf{x})$ is the differentiable objective with the constraints $g_i^{(v)}(\mathbf{x})$. Then, if $g_0^{(v)}(\mathbf{x})$ and $g_i^{(v)}(\mathbf{x})$ are suitable continuous functions and constraints, they must satisfy the following two properties:

1) $g_0^{(v)}(\mathbf{x}) \leq f_0(\mathbf{x}), g_i^{(v)}(\mathbf{x}) \leq f_i(\mathbf{x}) \quad \forall \mathbf{x}$;
2) $g_0^{(v)}((\mathbf{x}^*)^{(v-1)}) = f_0((\mathbf{x}^*)^{(v-1)}), g_i^{(v)}((\mathbf{x}^*)^{(v-1)}) \leq f_i((\mathbf{x}^*)^{(v-1)})$.

$(\mathbf{x}^*)^{(v-1)}$ is the optimal solution of the problem solved at iteration $(v-1)$-th. This means the solution sequence $\{(\mathbf{x}^*)\}^{(v)}$ of (14) monotonically increases the value of (13), i.e., $f_0((\mathbf{x}^*)^{(v)}) \geq f_0((\mathbf{x}^*)^{(v-1)})$ for all v, which guarantees the convergence of the sequential method. Next, if the following third property is also satisfied:

3) $\nabla g_0^{(v)}((\mathbf{x}^*)^{(v-1)}) = f_0((\mathbf{x}^*)^{(v-1)}), \nabla g_i^{(v)}((\mathbf{x}^*)^{(v-1)}) = \nabla f_i((\mathbf{x}^*)^{(v-1)})$.

then every limit point of $\{\mathbf{x}\}^{(v)}$ of (14) fulfills the Karush–Kuhn–Tucker (KKT) conditions of problem $\tilde{F}$ in (13).

Therefore, if a maximization problem finds suitable approximate problems which can fulfill the above three properties, its optimal value can be approximated by solving the monotonically increased sequential problems. The critical issue is that the suitable approximate problems are solved easier than the original problem. In the rest of this section, we will first find the sequential approximate problems to the numerator in problem (10).

4.2. Sequential Fractional Power Control Algorithm

Based on sequential optimization, we should find a sequence problem to approximate optimization problem (10). To circumvent this issue, we obtain the following main result with the first-order Taylor expansion at $\mathbf{p}_{fn}^{(v)}$ of $F^-(\mathbf{p}_{fn})$.

Proposition 1. *For any given $\mathbf{p}_{fn}^{(v)}$, the sequence approximation problem of (10), denoted by $G^{(v)}$ can be written as*

$$\max \quad \eta_n = \frac{F^+(\mathbf{p}_{fn}) - \tilde{F}(\mathbf{p}_{fn})}{\sum_{f \in \mathcal{F}} \sum_{k \in \mathcal{S}_{fn}} p_{kfn} + P_c} \tag{15a}$$

$$\text{s.t.} \quad C_{kfn}^+(\mathbf{p}_{fn}) - \tilde{C}(\mathbf{p}_{fn}) \geq 0, \tag{15b}$$

$$(7c), \tag{15c}$$

with optimal solution $\mathbf{p}_{fn}^{*(v)}$, where

$$\tilde{F}(\mathbf{p}_{fn}) = F^-\left(\mathbf{p}_{fn}^{(v)}\right) \\ - \left(\nabla F^-\left(\mathbf{p}_{fn}^{(v)}\right)\right)^T \left(\mathbf{p}_{fn} - \mathbf{p}_{fn}^{(v)}\right) \tag{16}$$

$$\tilde{C}(\mathbf{p}_{fn}) = C_{kfn}^-\left(\mathbf{p}_{fn}^{(v)}\right) \\ - \left(\nabla C^-\left(\mathbf{p}_{fn}^{(v)}\right)\right)^T \left(\mathbf{p}_{fn} - \mathbf{p}_{fn}^{(v)}\right) \tag{17}$$

If $\mathbf{p}_{fn}^{(v)} = \mathbf{p}_{fn}^{*(v-1)}, \forall v \geq 1$, then $\left\{\eta_n \mathbf{p}_{fn}^{*(v)}\right\}^{(v)}$ is monotonically increasing and converges to a value $\tilde{\eta}_n$. Furthermore, any limit point of sequence $\left\{\eta_n \mathbf{p}_{fn}^{*(v)}\right\}^{(v)}$ that achieves $\tilde{\eta}_n$ fulfills the KKT optimality conditions of (10a).

Proof of Proposition 1. As we know, any concave function is the upper-bounded of its first-order Taylor expansion at any point. Since $F^-(\mathbf{p}_{fn})$ and $C^-(\mathbf{p}_{fn})$ are concave functions, for any power vector $\mathbf{p}_{fn}^{(v)}$ we have

$$F^+(\mathbf{p}_{fn}) - F^-(\mathbf{p}_{fn}) \\ \geq F^+(\mathbf{p}_{fn}) - \tilde{F}(\mathbf{p}_{fn}) \\ = F^+(\mathbf{p}_{fn}) - F^-\left(\mathbf{p}_{fn}^{(v)}\right) \\ - \left(\nabla F^-\left(\mathbf{p}_{fn}^{(v)}\right)\right)^T \left(\mathbf{p}_{fn} - \mathbf{p}_{fn}^{(v)}\right), \tag{18}$$

$$C^+(\mathbf{p}_{fn}) - C^-(\mathbf{p}_{fn}) \\ \geq C^+(\mathbf{p}_{fn}) - \tilde{C}(\mathbf{p}_{fn}) \\ = C^+(\mathbf{p}_{fn}) - C_{kfn}^-\left(\mathbf{p}_{fn}^{(v)}\right) \\ - \left(\nabla C^-\left(\mathbf{p}_{fn}^{(v)}\right)\right)^T \left(\mathbf{p}_{fn} - \mathbf{p}_{fn}^{(v)}\right). \tag{19}$$

Hence, (15a) and (15b) are lower bounds of (10a) and (11), respectively. Since the lower bounds in (16) are tight when evaluated by $\mathbf{p}_{fn}^{(v)}$, it follows that (15a) and (15b) are equal to (10a) and (11), respectively, for $\mathbf{p}_{fn} = \mathbf{p}_{fn}^{(v)}$. Similarly, the gradients of (15a) and (15b) are equal to those of (10a) and (11), for $\mathbf{p}_{fn} = \mathbf{p}_{fn}^{(v)}$. Thus, (15) fulfills all the properties described in the above sub-section, which completes the proof of this proposition. □

For any $\mathbf{p}_{fn}^{(v)}$, problem (15) has a concave numerator and an affine denominator, while the constraint functions in (15b) and (15c) are both concave and affine. Therefore, (15) is a single-ratio problem, which can be solved by the generalized fractional programming. We adopt the widely used Dinkelbach's algorithm to solve it. According to Dinkelbach's method [33], we first introduce the following auxiliary function

$$T(\mathbf{p}_{fn}, \eta_n) = f_n(\mathbf{p}_{fn}) - \eta_n g_n(\mathbf{p}_{fn}), \tag{20}$$

with $f_n(\mathbf{p}_{fn}) = F^+(\mathbf{p}_{fn}) - \tilde{F}(\mathbf{p}_{fn})$, and $g_n(\mathbf{p}_{fn}) = \sum\limits_{f \in \mathcal{F}} \sum\limits_{k \in \mathcal{S}_{fn}} p_{kfn} + P_c$.

Theorem 2. Let η_n^*, $\mathbf{p}_{fn}^*$ and p_{kfn}^* denote the optimal value, optimal solution and its elements of problem (15), respectively. Then, we have

$$\eta_n^* = \frac{F^+(\mathbf{p}_{fn}^*) - \tilde{F}(\mathbf{p}_{fn}^*)}{\sum_{f \in \mathcal{F}} \sum_{k \in \mathcal{S}_{fn}} p_{kfn}^* + P_c} = \max \frac{F^+(\mathbf{p}_{fn}) - \tilde{F}(\mathbf{p}_{fn})}{\sum_{f \in \mathcal{F}} \sum_{k \in \mathcal{S}_{fn}} p_{kfn} + P_c}, \tag{21}$$

if and only if

$$\max \{ T(\mathbf{p}_{fn}, \eta_n^*) = f_n(\mathbf{p}_{fn}) - \eta_n^* g_n(\mathbf{p}_{fn}) \} \\
= f_n(\mathbf{p}_{fn}^*) - \eta_n^* g_n(\mathbf{p}_{fn}^*) = 0. \tag{22}$$

Proof of Theorem 2. Theorem 2 was proved in [33,36], and we omit it due to the limited space. □

The optimal η_n^* can be obtained by Dinkelbach's method, which is summarized in Algorithm 2. As shown in the algorithm, we need to solve the problem (23) for a given parameter $\eta_n^{(c)}$ in each iteration. In Algorithm 2, η_n has been updated as $\eta_n^{(c)}$ in each iteration until convergence. or reaching the maximum number of iterations. $\mathbf{p}_{fn}^{(c)}$ denotes the optimal power of the following problem in the c-th iteration, which can be obtained in Algorithm 3, as given by

$$\max_{\mathbf{p}_{fn}} T(\mathbf{p}_{fn}, \eta_n^{(c)}) = f_n(\mathbf{p}_{fn}) - \eta_n^{(c)} g_n(\mathbf{p}_{fn})$$
$$\text{s.t.} \quad C_{kfn}^+(\mathbf{p}_{fn}) - \tilde{C}(\mathbf{p}_{fn}) \geq 0, \tag{23}$$
$$(7c),$$

Algorithm 2 The Dinkelbach's algorithm.

Initialization phase:

Set iteration $c = 1$, $\eta_n^{(c)} > 0$, the maximum number of iterations C_{max}, and error tolerance $\tau > 0$.
1: **repeat**
2: Solve the equivalent problem (23) for a given $\eta_n^{(c)}$ to obtain the solution $\mathbf{p}_{fn}^{(c)}$.
3: $\eta_n^{(c)} = \frac{F^+(\mathbf{p}_{fn}^{(c)}) - \tilde{F}(\mathbf{p}_{fn}^{(c)})}{\sum_{f \in \mathcal{F}} \sum_{k \in \mathcal{S}_{fn}} p_{kfn}^{(c)} + P_c}$,
4: $c = c + 1$.
5: **until** $\left| T(\eta_n^{(c-1)}, \mathbf{p}_{fn}^{(c-1)}) \right| \leq \tau$ or $c > C_{max}$
6: $\eta_n^* = \eta_n^{(c-1)}$, $\mathbf{p}_{fn}^* = \mathbf{p}_{fn}^{(c-1)}$.

Algorithm 3 The algorithm for solving problem (23).

Initialization phase:

Set $\mathbf{p}_{fn}^{(0)}$, iteration index $v = 0$, the maximum iterations V_{max} and error tolerance μ. Calculate $f_n(\mathbf{p}_{fn}^{(0)}) - \eta_n^{(c)} g_n(\mathbf{p}_{fn}^{(0)})$.
1: **repeat**
2: Solve the problem (23) to obtain the optimal solution $\mathbf{p}_{fn}^*$ for given $\mathbf{p}_{fn}^{(v)}$ and $\eta_n^{(c)}$.
3: $v = v + 1$.
4: Set $\mathbf{p}_{fn}^{(v)} = \mathbf{p}_{fn}^*$ and cacluate $f_n(\mathbf{p}_{fn}^{(v)}) - \eta_n^{(c)} g_n(\mathbf{p}_{fn}^{(v)})$.
5: **until** $\left| f_n(\mathbf{p}_{fn}^{(v)}) - \eta_n^{(c)} g_n(\mathbf{p}_{fn}^{(v)}) - (f_n(\mathbf{p}_{fn}^{(v-1)}) - \eta_n^{(c)} g_n(\mathbf{p}_{fn}^{(v-1)})) \right| \leq \mu$
6: $\mathbf{p}_{fn}^* = \mathbf{p}_{fn}^{(v)}$.

4.3. Computational Complexity Analysis

In above sub-section, we have proposed the SFPCA including two steps, i.e., Algorithms 2 and 3. The computational complexity of them are separately discussed. First, we use C to denote the number of iterations for Algorithm 2, where C is bounded by C_{max}. From Section V, we can see that Algorithm 2 will converge after a few number of iterations. Then we discuss the computational complexity of Algorithm 3, the complexity of this algorithm is mainly caused by (23), and denoted by X. The computational complexity of (23) is $O(S_n{}^3)$ [37], where S_n is the number of MUs at SC_n. The complexity of Algorithm 3 is $X = VO(S_n{}^3)$, where V is the the number of iterations bounded by V_{max}. In summary, the computational complexity of the power control algorithm is $O(CX)$.

4.4. Sequential Exhaustive Algorithm

To evaluate the performance of the SFPCA, a sequential exhaustive algorithm (SEA) combined with sequential optimization and exhaustive search is proposed in this section. The detailed procedures of the compared algorithm is illustrated as follows. To solve the problem in an easier manner, we introduce the auxiliary variable y_n, as given by

$$y_n = B \sum_{f \in \mathcal{F}} \log_2 \left(1 + \frac{\sum_{k \in \mathcal{S}_{fn}} p_{kfn} H_{kfn}}{1 + \phi_{kfn}} \right). \tag{24}$$

if we fix y_n, the objective function (15a) can be recast as

$$\max_{\mathbf{p}_{fn}} \frac{y_n}{\sum_{f \in \mathcal{F}} \sum_{k \in \mathcal{S}_{fn}} p_{kfn}}$$

$$\text{s.t.} \sum_{f \in \mathcal{F}} \log_2 \left(1 + \frac{\sum_{k \in \mathcal{S}_{fn}} p_{kfn} H_{kfn}}{1 + \phi_{kfn}} \right) \geq y_n. \tag{25}$$

Due to the multi-user interference, we cannot solve problem (25) by standard convex optimization tools. Similar to SFPCA, sequential optimization is applied and the approximate problem can be shown as

$$\max_{\mathbf{p}_{fn}} \frac{F^+(\mathbf{p}_{fn}) - \tilde{F}(\mathbf{p}_{fn})}{\sum_{f \in \mathcal{F}} \sum_{k \in \mathcal{S}_{fn}} p_{kfn} + P_c} \tag{26a}$$

$$\text{s.t.} \quad F^+(\mathbf{p}_{fn}) - \tilde{F}(\mathbf{p}_{fn}) \geq y_n, \forall n \in \mathcal{N}, \tag{26b}$$

$$(7c), (11). \tag{26c}$$

It can be observed that since y_n is fixed, (26) is equivalent to minimize the linear function $\sum_{f \in \mathcal{F}} \sum_{k \in \mathcal{S}_{fn}} p_{kfn} + P_c$ in the denominator, subject to convex constraints. Then, problem (26) can be solved by plain convex programming. To implement an efficient line search for y_n, the bound of y_n is given by

$$\check{y}_n = FS_{fn} R_{req}$$

$$\leq \sum_{f \in \mathcal{F}} \log_2 \left(1 + \frac{\sum_{k \in \mathcal{S}_{fn}} p_{kfn} H_{kfn}}{1 + \phi_{kfn}} \right)$$

$$< \sum_{f \in \mathcal{F}} \log_2 \left(1 + \sum_{k \in \mathcal{S}_{fn}} P_{max} H_{kfn} \right) \tag{27}$$

$$= \hat{y}_n.$$

Then, the optimal $\mathbf{p}_{fn}$ can be obtained by searching an appropriate value of y_n with stepsize ε.

5. Numerical Results and Discussions

In this section, the effectiveness of our proposed MU association and power control algorithms in M2M-enabled HetNets with NOMA was demonstrated by Monte Carlo simulations. The HetNets included one MBS and two SBSs, and the radius of the cells for them were 200 m and 80 m, respectively. MUs were randomly and uniformly distributed. The values of the simulation parameters are summarized in Table 1.

We considered the EE performance obtained from EE maximization and sum-rate maximization with different P_{max} in Figure 2. The latter could be obtained in the first iteration of Algorithm 2 due to $q^{(1)} = 0$. In order to reflect the influence of q_{max}, we gave four schemes of different q_{max}. From the figure, we can see that all of the four schemes had a "green point", where EE and sum rate could both achieve their optimal values. Different from sum rate, EE became gradually flat while sum rate decreased after "green point" as P_{max} grew. The reason is when the maximum EE is achieved, no more transmit power is needed. For sum rate maximization, larger sum rate requires more transmit power, and its ratio (EE) may decrease, since the numerator (sum rate) and denominator (sum transmit power) both grow. We can also see that the EE decreased as q_{max} increased, because the increase of q_{max} means the data rate requirement of MUs increases. It is worth noting that even though MTCDs have lower data rate and transmit power, they can also have a strong influence on the overall uplink EE with their massive number. Furthermore, Algorithm 1 with higher EE had similar tendency as Algorithm 4, which proves the correctness of our algorithms.

Figure 3 shows the EE performance with respect to different data rate requirements of MUs with the different transmit power of MTCDs. The four curves all decreased as R_{req} increased. This is due to the fact that higher data rate will narrow the feasible value regions of the transmit power. Note that the four curves decreased slightly first, when $R_{req} = 150$ bps, the EE of the four schemes all declined distinctly, since higher data rate requirement may require more transmit power, destroying the balance of sum transmit power and sum rate. As explained above, the increase of q_{max} leads to the increase of data rate requirement, and the variation of EE is in line with the reason as the figure shown.

Table 1. simulation parameters.

Parameters	Meanings	Values
F	Number of BSs	3
B	The frequency bandwidth of each sub-channel	15 kHz
K	Number of MUs	40
U_k	Number of MTCDs of each MU	2
σ^2_{fn}	Noise variance	2 dBm
μ, τ	Error tolerance	10^{-3}
P_{max}	The maximum of transmit power of MU	0.2 W
q_{max}	The maximum of transmit power of MTCD	0.08 W
α	Path loss factor	3
P_c	The circuit power at each sub-channel	0.1 W
R_{req}	The data rate requirement of each MU	100 bps

Algorithm 4 Sequential exhaustive algorithm.

Initialization phase:

$\varepsilon > 0, \omega = \frac{\widehat{y}_n - \widecheck{y}_n}{\varepsilon}$

1: **for** $y_{fn} \in \left[\widecheck{y}_{fn} : \omega : \widehat{y}_{fn}\right)$ **do**
2: $\quad \mathbf{p}^*_{fn} = \arg\min \sum_{f \in \mathcal{F}} \sum_{k \in \mathcal{S}_{fn}} p_{kfn} + P_c$
3: **end for**
4: Obtain the optimal solution of $\mathbf{p}_{fn}$.

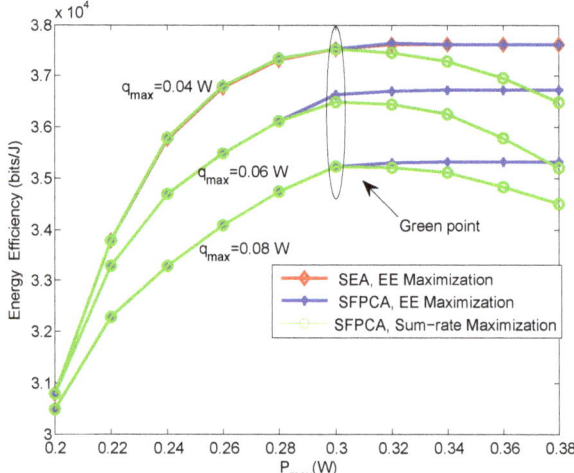

Figure 2. EE versus the maximum transmit power for different schemes.

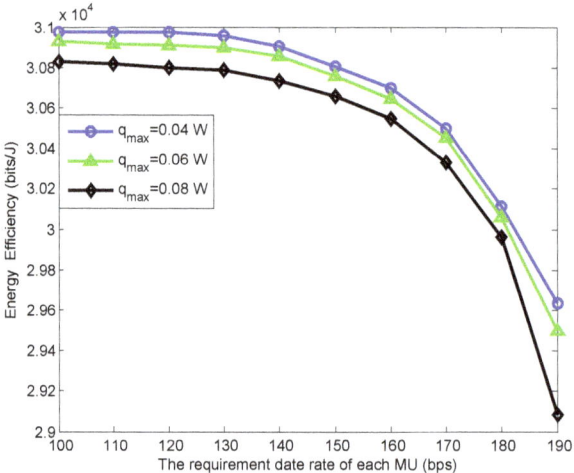

Figure 3. EE versus the date rate requirement of each MU for different schemes.

Figures 4 and 5 shows the convergence property of Algorithms 2 and 3. For simplicity, the numerical results in two figures are from a random chosen sub-channel, where $\eta_n^{(c)} = 1$. In Figure 4, we can see that the number of iterations are limited within four times. To show the influence of $\mathbf{p}_{fn}^{(0)}$, we give the different values of $\mathbf{p}_{fn}^{(0)}$ in Figure 5, where $\left| f_n(\mathbf{p}_{fn}^{(v)}) - \eta_n^{(c)} g_n(\mathbf{p}_{fn}^{(v)}) - (f_n(\mathbf{p}_{fn}^{(v-1)}) - \eta_n^{(c)} g_n(\mathbf{p}_{fn}^{(v-1)})) \right| = W(\mathbf{p}_{fn}^{(v)})$. It is shown that the initial values have an effect on the number of iterations. Specifically, when $\mathbf{p}_{fn}^{(0)} = 0 \times P_{max}$, less than 11 times is needed to reach the convergence. Although the initial values affect the number of iterations, it does not affect the final results.

To show the relationship between the different numbers of MUs and MTCDs and the EE, we have Figure 6. It is not surprise to see that the EE performance of all these schemes increases as P_{max} grows. From the four schemes, we can find out that the EE of $K = 40$ is much larger than that of $K = 15$, since the NOMA scheme can obtain much higher EE by supporting multiple MUs, and they can choose the suitable couples by swap operations for better EE. From the Algorithm 1, we know that power control algorithm needs to be executed after each swap operation, that is, the number of iterations increases with the increase of K and more process time are required. From the Figure 6, we can also see that the EE of $U_k = 2$ is larger than that of $U_k = 3$ under the same K, i.e., K has much greater impact on EE than U_k, since the NOMA scheme can obtain much higher EE by supporting multiple MUs and the increasing U_k represents the increase data rate requirement of MUs.

Figure 7 presents the cumulative distribution function (CDF) of the number of swap operations of different scenarios when the matching algorithms reached convergence. From the figure we can see that more swap operations were needed for a larger number of MUs and sub-channels, such as, $K = 40, N = 3$ needed more swap operations than that $K = 40, N = 2$ and $K = 15, N = 3$. Especially, less than 70 swap operations were needed for $K = 40$ and $N = 3$.

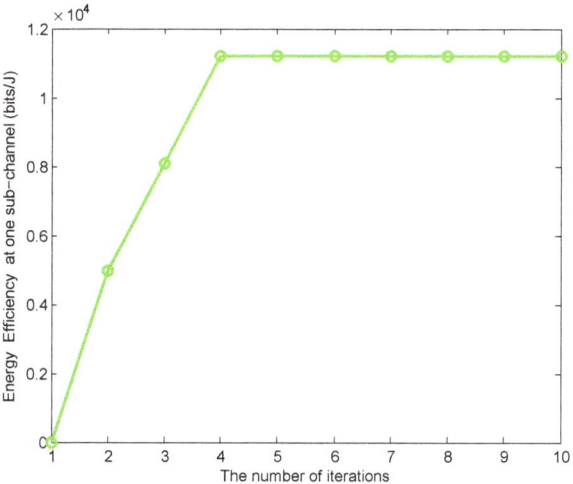

Figure 4. Convergence property of Algorithm 2.

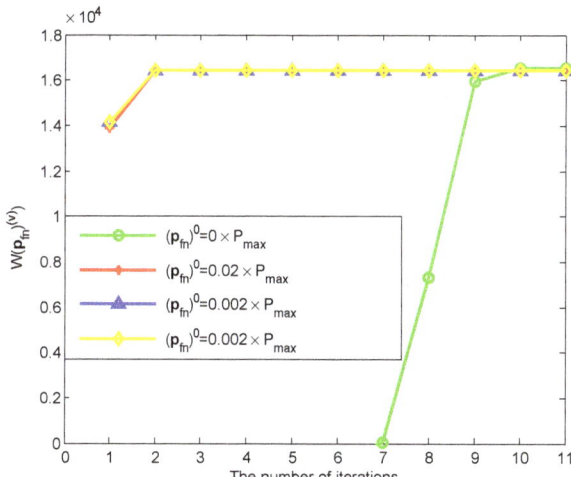

Figure 5. Convergence property of Algorithm 3.

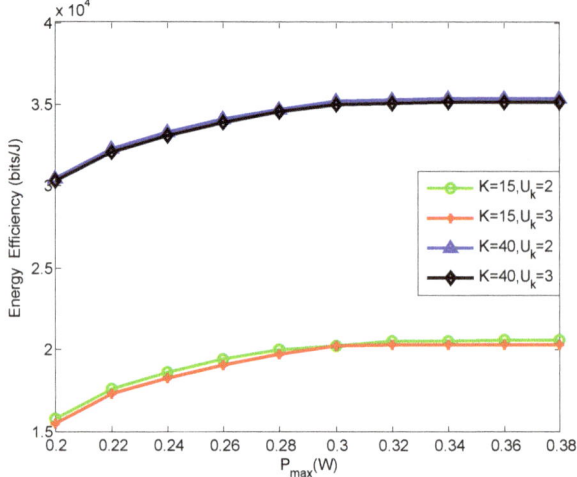

Figure 6. EE versus the maximum transmit power for the different number of MUs and MTCDs.

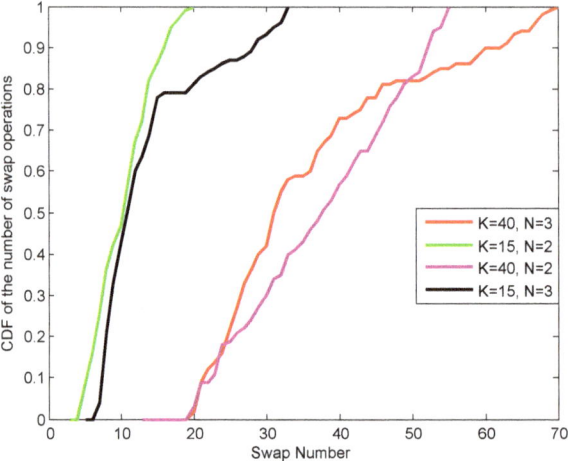

Figure 7. CDF of the number of swap operations for convergence.

6. Conclusions

This work investigated the uplink EE maximization problem in M2M-enabled HetNets with NOMA, where a MU acting as an MTCG can decode and forward both the information of MTCDs and its own data to the BS directly. Due to the limited spectrum resource, each BS shared the same sub-channels and NOMA was adopted between MUs in the same BS and sub-channel. The EE maximization problem was formulated, where MU association and power control were combined with each other. To solve it, a MU association matching algorithm was proposed based on the matching game. Under a given MU association, the uplink EE maximization was transformed into the EE maximization of each sub-channel. Two power control algorithms were provided to obtain the suboptimal power solutions based on sequential optimization. Simulation results showed that our proposed algorithms performed better than EE performance. It is known that cellular network is a key way to connect the M2M communications to the core network; our proposed scheme provided a new strategy for MTCDs to connect the cellular network with regard to MUs as their MTCGs based NOMA, and the power control of MTCDs was also considered as the constraints for the EE optimization problem. In fact, large scale devices are a more realistic scenario for 5G and next generation network, and since the number of MTCDs is considered on a small scale in this paper, the extension of our algorithms for large scale devices is one of the future works. Furthermore, the research of high computation complexity of the proposed algorithms for large scale devices is also a significant problem.

Author Contributions: The main contributions of S.Z. were to create the main ideas and execute performance evaluation by extensive simulation, while G.K. worked as the advisor to discuss, create, and advise the main ideas and performance evaluations together. All authors read and approved the final version of the paper.

Funding: This work was supported by the National Science and Technology Major Project of China (No. 2017ZX03001022) and National Natural Science Foundation of China (61471064).

Conflicts of Interest: The authors declare no conflict of interest.

References

1. Ghosh, A.; Mangalvedhe, N.; Ratasuk, R.; Mondal, B.; Cudak, M.; Visotsky, E.; Thomas, T.A.; Andrews, J.G.; Xia, P.; Jo, H.S.; et al. Heterogeneous cellular networks: From theory to practice. *IEEE Commun. Mag.* **2012**, *6*, 54–64. [CrossRef]
2. Damnjanovic, A.; Montojo, J.; Wei, Y.; Ji, T.; Luo, T.; Vajapeyam, M.; Yoo, T.; Song, O.; Malladi, D. A survey on 3GPP heterogeneous networks. *IEEE Wirel. Commun.* **2011**, *3*, 10–21. [CrossRef]
3. Ericsson, A.B. *5G Radio Access, Research and Vision*; Saito, Y., Benjebbour, A., Eds.; June 2013.
4. Shin, W.; Vaezi, M.; Lee, B.; Love, D.J.; Lee, J.; Poor, H.V. Non-orthogonal multiple access in multi-cell networks: Theory, performance, and practical challenges. *IEEE Commun. Mag.* **2017**, *10*, 2–9. [CrossRef]
5. Ding, Z.; Yang, Z.; Fan, P.; Poor, H.V. On the performance of non-orthogonal multiple access in 5G systems with randomly deployed users. *IEEE Signal Process. Lett.* **2014**, *12*, 1501–1505. [CrossRef]
6. Ding, Z.; Peng, M.; Poor, H.V. Cooperative non-orthogonal multiple access in 5G systems. *IEEE Commun. Lett.* **2015**, *8*, 1462–1465. [CrossRef]
7. Zhang, N.; Wang, J.; Kang, G.; Liu, Y. Uplink non-orthogonal multiple access in 5G systems. *IEEE Commun. Lett.* **2016**, *3*, 458–461. [CrossRef]
8. Zhang, H.; Fang, F.; Cheng, J.; Long, K.; Wang, W.; Leung, V.C.M. Energy-efficient resource allocation in NOMA heterogeneous networks. *IEEE Wirel. Commun.* **2018**, *2*, 48–53. [CrossRef]
9. Mokdad, A.; Azmi, P.; Mokari, N. Radio resource allocation for heterogeneous traffic in GFDM-NOMA heterogeneous cellular networks. *IET Commun.* **2016**, *12*, 1444–1455. [CrossRef]
10. Zhao, J.; Liu, Y.; Chai, K.K.; Nallanathan, A.; Chen, Y.; Han, Z. Spectrum allocation and power control for non-orthogonal multiple access in HetNets. *IEEE Trans. Wirel. Commun.* **2017**, *9*, 5825–5837. [CrossRef]
11. Ni, D.; Hao, L.; Tran, Q.T.; Qian, X. Power allocation for downlink NOMA heterogeneous networks. *IEEE Access* **2018**, *6*, 26742–26752. [CrossRef]
12. Ni, D.; Hao, L.; Tran, Q.T.; Qian, X. Transmit power minimization for downlink multi-cell multi-carrier NOMA networks. *IEEE Commun. Lett.* **2018**, *12*, 2459–2462. [CrossRef]
13. Fang, F.; Cheng, J.; Ding, Z. Joint energy efficient subchannel and power optimization for a downlink NOMA heterogeneous network. *IEEE Trans. Veh. Technol.* **2019**, *68*, 1351–1364. [CrossRef]
14. Moltafet, M.; Azmi, P.; Mokari, N.; Javan, M.R.; Mokdad, A. Optimal and fair energy efficient resource allocation for energy harvesting-enabled-PD-NOMA-based HetNets. *IEEE Trans. Wirel. Commun.* **2018**, *3*, 2054–2067. [CrossRef]
15. Pentikousis, K. In search of energy-efficient mobile networking. *IEEE Commun. Mag.* **2010**, *1*, 95–103. [CrossRef]
16. IMT-2020(5G) Promotion Group. *5G Vision and Requirements*; May 2014.
17. Buzzi, S.; Chih-Lin, I.; Klein, T.E.; Poor, H.V.; Yang, C.; Zappone, A. A survey of energy-efficient techniques for 5G networks and challenges ahead. *IEEE J. Sel. Areas Commun.* **2016**, *4*, 697–709. [CrossRef]
18. Qin, Z.; Yue, X.; Liu, Y.; Ding, Z.; Nallanathan, A. User association and resource allocation in unified NOMA enabled heterogeneous ultra dense networks. *IEEE Commun. Mag.* **2018**, *6*, 86–92. [CrossRef]
19. Yang, Z.; Xu, W.; Xu, H.; Shi, J.; Chen, M. Energy efficient non-orthogonal multiple access for machine-to-machine communications. *IEEE Commun. Lett.* **2017**, *4*, 817–820. [CrossRef]

20. Shirvanimoghaddam, M.M.; Dohler, M.; Johnson, S.J. Massive non-orthogonal multiple access for cellular IoT: potentials and limitations. *IEEE Commun. Mag.* **2017**, *9*, 55–61. [CrossRef]
21. Chen, S.; Ma, R.; Chen, H.; Zhang, H.; Meng, W.; Liu, J. Machine-to-machine communications in ultra-dense networks-a survey. *IEEE Commun. Surv. Tutor.* **2017**, *3*, 1478–1503. [CrossRef]
22. Yang, Z.; Pan, Y.; Xu, W.; Guan, R.; Wang, Y.; Chen, M. Energy efficient resource allocation for machine-to-machine communications with NOMA and energy harvesting. In Proceedings of the 2017 IEEE Conference on Computer Communications Workshops (INFOCOM WKSHPS), Atlanta, GA, USA, 1–4 May 2017; pp. 1–6. [CrossRef]
23. Xu, B.; Chen, Y.; Carrion, J.R.; Zhang, T. Resource allocation in energy-cooperation enabled two-tier NOMA HetNets towards green 5G. *IEEE J. Sel. Areas Commun.* **2017**, *12*, 2758–2770. [CrossRef]
24. Qian, L.P.; Wu, Y.; Zhou, H.; Shen, X.S. Joint uplink base station sssociation and power control for small-cell networks with non-orthogonal multiple access. *IEEE Trans. Wirel. Commun.* **2017**, *19*, 5567–5582. [CrossRef]
25. Yang, Z.; Xu, W.; Pan, Y.; Pan, C.; Chen, M. Energy efficient resource allocation in machine-to-machine communications with multiple access and energy harvesting for IoT. *IEEE Internet Things J.* **2018**, *5*, 229–245. [CrossRef]
26. Li, Z.; Gui, J. Energy-efficient resource allocation with hybrid TDMA–NOMA for cellular-enabled machine-to-machine communications. *IEEE Access* **2019**, *7*, 105800–105815. [CrossRef]
27. Lv, T.; Ma, Y.; Zeng, J.; Mathiopoulos, P.T. Millimeter-wave NOMA transmission in cellular M2M communications for Internet of Things. *IEEE Internet Things J.* **2018**, *5*, 1989–2000. [CrossRef]
28. Shahini, A.; Ansari, N. NOMA aided narrowband IoT for machine type communications with user clustering. *IEEE Internet Things J.* **2019**, *6*, 7183–7191. [CrossRef]
29. Han, S.; Xu, X.; Tao, X.; Zhang, P. Joint power and sub-channel allocation for secure transmission in NOMA-based mMTC networks. *IEEE Syst. J.* **2019**, *13*, 2476–2487. [CrossRef]
30. Mohamed, A.; Onireti, O.; Imran, M.A.; Imran, A.; Tafazolli, R. Control-data separation architecture for cellular radio access networks: A survey and outlook. *IEEE Commun. Surv. Tutor.* **2016**, *18*, 446–465. [CrossRef]
31. Bodine-Baron, E.; Lee, C.; Chong, A.; Hassibi, B.; Wierman, A. Peer effects and stability in matching markets. In Proceedings of the 4th Symposium on Algorithmic Game Theory (SAGT), Amalfi, Italy, 17–19 October 2011; pp. 117–129. [CrossRef]
32. Aydin, O.; Jorswieck, E.A.; Aziz, D.; Zappone, A. Energy-spectral efficiency tradeoffs in 5G multi-operator networks with heterogeneous constraints. *IEEE Trans. Wirel. Commun.* **2017**, *9*, 5869–5881. [CrossRef]
33. Zappone, A.; Bjornson, E.; Sanguinetti, L.; Jorswieck, E. Globally optimal energy-efficient power control and receiver design in wireless networks. *IEEE Trans. Signal Process.* **2017**, *11*, 2844–2859. [CrossRef]
34. Dinkelbach, W. On nonlinear fractional programming. *Manag. Sci.* **1967**, *7*, 492–498. [CrossRef]
35. Ali, M.S.; Tabassum, H.; Hossain, E. Dynamic user clustering and power allocation for uplink and downlink non-orthogonal multiple access (NOMA) systems. *IEEE Access* **2016**, *4*, 6325–6343. [CrossRef]
36. Li, Y.; Sheng, M.; Yang, C.; Wang, X. Energy efficiency and spectral efficiency tradeoff in interference-limited wireless networks. *IEEE Commun. Lett.* **2013**, *10*, 1924–1927. [CrossRef]
37. Kha, H.H.; Tuan, H.D.; Nguyen, H.H. Fast global optimal power allocation in wireless networks by local D.C. programming. *IEEE Trans. Wirel. Commun.* **2012**, *2*, 510–515. [CrossRef]

© 2019 by the authors. Licensee MDPI, Basel, Switzerland. This article is an open access article distributed under the terms and conditions of the Creative Commons Attribution (CC BY) license (http://creativecommons.org/licenses/by/4.0/).

Article

A Dynamic Access Probability Adjustment Strategy for Coded Random Access Schemes

Jingyun Sun [1], Rongke Liu [1,*] and Enrico Paolini [2,*]

[1] School of Electronics and Information Engineering, Beihang University, 37 Xueyuan Road, Haidian District, Beijing 100191, China; sunjingyun@buaa.edu.cn
[2] Department of Electrical, Electronic, and Information Engineering, University of Bologna, via Dell'Universitá 50, 47522 Cesena (FC), Italy
* Correspondence: rongke_liu@buaa.edu.cn (R.L.); e.paolini@unibo.it (E.P.); Tel.: +86-10-8233-9475 (R.L.); +39-0547-339137 (E.P.)

Received: 13 August 2019; Accepted: 25 September 2019; Published: 27 September 2019

Abstract: In this paper, a dynamic access probability adjustment strategy for coded random access schemes based on successive interference cancellation (SIC) is proposed. The developed protocol consists of judiciously tuning the access probability, therefore controlling the number of transmitting users, in order to resolve medium access control (MAC) layer congestion states in high load conditions. The protocol is comprised of two steps: Estimation of the number of transmitting users during the current MAC frame and adjustment of the access probability to the subsequent MAC frame, based on the performed estimation. The estimation algorithm exploits a posteriori information, i.e., available information at the end of the SIC process, in particular it relies on both the frame configuration (residual number of collision slots) and the recovered users configuration (vector of recovered users) to effectively reduce mean-square error (MSE). During the access probability adjustment phase, a target load threshold is employed, tailored to the packet loss rate in the finite frame length case. Simulation results revealed that the developed estimator was able to achieve remarkable performance owing to the information gathered from the SIC procedure. It also illustrated how the proposed dynamic access probability strategy can resolve congestion states efficiently.

Keywords: congestion; estimation; irregular repetition slotted ALOHA; medium access control; random access; successive interference cancellation

---

1. Introduction

In machine-type and Internet-of-Things (IoT) communications, users generate a large amount of bursty traffic to transmit over a shared communication medium. Coordinated multiple access schemes turn to be impractical and generally inefficient in such scenarios. For this reason, random access schemes have attracted a renewed interest, as they provide a practical way for uncoordinated users to contend for channel resource.

Pure ALOHA scheme [1] was proposed in 1968 to share a channel among a number of users sending packets as soon as they have data to transmit. Classical slotted ALOHA [2] is a distributed random access scheme in which time is divided into slots of equal duration with each transmission starting only at the beginning of a time slot. In both variants, an absence of coordination among users may lead to collisions (two or more packets are received in overlapping time windows). All packets involved in a collision are often reported as useless and are retransmitted after a random delay, according to some probability distribution, or (in the framed case) in the next frame. As a result, pure ALOHA and slotted ALOHA suffer from a throughput penalty and an under-utilization of channel resource. The optimal normalized throughput of pure ALOHA is 0.18 and the throughput of slotted ALOHA is increased to 0.37.

The expression of "coded random access" refers to a set of random access schemes that combine the packet repetition of users with successive interference cancellation (SIC) at the receiver. The first coded random access scheme is collision resolution diversity slotted ALOHA (CRDSA) [3], where each user sends two packet replicas in two random slots of the frame, and then SIC is applied to recover the collided packets in an iterative fashion. After CRDSA, CRDSA++ [4] was proposed to further improve throughput by increasing the number of packet replicas. In [5], where irregular repetition slotted ALOHA (IRSA) was proposed, the SIC-based random access process is conveniently described by a bipartite graph, establishing a bridge between the SIC procedure and the iterative erasure decoding of graph-based codes. In IRSA, the packet repetition rate is irregular from user to user and is chosen independently by each active user according to a suitably designed probability distribution. Since then, coded random access emerged as a new paradigm and has been the subject of several investigations over the past few years (e.g., [6–12] and references therein). As a result, the throughput has substantially increased which makes it a practical and efficient solution to support uncoordinated access.

Despite their numerous advantages, coded random access schemes exhibit lower critical points in traffic load. In other words, the throughput of these schemes is maximized for load values less than 1 and, for larger values of the load, it decreases very rapidly. Congestion occurs when the number of active users is greater than the receiver processing capacity. Several control methods for random access schemes have been investigated, which may be classified into two kinds: Dynamic frame length based methods and dynamic access probability based methods. In dynamic framed slotted ALOHA (DFSA) systems, the frame size is adjusted dynamically according to the estimated number of active users in order to maximize the system efficiency [13–17]. In dynamic access probability based schemes, on the other hand, an access controller is required to adjust the users access probability under high traffic loads in order to limit the number of transmitting users [18–21]. However, in [18–20], the estimation process was simply based on the status of frame slots before the application of SIC and in [21], the estimation is assumed to be ideal at the receiver. Furthermore, the proposed random access control mechanisms in [18] are based on random access schemes without SIC at the receiver, which is not applicable for coded random access schemes. In both [19,20], users directly employ the load threshold from [5], which is obtained via asymptotic analysis (frame length and user population size tending to infinity, their ratio remaining constant). When applied to the finite frame length case, asymptotic load thresholds tend to be beyond the actual critical point, which may yield considerable throughput losses.

In this paper, a dynamic access probability based strategy for coded random access schemes is proposed to resolve congestion. The proposed strategy performs two main tasks: Estimation of number of transmitting users in the current frame and the adjustment of access probability in the next frame based on the estimation results. In our previous work [22–24] techniques for a more reliable estimation of the number of transmitting users in coded random access schemes were developed and more specially, the number of transmitting users in the current frame was estimated using a posteriori information gathered throughout the SIC process. A posteriori estimation was considered for CRDSA in [22], for IRSA in [23] and for CRDSA over a packet and slot erasure channel in [24]. Notably, [22–24] were entirely focused on the estimation process, without any attempt to exploit it within a dynamic access probability adjustment protocol. The usage of a target load threshold tailored to the finite frame length case and the introduction of a state judgment to avoid not fully reliable estimation in high traffic load conditions are other original features of this manuscript.

The system model and some preliminary definitions are provided in Section 2. The estimation algorithm for the number of transmitting users in the current frame is addressed in Section 3.1, while the access probability adjustment strategy is proposed in Section 3.2. Numerical results are illustrated in Section 4 and concluding remarks are given in Section 5.

2. Preliminaries

2.1. System Model

We consider a scenario where multiple users contend for access to a single central receiver. The medium access control (MAC) layer is organized into frames and the random access scheme is a slotted one. We denote random variables by capital letters, while their realizations and deterministic quantities are denoted by lower case letters. The frame length is fixed and divided into m time slots with equal duration.

Active users are the ones who have packets to transmit. Congestion occurs when the number of active users is too large in comparison to the available resources (a more precise definition of congestion is given in Section 3.2). We use the subscript (k) to represent the index of the MAC frame. If there is no congestion or the congestion is resolved, the index (k) is re-initialized to (0) in the next frame, otherwise, it keeps counting. As such, a frame index $k \geq 1$ indicates that we are in the k-th frame of the current congestion event.

User population size is n_{pop}. The number of active users is unknown to the receiver and is modeled by a random variable N_a; the number of active users at the beginning of frame k is $N_a^{(k)}$. No new user activates before the current congestion has been resolved. Denoting by $\Delta^{(k)}$, the number of users that are recovered while processing frame k, for $k \geq 1$ we have:

$$N_a^{(k)} = N_a^{(k-1)} - \Delta^{(k-1)}. \tag{1}$$

Transmitting users are the ones who are allowed to transmit their packets in the frame. Let $T_a^{(k)}$ be the number of transmitting users during frame k. Moreover, denote by $p_{ac}^{(k)}$ the access probability of the active users during frame k. At the beginning of the k-th frame, each active user becomes a transmitting one with probability $p_{ac}^{(k)}$, independently of other active users. Hence, the conditional expected value of $T_a^{(k)}$ is:

$$\mathbb{E}[T_a^{(k)} | n_a^{(k)}] = n_a^{(k)} p_{ac}^{(k)}. \tag{2}$$

Each transmitting user is frame- and slot- synchronous and attempts at most one packet transmission per frame.

In every frame corresponding to $k = 0$, all active users transmit their packets to the receiver, i.e., we have $p_{ac}^{(0)} = 1$ and $t_a^{(0)} = n_a^{(0)}$. The instantaneous channel load over frame k is defined as:

$$G^{(k)} = \frac{t_a^{(k)}}{m} \tag{3}$$

and represents the average number of packet transmissions per slot. The throughput over frame k is defined as:

$$T_h^{(k)} = \frac{t_a^{(k)}}{m}(1 - P_L) \tag{4}$$

representing the average number of successfully recovered packets per slot by the receiver. The quantity P_L in Equation (4) is the packet loss rate over the frame, which is expressed as:

$$P_L = 1 - \frac{\delta^{(k)}}{t_a^{(k)}}. \tag{5}$$

In IRSA, each transmitting user sends L packet replicas to slots picked uniformly at random. The number of replicas, named user degree, is a discrete random variable probability mass function (p.m.f.) $\{\Lambda_l\}$, where $\Lambda_l = P(L = l)$ is the probability that a user generates l packet replicas. Users choose their replica factor (i.e., user degree) L independently of each other, with no coordination, and the values of user degree are according to distribution $\{\Lambda_l\}$. We also represent $\{\Lambda_l\}$ in polynomial form, as $\Lambda(x) = \sum_{l=1}^{d_{max}} \Lambda_l x^l$, where d_{max} is the maximum number of packet replicas per user. Both information about the transmitting user index (assuming users are indexed from 1 to n_{pop}) and pointers to the slots where the other replicas have been transmitted are included in the header of each packet replica. CRDSA can be seen as IRSA with $\Lambda(x) = x^{d_{max}}$.

In this paper, a classical collision channel model is adopted. After packet replica transmissions, each slot takes one of the following three states: Empty slot (no packet replica transmitted in that slot), singleton slot (only one packet replica transmitted in that slot), and collision slot (two or more than two packet replicas transmitted in that slot). The receiver can always correctly classify the state of each slot. Collision slots provide no information to the receiver about the number and content of collided packet replicas directly. However, as soon as the contribution of interference, generated by some transmitting users on the slot, is canceled and only one packet replica is left in it, the slot status is updated to singleton. Similarly, if all of the packet replicas transmitted in the slot are recovered by the receiver, the slot status (singleton slot or collision) is updated to empty. Packet replicas from singleton slots are always correctly received, which means that packet losses may only be generated by unresolved collisions.

After transmissions, the pointers to twin replicas in the header of the packet enable SIC at the receiver. At first, the receiver stores the content of the frame. Then, the receiver performs iteratively the following procedure, consisting of two subsequent steps:

1. Pick out the singleton slots in the frame. For each singleton slot, extract the transmitting user index, the content of the packet replica, and positions of other twin replicas. Identified users in this step become recovered users;

2. For each user recovered at step 1, remove the user's contribution of interference in the slots where the packet replicas have been transmitted. A new singleton slot will appear if, after interference cancellation, they contain only one replica.

The iterative SIC procedure terminates when all slots are empty ones, in which case SIC succeeds, or when no singleton slot can be found but collision slots still exist, in which case it fails. At the end of the SIC procedure, the residual number of empty slots in the frame is denoted by M_e, and the residual number of collided slots per frame by M_c. Obviously, we have $M_e + M_c = m$.

Example 1. *With reference to Figure 1, $t_a = 4$ users transmit their packets to a frame with $m = 5$ slots. User u_1 generates three replicas of his packet, and sends them to s_1, s_3, and s_4, respectively. Each of the other users generate two replicas of the corresponding packets and transmit them as illustrated in the figure. At the receiver, slots s_1 and s_4 are singleton slots and the left s_2, s_3, and s_5 are collison slots.*

Figure 2 provides a graphical interpretation (first proposed in [5]) of the iterative SIC procedure performed on the frame of Figure 1. In the presented graph, "slot nodes" represent slots and "user nodes" represent users. In the first SIC iteration, s_1 and s_4 are singleton slots and the corresponding packet replicas are correctly received, making u_1 a recovered user. The pointer to slot s_3, where the twin of the replica in s_1 has been transmitted, is extracted (step 1). After the interference from recovered user u_1 in slot s_3 is canceled and only one packet replica is left in s_3, making s_3 a new singleton slot (step 2). Then a second iteration is triggered. After three SIC iterations, users u_2 and u_3 remain unrecovered, there are no singleton slots in the frame, and SIC terminates with failure.

The feedback frame configuration signal is $\{0, 1, 0, 0, 1\}$ which indicates that s_1, s_3 and s_4 are empty slots and that s_2 and s_5 are unresolved collision slots. Receiving this feedback signal, u_2 and u_3 become aware that their packets have not been successfully received.

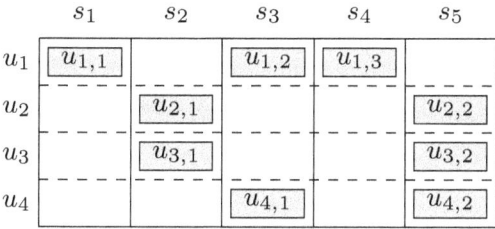

Figure 1. Example of a MAC frame with $t_a = 4$ transmitting users and $m = 5$ slots. User u_1 sends three packet replicas and the other users each send two packet replicas. Slots s_1 and s_4 are singleton slots and the left s_2, s_3, and s_5 are collision slots.

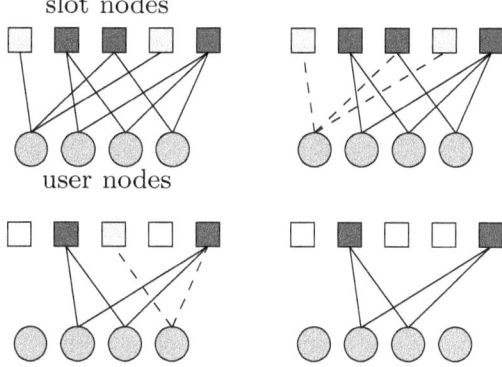

Figure 2. Example of successive interferece cancellation (SIC) procedure corresponding to Figure 1. Squares correspond to slots and circles correspond to users.

2.2. Threshold Definition and Notation

Throughout the paper we define a load threshold G° as the maximum load such that the packet loss rate falls below a given target value P_L°. In other words, when the instantaneous load G is below G°, we have $P_L \leq P_L^\circ$, otherwise we have $P_L > P_L^\circ$.

In Table 1, some examples of probability distributions $\Lambda(x)$ are shown with the corresponding target load threshold values. The first two rows in the table represent CRDSA schemes, where each user transmits the same number of replicas. The last two rows represent IRSA schemes, where the number of replicas per user is irregular. The values of G° have been obtained via a Monte Carlo simulation, for MAC frame length $m = 200$ and target packet loss rate $P_L^\circ = 0.01$.

Table 1. Load threshold G° for different probability distributions $\Lambda(x)$, for MAC frame length $m = 200$, and packet loss rate target $P_L^\circ = 0.01$.

Distribution, $\Lambda(x)$	G°
$\Lambda_1(x) = x^2$	0.35
$\Lambda_2(x) = x^4$	0.69
$\Lambda_3(x) = 0.5x^2 + 0.28x^3 + 0.22x^8$	0.705
$\Lambda_4(x) = 0.25x^2 + 0.6x^3 + 0.15x^8$	0.76

Figure 3 shows the packet loss rate P_L versus instantaneous load G for the distributions in Table 1 and frame length $m = 200$. As previously remarked, the SIC process in IRSA can be described by a bipartite graph, where unresolved collisions are associated with graphical structures known, in the low-density parity-check (LDPC) coding jargon, as stopping sets. It is well known that the impact of small stopping sets on the finite-length performance is strictly related to the fraction of degree-2 variable nodes in its bipartite graph and a similar role is played by degree-2 users in IRSA. As observed in the figure, the limitation of degree-2 repetition has a better error floor performance, but a poorer waterfall performance. The detailed packet loss rate performance analysis for IRSA schemes have been addressed in [5].

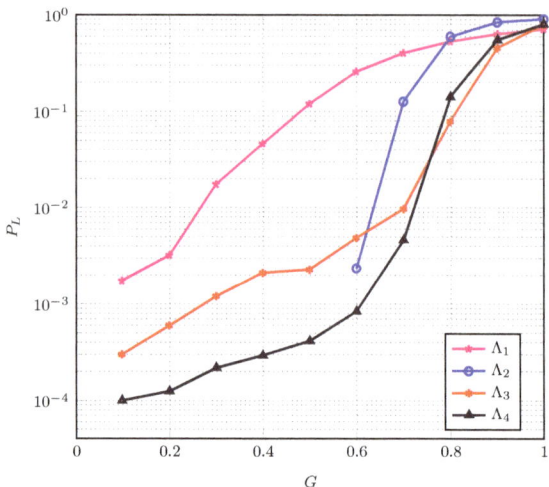

Figure 3. Packet loss rate P_L versus instantaneous load G for frame length $m = 200$ and the distributions in Table 1.

2.3. Combinatorial Parameters

We denote by $|\vec{v}| = \sum_{i=1}^{n} |v_i|$ the ℓ_1 norm of a real-valued vector $\vec{v} = (v_1, \ldots, v_n)$. Moreover, given a second vector $\vec{w} = (w_1, \ldots, w_n)$ whose elements are nonnegative integers, we use the compact notation $\vec{v}^{\vec{w}}$ for $v_1^{w_1} \cdots v_n^{w_n}$.

Let $\vec{o} = (o_1, \ldots, o_{d_{\max}})$ be a vector whose elements are all nonnegative integers. Let $\mathcal{M}(\vec{o}, b)$ be the set of all $|\vec{o}| \times b$ binary matrices $\mathbf{M}$, with rows and columns indexed from 1 to $|\vec{o}|$ and from 1 to b, respectively, that fulfill the following properties: 1. The matrix $\mathbf{M}$ has the structure:

$$\mathbf{M} = [\mathbf{M}_1^T \ \mathbf{M}_2^T \ \cdots \ \mathbf{M}_{d_{\max}}^T]^T$$

where $\mathbf{M}_i$ has dimension $o_i \times b$ and all of its rows have Hamming weight i. 2. Every column of $\mathbf{M}$ has Hamming weight at least 2.

Example 2. Let $\vec{o} = (o_1, o_2, o_3) = (0, 1, 3)$ and $b = 5$. Each matrix in $\mathbf{M} \in \mathcal{M}(\vec{o}, b)$ has dimension 4×5. Its row indexes should be thought as partitioned into the two subsets $\{1\}$ and $\{2, 3, 4\}$. The row of index 1 has weight 2, and the rows of indexes 2, 3, and 4 have weight 3. Every column of $\mathbf{M}$ has weight of at least 2. An example of matrix $\mathbf{M} \in \mathcal{M}(\vec{o}, b)$ is:

$$M = \begin{pmatrix} 1 & 0 & 0 & 1 & 0 \\ 0 & 1 & 0 & 1 & 1 \\ 1 & 1 & 1 & 0 & 0 \\ 0 & 0 & 1 & 1 & 1 \end{pmatrix}.$$

The following lemma provides a formal expression for the cardinality of the set $\mathcal{M}(\vec{o}, b)$.

Lemma 1. *For given $\vec{o}$ and b, let $h(\vec{o}, b)$ be the cardinality of the set $\mathcal{M}(\vec{o}, b)$. Moreover, let $\vec{x} = (x_1, x_2, \ldots, x_{|\vec{o}|})$ and:*

$$\vec{q} = (\underbrace{1, \ldots, 1}_{o_1}, \underbrace{2, \ldots, 2}_{o_2}, \ldots, \underbrace{d_{max}, \ldots, d_{max}}_{o_{d_{max}}}). \tag{6}$$

Define the multivariate polynomials $A(\vec{x})$ and $B_{j,l}(\vec{x})$ as:

$$A(\vec{x}) = \prod_{i=1}^{|\vec{o}|}(1 + x_i) - \left(1 + \sum_{i=1}^{|\vec{o}|} x_i\right) \tag{7}$$

and:

$$B_{j,l}(\vec{x}) = \left(\sum_{i=1}^{|\vec{o}|} x_i\right)^l \left(\prod_{i=1}^{|\vec{o}|}(1 + x_i)\right)^j. \tag{8}$$

Then, we have:

$$h(\vec{o}, b) = \text{coeff}((A(\vec{x}))^b, \vec{x}^{\vec{q}}) \tag{9}$$

$$= \sum_{j=0}^{b} \sum_{l=0}^{b-j} \binom{b}{j} \binom{b-j}{l} (-1)^{b-j} \text{coeff}(B_{j,l}(\vec{x}), \vec{x}^{\vec{q}}) \tag{10}$$

where $\text{coeff}(P(\vec{x}), \vec{x}^{\vec{r}})$ is the coefficient of $\vec{x}^{\vec{r}}$ in the multivariate polynomial $P(\vec{x})$.

Proof. Let $\vec{c}^T = (c_1, \ldots, c_{|\vec{o}|})^T$ be the generic column and define a multivariate enumerating function for valid columns (i.e., columns with weight of at least 2):

$$A(\vec{x}) = \sum_{\vec{c}: |\vec{c}| \geq 2} \vec{x}^{\vec{c}}. \tag{11}$$

It is easy to recognize that an equivalent expression for $A(\vec{x})$ is the one shown in Equation (7). This is because $(1 + x_1) \cdots (1 + x_{|\vec{o}|})$ provides the sum of all monomials in the variables $x_1, \ldots x_{|\vec{o}|}$ with a unitary coefficient, to which we subtract all monomials of degrees 0 and 1 as required by the condition of validity.

Considering now b columns and applying properties of generating functions, $\text{coeff}((A(\vec{x}))^b, \vec{x}^{\vec{w}})$ is the number of $|\vec{o}| \times b$ binary matrices such that all matrix columns are valid and such that the weight of row i is w_i. This immediately leads to Equation (9). The equivalent expression of Equation (10) is obtained by simple algebraic manipulation of the multivariate polynomial $(A(\vec{x}))^b$.

In particular, it is obtained by applying Newton's binomial formula twice and by exploiting the identity $\text{coeff}(\sum_i \alpha_i P_i(\vec{x}), \vec{x}^{\vec{w}}) = \sum_i \alpha_i \text{coeff}(P_i(\vec{x}), \vec{x}^{\vec{w}})$. □

3. Dynamic Access Probability Algorithm

In this section we introduce the proposed multiple access strategy based on a dynamic adjustment of the users access probability. Section 3.1 addresses estimation of the number of transmitting users; Section 3.2 exploits the developed estimator to perform congestion detection and resolution via dynamic access probability adjustment.

3.1. Number of Transmitting Users Estimation

In this subsection, we exploit frame configuration information at the end of SIC to estimate the number $t_a^{(k)}$ of transmitting users in the k-th frame, when an SIC failure occurs. For the sake of notational simplicity, the superscript (k) is temporarily omitted.

The total number of transmitting users is denoted by t_a. We also denote by $t_{a,l}$ the number of such users that employ the replica factor l. Clearly, we have $t_a = \sum_{l=1}^{d_{\max}} t_{a,l}$. The vector $\vec{t}_a = (t_{a,1}, t_{a,2}, \ldots, t_{a,d_{\max}})$ is referred to as transmitting users configuration at the beginning of the frame. The number of transmitted users that are recovered at the end of the SIC process is denoted by $\delta \leq t_a$. Out of these δ recovered users, $\delta_l \leq t_{a,l}$ are the ones using replica factor l, so that $\delta = \sum_{l=1}^{d_{\max}} \delta_l$. The vector $\vec{\delta} = (\delta_1, \delta_2, \ldots, \delta_{d_{\max}})$ is referred to as the recovered users configuration at the end of SIC.

Hereafter we develop a compact expression for the a posteriori probability distribution of the configuration $\vec{t}_a$ of transmitting users, given the number m_c of residual collision slots and the configuration $\vec{\delta}$ of recovered users observed at the end of SIC. This probability is denoted by $P(\vec{t}_a | m_c, \vec{\delta})$. Note that, as transmitting users pick their slots uniformly at random, it is sufficient to condition to the number of collision slots (and not to their positions in the frame). The corresponding probability distribution of the number t_a of transmitting users is given by:

$$P(t_a | m_c, \vec{\delta}) = \sum_{\vec{t}_a : |\vec{t}_a| = t_a} P(\vec{t}_a | m_c, \vec{\delta}). \tag{12}$$

A maximum a posteriori (MAP) estimator for the number of transmitting users then returns the value:

$$\hat{t}_a = \underset{t_a}{\operatorname{argmax}} \, P(t_a | m_c, \vec{\delta}). \tag{13}$$

Theorem 1. *The a posteriori probability distribution of the configuration $\vec{t}_a$ of the transmitting users fulfills:*

$$P(\vec{t}_a | m_c, \vec{\delta}) \propto \binom{\vec{t}_a}{\vec{\delta}} \frac{h(\vec{t}_a - \vec{\delta}, m_c)}{\prod_{l=1}^{d_{\max}} \binom{m}{l}^{t_{a,l}}} P(\vec{t}_a) \tag{14}$$

where $h(\vec{o}, b)$ is given by Lemma 1, $\binom{\vec{t}_a}{\vec{\delta}} = \prod_l \binom{t_{a,l}}{\delta_l}$, and $P(\vec{t}_a)$ is the a priori probability that the transmitting users configuration equals $\vec{t}_a$.

Proof. From Bayes' rule we have:

$$P(\vec{t}_a | m_c, \vec{\delta}) = \frac{P(m_c, \vec{\delta} | \vec{t}_a) P(\vec{t}_a)}{P(m_c, \vec{\delta})}$$

$$\propto P(m_c, \vec{\delta} | \vec{t}_a) P(\vec{t}_a). \tag{15}$$

Let $T(\vec{t}_a, m_c, \vec{\delta})$ be the number of ways in which $|\vec{t}_a|$ transmitting users with configuration $\vec{t}_a$ can transmit their packet replicas in the frame so that, at the end of SIC, there are m_c unresolved collision slots and a recovered users configuration $\vec{\delta}$. Moreover, let $T(\vec{t}_a)$ be the number of ways in which $|\vec{t}_a|$ transmitting users with configuration $\vec{t}_a$ can place their packet replicas in the frame. The conditional probability $P(m_c, \vec{\delta}|\vec{t}_a)$ can be expressed as:

$$P(m_c, \vec{\delta}|\vec{t}_a) = \frac{T(\vec{t}_a, m_c, \vec{\delta})}{T(\vec{t}_a)}. \tag{16}$$

The quantity $T(\vec{t}_a)$ is readility shown to be given by:

$$T(\vec{t}_a) = \prod_{l=1}^{d_{max}} \binom{m}{l}^{t_{a,l}}. \tag{17}$$

To develop an expression for $T(\vec{t}_a, m_c, \vec{\delta})$, we proceed as follows. At the end of SIC, $|\vec{t}_a - \vec{\delta}|$ transmitting users with configuration $\vec{t}_a - \vec{\delta}$ remain unrecovered. The number of ways in which these users transmit their packet replicas to m_c slots, forming m_c collisions (at least two replicas per slot) is $h(\vec{t}_a - \vec{\delta}, m_c)$. If we let $g(\vec{\delta}, m_c)$ be the number of ways in which $|\vec{\delta}|$ transmitting users with configuration $\vec{\delta}$ can place their packet replicas in a frame with $m - m_c$ free slots and m_c unresolvable collision slots, so that SIC can recover all of them, we can write (no formal expression for $g(\vec{\delta}, m_c)$ is provided because this parameter, not depending on $\vec{t}_a$ does not play any role in the estimation process of Equation(13)):

$$T(\vec{t}_a, m_c, \vec{\delta}) = \binom{\vec{t}_a}{\vec{\delta}} \binom{m}{m_c} h(\vec{t}_a - \vec{\delta}, m_c) g(\vec{\delta}, m_c). \tag{18}$$

Incorporating Equation (17) and Equation (18) into Equation (16) and then Equation (16) into Equation (15), and omitting all terms not depending on $\vec{t}_a$, we obtain Equation (14). □

Although Equations (13) and (14) define an exact MAP estimator, computing $h(\vec{t}_a - \vec{\delta}, m_c)$ turns out to be a complex task, becoming already intractable for frame sizes in the order of a few tens. For this reason we employ an approximated MAP estimator. In the approximation, all packet replicas, even from the same user, are regarded as distinguishable packets. Equivalently, each user chooses l slots with replacement. In this approximate setting, we have

$$P(\vec{t}_a|m_c, \vec{\delta}) \propto \binom{\vec{t}_a}{\vec{\delta}} \frac{h((\sum_{l=1}^{d_{max}}(t_l - \delta_l)l), m_c)}{m^{\sum_{l=1}^{d_{max}} t_l l}} P(\vec{t}_a) \tag{19}$$

where $(\sum_{l=1}^{d_{max}}(t_l - \delta_l)l)$ represents a vector with only one element, corresponding to $\vec{o} = (o_1)$ in $h(\vec{o}, b)$. The value of $h((\sum_{l=1}^{d_{max}}(t_l - \delta_l)l), m_c)$ is the number of ways in which $\sum_{l=1}^{d_{max}}(t_l - \delta_l)l$ packet replicas are sent to m_c slots, such that each slot receives not less than 2 packet replicas [25]; $m^{\sum_{l=1}^{d_{max}} t_l l}$ is the total number of ways in which $\sum_{l=1}^{d_{max}} t_l l$ packet replicas can be accommodated into the m slots.

As an estimator performance measure we consider the MSE, defined as:

$$M_\epsilon = \mathbb{E}[\epsilon^2] \tag{20}$$

where $\epsilon = \hat{t}_a^{(k)} - t_a^{(k)}$ is the estimation error.

3.2. Access Probability Adjustment Strategy

We say that we have a congestion on frame k whenever

$$n_a^{(k)} > G°m, \qquad (21)$$

where we recall that $n_a^{(k)}$ is the number users that are active on frame k. Our purpose is to exploit the developed estimator to detect congestion states and dynamically adjust the users access probability to improve overall efficiency. Congestion states are resolved by tuning the access probability to control the number of transmitting users in the next frame.

The proposed scheme is based on the definition of three possible states for a frame, namely:

- **Not fully reliable estimate.** In high load conditions, SIC typically stops prematurely with a relatively small number of recovered users. We say that the estimate $\hat{t}_a^{(k)}$ is not fully reliable when the number of users recovered by processing the frame is smaller than the number of users that could not be recovered:

$$\delta^{(k)} < t_a^{(k)} - \delta^{(k)} \qquad (22)$$

or, equivalently,

$$\delta^{(k)} < t_a^{(k)}/2. \qquad (23)$$

- **Congestion with reliable estimate.** The number of active users is above threshold $G°m$, but the number of users recovered by processing the frame is not less than the number of users that could not be recovered:

$$n_a^{(k)} > G°m \quad \text{and} \quad \delta^{(k)} \geq t_a^{(k)} - \delta^{(k)}. \qquad (24)$$

- **No congestion.** The number of active users is below threshold $G°m$:

$$n_a^{(k)} \leq G°m. \qquad (25)$$

In the first case, a large number of transmitting users is unrecovered, and the packet loss rate is larger than 0.5. As illustrated in the numerical results section, the estimation MSE increases with the number of transmitting users and the estimate is therefore regarded as not suitable to design the access probability p_{ac} in the subsequent frame. In contrast, in the last two cases the access probability in the next frame is calculated directly by employing the estimate of the number of transmitting users.

In the generic frame k, after all transmitting users have performed the transmission of their packet replicas, the receiver performs the SIC procedure. At the end of SIC, the receiver executes the procedure described in Algorithm 1. This procedure is executed regardless of the SIC termination status (success or failure). An explanation of Algorithm 1 is provided in the following.

Algorithm 1: Receiver procedure

1. **if** $m_c^{(k)} > 0$ && $\dfrac{P(t_a^{(k)} > 2\delta^{(k)} | m_c^{(k)}, \vec{\delta}^{(k)})}{P(t_a^{(k)} \leq 2\delta^{(k)} | m_c^{(k)}, \vec{\delta}^{(k)})} > 1$ **then**
2. $\quad p_{ac}^{(k+1)} \leftarrow \dfrac{G^\circ m}{n_{\text{pop}} - \sum_{i=0}^{k} \delta^{(i)}}$;
3. $\quad k \leftarrow k + 1$;
4. **else**
5. $\quad$ **if** $m_c^{(k)} == 0$ **then**
6. $\quad\quad \hat{t}_a^{(k)} = \delta^{(k)}$;
7. $\quad$ **else**
8. $\quad\quad$ calculate $\hat{t}_a^{(k)}$ according to Equation (19)
9. $\quad$ **end**
10. $\quad \hat{n}_a^{(k)} = \hat{t}_a^{(k)} / p_{ac}^{(k)}$;
11. $\quad$ **if** $\hat{n}_a^{(k)} > G^\circ m$ **then**
12. $\quad\quad p_{ac}^{(k+1)} \leftarrow \dfrac{G^\circ m}{\hat{n}_a^{(k)} - \delta^{(k)}}$;
13. $\quad\quad k \leftarrow k + 1$;
14. $\quad$ **else**
15. $\quad\quad p_{ac}^{(1)} \leftarrow 1$;
16. $\quad\quad k \leftarrow 0$;
17. $\quad$ **end**
18. **end**
19. broadcast k, $p_{ac}^{(k)}$, and $C^{(k)}$ to the users;

The first step (line 1) consists of detecting whether Equation (23) is fulfilled or not. When SIC succeeds ($m_c^{(k)} = 0$), the estimation is perfect. The algorithm jumps to line 6 and simply sets $\hat{t}_a^{(k)} = \delta^{(k)}$. In case of an SIC failure ($m_c^{(k)} > 0$), the algorithm applies a two-hypotheses MAP detector, whose development is presented in Appendix A, to decide whether Equation (23) holds (in which case estimation is considered unreliable) or not. Concretely, if

$$\frac{P(t_a^{(k)} > 2\delta^{(k)} | m_c^{(k)}, \vec{\delta}^{(k)})}{P(t_a^{(k)} \leq 2\delta^{(k)} | m_c^{(k)}, \vec{\delta}^{(k)})} > 1, \qquad (26)$$

then Equation (23) is assumed to hold and the estimation Equation (19) is regarded as not reliable enough. Otherwise, the estimate $\hat{t}_a^{(k)}$ is employed to design the access probability in the next frame.

When Equation (26) is satisfied, a 'not fully reliable estimate' state is detected and the number of transmitting users is detected to be large enough to create a congestion but the relatively large estimation MSE prevents from relying on $\hat{t}_a^{(k)}$ to reliably adjust the access probability in the next frame. At the end of frame k a number $\sum_{i=0}^{k} \delta^{(i)}$ of active users have been recovered since the beginning of congestion. Therefore, at the beginning of the subsequent frame, the number of unresolved active users fulfills $n_a^{(k+1)} \leq n_{\text{pop}} - \sum_{i=0}^{k} \delta^{(i)}$. To make the expected number of transmitting users in the subsequent frame below the target number $G^\circ m$, we set the access probability according to (line 2)

$$p_{ac}^{(k+1)} = \frac{G^\circ m}{n_{\text{pop}} - \sum_{i=0}^{k} \delta^{(i)}}. \qquad (27)$$

This way, the conditional expected number of transmitting users in the next frame is

$$\mathbb{E}[T_a^{(k+1)}|n_a^{(k+1)}] = n_a^{(k+1)} p_{ac}^{(k+1)}$$
$$= \frac{n_a^{(0)} - \sum_{i=0}^{k}\delta^{(i)}}{n_{\text{pop}} - \sum_{i=0}^{k}\delta^{(i)}} G^\circ m, \tag{28}$$

where $n_a^{(0)} - \sum_{i=1}^{k}\delta^{(i)}$ represents the actual number of unrecovered active users at the beginning of frame $k+1$.

When the estimation result is detected to be reliable, an acceptable estimation MSE is assumed by the receiver, which exploits $\hat{t}_a^{(k)}$ (equal to $\delta^{(k)}$ in case of a SIC success or by Equation (19) in case of a failure) to obtain an estimate of the number of active users on frame k. Specifically, the receiver performs (line 10):

$$\hat{n}_a^{(k)} = \hat{t}_a^{(k)} / p_{ac}^{(k)}. \tag{29}$$

The estimate $\hat{n}_a^{(k)}$ is compared with the threshold $G^\circ m$ (line 11). If $\hat{n}_a^{(k)} > G^\circ m$ the receiver declares a congestion with a reliable estimate state. The system is suffering from congestion, but most of (or all of) transmitting users have been recovered by SIC. The relatively low estimation MSE allows confidently using $\hat{n}_a^{(k)}$ to set the access probability for the next frame. If the access probability is kept unchanged in the subsequent frames, the number of transmitting users will deviate progressively from the target $G^\circ m$, leading to a low throughput. To make efficient use of channel resources, we increase the access probability in such a way as to maintain the number of transmitting users close to the target $G^\circ m$ in the next frame. From Equation (1), the estimated number of unrecovered active users at the beginning of the frame $k+1$ is $\hat{n}_a^{(k+1)} = \hat{n}_a^{(k)} - \delta^{(k)}$. The target conditional expected number of transmitting users in frame $k+1$ is

$$\mathbb{E}[T_a^{(k+1)}|n_a^{(k+1)}] = G^\circ m. \tag{30}$$

Thus, the access probability over frame $k+1$ is set to (line 12):

$$p_{ac}^{(k+1)} = \frac{G^\circ m}{\hat{n}_a^{(k)} - \delta^{(k)}}. \tag{31}$$

If $\hat{n}_a^{(k)} < G^\circ m$, a no congestion state is detected. The frame index k is re-initialized to 0 and the users access probability is set to be 1 (lines 15 and 16).

As a last step (line 19), the receiver broadcasts to the users the index of the next frame (index of the current frame increased by 1 if a congestion is detected and 0 otherwise), the access probability to be employed by active users in the next frame, and the list of indexes of collision slots at the end of SIC in the current frame. Upon receiving feedback from the receiver, users behave as follows:

- If $k > 0$ (congestion), in the next frame each backlogged user attempts access to the frame with probability equal to the new access probability. Each non-backlogged user is prevented from transmitting new packets;
- If $k = 0$ (no congestion), users that are in a backlog state retransmit their packet. Users that are not backlogged take their normal access activity.

Users are updated by the receiver about congestion or no congestion simply through the index k. Moreover, each of them knows whether or not it is in a backlog state simply by looking at the list of collision slot indexes $\mathcal{C}^{(k)}$. Note that, if $k = 0$ (no congestion) is broadcasted by the receiver, this does not necessarily mean that SIC has succeeded as there may be a small number of users unrecovered even though the system is not suffering from congestion. In this case, we simply let backlogged users retransmit packet replicas with probability 1 in the subsequent frame, together with possible fresh replicas from newly activated users.

4. Numerical Results

This section is organized into two subsections. In Section 4.1 we show results on the estimation of the number of transmitting users, while in Section 4.2 we address the performance achieved by the proposed scheme.

4.1. Estimation of Transmitting Users

In this section, we present Monte Carlo simulation results using the approximated estimator discussed at the end of Section 3.1. Let the frame length be $m = 200$ and the user population size be $n_{pop} = 400$. Users are assumed to activate independently of each other at the beginning of every new frame. In each run, after users transmissions, SIC is applied at the receiver and then the developed approximated estimator is applied.

Figure 4 shows the average throughput and throughput standard deviation versus the instantaneous load G for IRSA with $\Lambda(x) = 0.5x^2 + 0.28x^3 + 0.22x^8$ [5]. The maximum average throughput is achieved at a value of G that is approximately equal to 0.8. However, the realizations of the per-frame throughput fluctuate around its statistical mean, the throughput standard deviation representing a reliable measure of the bobbing range (i.e., dispersion). A large standard deviation makes the average throughput a not fully meaningful parameter since, due to the per-frame throughput fluctuations, we have a higher probability that the system falls into a not fully reliable estimate state. In this respect, the peak average throughput is not necessarily a good working point, as the statistical mean alone is not able to capture the probability of falling into such an "outage" state.

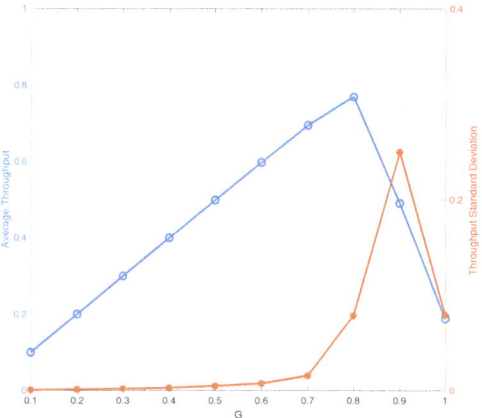

Figure 4. Average throughout and throughput standard deviation versus the instantaneous load G for IRSA with $\Lambda(x) = 0.5x^2 + 0.28x^3 + 0.22x^8$.

Figure 5 shows the estimation performance after SIC iterations, letting the SIC-based receiver run until no active user is recovered. As a comparison, we also consider the estimation performance using the frame configuration before SIC iterations, which is reviewed in Appendix B. In the figure, the solid line is relevant to the proposed estimation making use of the frame configuration and recovered users configuration after SIC iterations. Moreover, a dashed line corresponds to the estimation based on the initially received frame, before SIC is applied. As observed in the figure, the proposed estimation algorithm is able to reduce the MSE effectively over the whole range of G values. It is also worth noting that the performance of the proposed estimator relies on the SIC performance. In low load conditions, the SIC procedure stops with a large number of users recovered, so in this region the proposed estimation algorithm is more effective. In contrast, in high load conditions SIC almost always stops prematurely, recovering a small number of users, leading the proposed estimation algorithm to be less effectively.

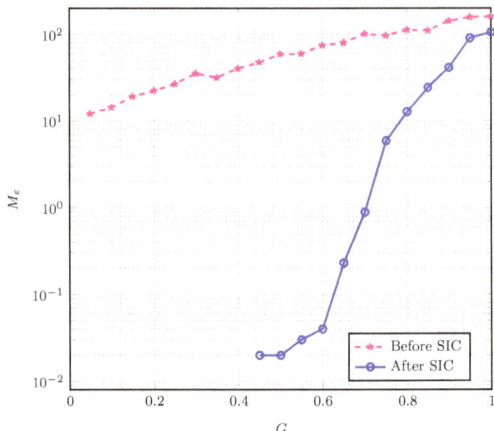

Figure 5. Mean squared error (MSE) versus G for IRSA with $\Lambda(x) = 0.5x^2 + 0.28x^3 + 0.22x^8$.

4.2. Dynamic Access Probability Simulation Results

In this subsection, we present the simulation results for dynamic access probability based coded random access schemes using the mentioned estimation methods. The frame length is $m = 200$ and the user population size is $n_{\text{pop}} = 2000$. Moreover, the considered IRSA distribution is $\Lambda(x) = 0.5x^2 + 0.28x^3 + 0.22x^8$. Each non-backlogged user activates, independently of the other users, with probability $\pi = 0.8$ at the beginning of every frame with $k = 0$. At the first frame, there are no backlogged users. The target traffic threshold G° is set to 0.65, 0.705, 0.80, and 0.938 respectively, of which $G^\circ = 0.708$ is associated with $P_L^\circ = 0.01$ and $G^\circ = 0.938$ is the asymptotic threshold of the considered IRSA distribution [5]. The initial access probability is $p_{ac}^{(0)} = 1$. We analyzed the system performance, during congestion resolution periods, through numerical simulations. Every simulation consisted of a sufficiently large number of runs and, in each run, the simulation was stopped when the congestion was resolved.

As a benchmark, consider transmission without any dynamic access probability adjustment process. The expected number of active users (transmitting users) in the initial frame is 1600. The average repetition rate is 3.6, corresponding to an expected number of 6480 packet replicas transmitted over the 200 slots. At the receiver, we have a vanishing probability to find singleton slots capable of triggering the SIC process. Without dynamic access probability adjustment, the packet loss rate becomes very close to 1 and the throughput very close to 0, meaning that almost no users are recovered in the subsequent frames, making system congestion unresolvable.

Figures 6–8 show that the proposed access probability algorithm works well to resolve congestion. The users access probability is adjusted dynamically to track the number of active users. At frame 1, the access probability is decreased quickly to avoid working in the high load region. In this way, the estimator can provide a reliable estimate at the end of the frame and the receiver is able to perform an accurate access probability design for the users in the next frame. Then the access probability is adjusted dynamically to make the number of transmitting users around the target $G^\circ m$. It is increased slowly as some users are recovered by the receiver in each transmission. Each curve is plotted up to the maximum value of k for which congestions remain unresolved, which is different for the different choices of the target load threshold.

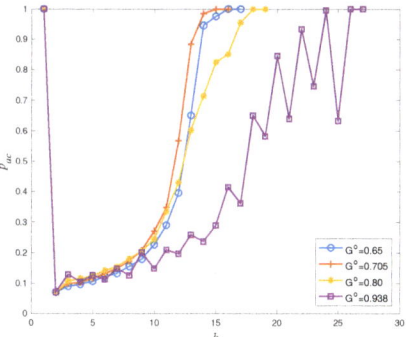

Figure 6. Access probability p_{ac} versus frame index k for IRSA with $\Lambda(x) = 0.5x^2 + 0.28x^3 + 0.22x^8$.

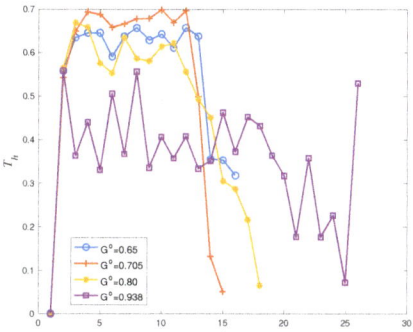

Figure 7. Throughput performance T_h versus frame index k for IRSA with $\Lambda(x) = 0.5x^2 + 0.28x^3 + 0.22x^8$.

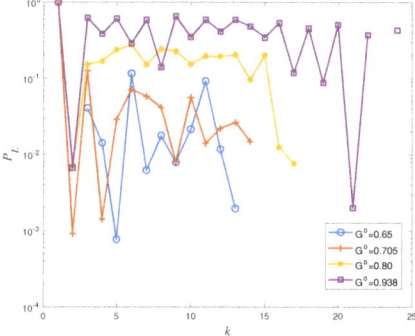

Figure 8. Packet loss rate P_L versus frame index k for IRSA with $\Lambda(x) = 0.5x^2 + 0.28x^3 + 0.22x^8$.

Back to Figure 3, we have seen that the IRSA scheme tends to show a packet loss rate floor at low offered traffic regimes, the floor appearing around $P_L = 10^{-2}$ (corresponding to $G = 0.705$) for

$\Lambda(x) = 0.5x^2 + 0.28x^3 + 0.22x^8$. For larger values of G (corresponding to the waterfall packet loss rate region), the packet loss rate increases rapidly: A $P_L \simeq 0.08$ is achieved at $G = 0.80$ and a $P_L \simeq 0.5$ is achieved at $G = 0.938$. Consequently, in Figure 8, systems with target $G° = 0.65$ and $G° = 0.705$ have a similar packet loss rate performance, and they perform better than those with target $G° = 0.80$ and $G° = 0.938$. Furthermore, due to estimation errors and to fluctuations of the actual number of transmitting users, we observe a minor packet loss rate deviation between Figure 8 and Figure 3. For example, in Figure 8, the packet loss rate with target $G° = 0.80$ is around 0.2, while in Figure 3, the packet loss rate at $G° = 0.80$ is approximately equal to 0.008.

As a final remark, recall that the throughput is defined as $G(1 - P_L)$. The influence of P_L at $G \leq 0.705$ is small, so that the per-frame throughput $T_h^{(k)}$ is approximately equal to the instantaneous load G. That is why in Figure 7, the throughput performance with target $G° = 0.705$ is better than that with target $G° = 0.65$. However, for the cases $G° = 0.8$ and $G° = 0.938$, the influence of P_L can not be ignored any more. The system performance is worse even though the load target $G°$ is higher, since the packet loss rate is now considerably higher.

5. Conclusion

In this paper, we proposed a technique to estimate the number of transmitting users in each frame of an IRSA-based coded random access system. The estimated number of transmitting users in the current frame was exploited to adjust the users access probability in the next frame. Frame configuration information as well as recovered users configuration information at the end of the SIC procedure were employed to make the estimation more accurate. Numerical results revealed how the derived dynamic access probability strategy could resolve congestion efficiently, with a stable throughput and a target packet loss rate performance for a proper choice of the parameter $G°$. Interesting directions of investigation include the exact efficient evaluation of the $h(\vec{o}, b)$ function (addressed in Lemma 1), to make the optimum estimator applicable to large communication networks. Adjusting the frame length dynamically in situations of slowly varying traffic load over a large scale is another direction of investigation that, to the best of the authors' knowledge, has not been so far addressed in the coded random access context.

Author Contributions: All the authors contributed extensively to the work presented in this paper and to writing the paper itself. J.S. and E.P. conceived the idea and developed the proposed approaches. E.P. gave advice on the research and helped in editing the paper. E.P. and R.L. improved the quality of the manuscript and completed revision.

Funding: This research was funded by the China Scholarship Council (grant No. 201706020024).

Conflicts of Interest: The authors declare no conflict of interest.

Appendix A. Justification and Implementation of Equation (26)

This appendix justifies and addresses the implementation of the two-hypotheses MAP detection rule Equation (26). Let the two hypotheses be H_0 and H_1. Moreover, let r represent the observation. The optimum detection rule consists of making the decision $\hat{H}_0$ when $P(H_0|r) > P(H_1|r)$ and of making the decision $\hat{H}_1$ otherwise.

In our case, H_0 corresponds to a 'not fully reliable estimate' state (satisfied Equation (23)); H_1 corresponds to a 'reliable estimate' state (Equation (23) not holding). The observation is $(m_c^{(k)}, \vec{\delta}^{(k)})$. Hence, we have:

$$\frac{P(t_a^{(k)} > 2\delta^{(k)} | m_c^{(k)}, \vec{\delta}^{(k)})}{P(t_a^{(k)} \leq 2\delta^{(k)} | m_c^{(k)}, \vec{\delta}^{(k)})} \begin{array}{c} \hat{H}_0 \\ \gtreqless \\ \hat{H}_1 \end{array} 1, \tag{A1}$$

where $P(t_a^{(k)} > 2\delta^{(k)}|m_c^{(k)}, \vec{\delta}^{(k)}) = \sum_{t_a^{(k)}: t_a^{(k)} > 2\delta^{(k)}} P(t_a^{(k)}|m_c^{(k)}, \vec{\delta}^{(k)})$ and $P(t_a^{(k)} \leq 2\delta^{(k)}|m_c^{(k)}, \vec{\delta}^{(k)}) = \sum_{t_a^{(k)}: t_a^{(k)} \leq 2\delta^{(k)}} P(t_a^{(k)}|m_c^{(k)}, \vec{\delta}^{(k)})$. The probability $P(t_a^{(k)}|m_c^{(k)}, \vec{\delta}^{(k)})$ can be expressed as:

$$P(t_a^{(k)}|m_c^{(k)}, \vec{\delta}^{(k)}) = \sum_{\vec{t}_a: |\vec{t}_a^{(k)}| = t_a^{(k)}} P(\vec{t}_a^{(k)}|m_c, \vec{\delta}^{(k)}), \tag{A2}$$

where $P(\vec{t}_a^{(k)}|m_c, \vec{\delta}^{(k)})$ comes from the estimator Equation (14).

Appendix B. Estimation Using Collision Slots before SIC

The user's repetition rate is $\Lambda(x)$. Define $\Lambda'(1)$ as the average user repetition rate given by $\Lambda'(1) = \sum_l l \Lambda_l$. It is easy to verify that the probability that a generic user sends a packet replica within a given slot is $\Lambda'(1)/m$. As the users send packet replicas randomly, the slot degree distribution is binomially distributed. The probability that a slot has l collided users is given by:

$$\Psi_l = \binom{t_a}{l} \left(\frac{\Lambda'(1)}{m}\right)^l \left(1 - \frac{\Lambda'(1)}{m}\right)^{t_a - l}. \tag{A3}$$

Before SIC iterations, the probability p_e that a given slot is empty, the probability p_s that a given slot is singleton and the probability p_c that a given slot is a collision one can be expressed respectively as:

$$p_e = \left(1 - \frac{\Lambda'(1)}{m}\right)^{t_a}, \tag{A4}$$

$$p_s = t_a \frac{\Lambda'(1)}{m} \left(1 - \frac{\Lambda'(1)}{m}\right)^{t_a - 1} \tag{A5}$$

and:

$$p_c = 1 - p_e - p_s. \tag{A6}$$

An estimation for T_a using frame configuration before SIC performs:

$$\hat{t}_a = \underset{t_a}{\operatorname{argmax}} P(t_a|w_c), \tag{A7}$$

where w_c is the number of collision slots before SIC iterations. Following Bayes' rule, $P(t_a|w_c)$ may be developed as:

$$P(t_a|w_c) \propto P(w_c|t_a)P(t_a)$$
$$\propto p_c^{w_c}(1 - p_c)^{m - w_c}. \tag{A8}$$

References

1. Abramson, N. The ALOHA system: Another alternative for computer communications. In Proceedings of the Fall Joint Computer Confonference, New York, NY, USA, 17–19 November 1970; pp. 281–285.
2. Roberts, L.G. ALOHA packet systems with and without slots and capture. *ACM SIGCOM Comput. Commun. Rev.* **1975**, *5*, 28–42. [CrossRef]
3. Casini, E.; De Gaudenzi, R.; Herrero, O.D.R. Contention resolution diversity slotted ALOHA (CRDSA): An enhanced random access scheme for satellite access packet networks. *IEEE Trans. Wirel. Commun.* **2007**, *6*, 1408–1419. [CrossRef]

4. De Gaudenzi, R.; Herrero, O.D.R. Advances in random access protocols for satellite networks. In Proceedings of the 2009 International Workshop on Satellite and Space Communications, Tuscany, Italy, 9–11 September 2009; pp. 331–336.
5. Liva, G. Graph-based analysis and optimization of contention resolution diversity slotted ALOHA. *IEEE Trans. Commun.* **2011**, *59*, 477–487. [CrossRef]
6. Amat, A.G.; Liva, G. Finite Length Analysis of Irregular Repetition Slotted ALOHA in the Waterfall Region. *IEEE Commun. Lett.* **2018**, *22*, 886–889. [CrossRef]
7. De Gaudenzi, R.; Herrero, O.D.R.; Gallinaro, G.; Cioni, S.; Arapoglou, P.D. Random access schemes for satellite networks, from VSAT to M2M: A survey. *Int. J. Satellite Commun. Netw.* **2018**, *36*, 66–107. [CrossRef]
8. Alvi, S.; Durrani, S.; Zhou, X. Enhancing CRDSA with Transmit Power Diversity for Machine-Type Communication. *IEEE Trans. On Vehic. Tech.* **2018**, *67*, 7790–7794. [CrossRef]
9. Paolini, E.; Liva, G.; Chiani, M. Coded slotted ALOHA: A graph-based method for uncoordinated multiple access. *IEEE Trans. Inf. Theory* **2015**, *61*, 6815–6832. [CrossRef]
10. Paolini, E.; Stefanović, Č.; Liva, G.; Popovski, P. Coded random access: Applying codes on graphs to design random access protocols. *IEEE Commun. Mag.* **2015**, *53*, 144–150. [CrossRef]
11. Taghavi, A.; Vem, A.; Chamberland, J.-F.; Narayanan, K.R. On the design of universal schemes for massive uncoordinated multiple access. In Proceedings of the 2016 IEEE International Symposium on Information Theory (ISIT), Barcelona, Spain, 10–15 July 2016; pp. 345–349.
12. Sandgren, E.; i Amat, A.G.; Brännström, F. On frame asynchronous coded slotted ALOHA: Asymptotic, finite length, and delay analysis. *IEEE Trans. Commun.* **2017**, *65*, 691–704. [CrossRef]
13. Schoute, F. Dynamic frame length ALOHA. *IEEE Trans. Commun.* **1983**, *31*, 565–568. [CrossRef]
14. Chen, W.-T. An accurate tag estimate method for improving the performance of an RFID anticollision algorithm based on dynamic frame length ALOHA. *IEEE Trans. Autom. Sci. Eng.* **2009**, *6*, 9–15. [CrossRef]
15. Eom, J.-B.; Lee, T.-J. Accurate tag estimation for dynamic framed-slotted ALOHA in RFID systems. *IEEE Commun. Lett.* **2010**, *14*, 60–62. [CrossRef]
16. Wu, H.; Zeng, Y. Bayesian tag estimate and optimal frame length for anti-collision ALOHA RFID system. *IEEE Trans. Autom. Sci. Eng.* **2010**, *7*, 963–969. [CrossRef]
17. Zanella, A. Estimating collision set size in framed slotted ALOHA wireless networks and RFID systems. *IEEE Commun. Lett.* **2012**, *16*, 300–303. [CrossRef]
18. Rivero Angeles, M.E.; Lara Rodriguez, D.; Cruz-Perez, F.A. Random-access control mechanisms using adaptive traffic load in ALOHA and CSMA strategies for EDGE. *IEEE Trans. Vehic. Tech.* **2005**, *54*, 1160–1186. [CrossRef]
19. Lee, M.W.; Lee, J.K.; Lim, J.S. R-CRDSA: Reservation-Contention Resolution Diversity Slotted ALOHA for Satellite Networks. *IEEE Commun. Lett.* **2012**, *16*, 1576–1579. [CrossRef]
20. Noh, H.J.; Lee, J.K.; Lim, J.S. Performance evaluation of access control for CRDSA and R-CRDSA under high traffic load. In Proceedings of the 2013 IEEE Military Communications Conference, San Diego, CA, USA, 18–20 November 2013; pp. 1365–1370.
21. Sun, J.; Liu, R. Irregular Repetition Slotted ALOHA with Priority (P-IRSA). In Proceedings of the 2016 IEEE Vehicular Technology Conference (VTC Spring), Nanjing, China, 15–18 May 2016; pp. 1–5.
22. Sun, J.; Liu, R.; Paolini, E. Detecting the Number of Active Users in Coded Random Access Systems. In Proceedings of the 2018 IEEE 29th Annual International Symposium on Personal, Indoor and Mobile Radio Communications (PIMRC), Bologna, Italy, 9–12 September 2018; pp. 1–7.
23. Sun, J.; Liu, R.; Paolini, E. Detecting the Number of Active Users in IRSA Access Protocols. In Proceedings of the 2018 IEEE 29th Annual International Symposium on Personal, Indoor and Mobile Radio Communications (PIMRC), Bologna, Italy, 9–12 September 2018; pp. 1972–1976.
24. Sun, J.; Liu, R.; Paolini, E. Unrecovered Users Distribution in Coded Random Access Systems with Erasures. In Proceedings of the 2019 IEEE International Conference on Communications (ICC), Shanghai, China, 20–24 May 2019; pp. 1–6.
25. Schmetterer, L. An introduction to combinatorial analysis by J. Riordan. *Phys. Today* **1959**, *12*, 158.

© 2019 by the authors. Licensee MDPI, Basel, Switzerland. This article is an open access article distributed under the terms and conditions of the Creative Commons Attribution (CC BY) license (http://creativecommons.org/licenses/by/4.0/).

Article

On the Capacity of 5G NR Grant-Free Scheduling with Shared Radio Resources to Support Ultra-Reliable and Low-Latency Communications

M. Carmen Lucas-Estañ *, Javier Gozalvez and Miguel Sepulcre

Department of Communications Engineering, Universidad Miguel Hernández de Elche (UMH), Avda. de la Universidad s/n, 03202 Elche, Spain
* Correspondence: m.lucas@umh.es; Tel.: +34-965-222-424

Received: 14 June 2019; Accepted: 10 August 2019; Published: 16 August 2019

Abstract: 5G and beyond networks are being designed to support the future digital society, where numerous sensors, machinery, vehicles and humans will be connected in the so-called Internet of Things (IoT). The support of time-critical verticals such as Industry 4.0 will be especially challenging, due to the demanding communication requirements of manufacturing applications such as motion control, control-to-control applications and factory automation, which will require the exchange of critical sensing and control information among the factory nodes. To this aim, important changes have been introduced in 5G for Ultra-Reliable and Low-Latency Communications (URLLC). One of these changes is the introduction of grant-free scheduling for uplink transmissions. The objective is to reduce latency by eliminating the need for User Equipments (UEs—sensors, devices or machinery) to request resources and wait until the network grants them. Grant-free scheduling can reserve radio resources for dedicated UEs or for groups of UEs. The latter option is particularly relevant to support applications with aperiodic or sporadic traffic and deterministic low latency requirements. In this case, when a UE has information to transmit, it must contend for the usage of radio resources. This can lead to potential packet collisions between UEs. 5G introduces the possibility of transmitting K replicas of the same packet to combat such collisions. Previous studies have shown that grant-free scheduling with K replicas and shared resources increases the packet delivery. However, relying upon the transmission of K replicas to achieve a target reliability level can result in additional delays, and it is yet unknown whether grant-free scheduling with K replicas and shared resources can guarantee very high reliability levels with very low latency. This is the objective of this study, that identifies the reliability and latency levels that can be achieved by 5G grant-free scheduling with K replicas and shared resources in the presence of aperiodic traffic, and as a function of the number of UEs, reserved radio resources and replicas K. The study demonstrates that current Fifth Generation New Radio (5G NR) grant-free scheduling has limitations to sustain stringent reliability and latency levels for aperiodic traffic.

Keywords: grant-free; scheduling; URLLC; ultra-reliable and low-latency communications; 5G; deterministic; time-critical; reliability; latency; aperiodic traffic; Industry 4.0

1. Introduction

5G networks are being designed with the objective to support a broad range of verticals such as manufacturing, transport, health, energy and entertainment. To this aim, important changes have been introduced to increase data rates (enhanced mobile broadband, or eMBB), efficiently support large amounts of devices (massive machine type communications, or mMTC) and guarantee unprecedented reliability and latency levels (Ultra-Reliable and Low-Latency Communications or URLLC) [1]. Supporting URLLC is particularly relevant for many Industry 4.0 manufacturing applications, such as motion control (requires a maximum latency of 1 ms and a reliability of 1–10^{-6} [2]), control-to-control applications (maximum latency of 4 ms and a reliability of 1–10^{-8} [1]) and factory automation (maximum latency between 0.25 ms and 2.5 ms and reliability requirements up to 1–10^{-9} [3]). These applications require the exchange of information between sensors, actuators and controllers through an industrial sensor and control network. 5G has the potential to provide the connectivity required by the Industry 4.0 to digitalize factories and to support data-intensive services while ubiquitously guaranteeing low latency and reliable connections. This has actually been acknowledged through the establishment of the 5G Alliance for Connected Industries and Automation (5G-ACIA) [4].

5G has introduced significant changes to support URLLC [5]. Some of these changes focus at the Radio Access Network level, since the medium access mechanisms account for an important part of the total end-to-end transmission delay [6]. This is for example the case of the grant-based scheduling process for uplink (UL) transmissions in legacy LTE (Long Term Evolution) 4G networks. Grant-based scheduling requires a User Equipment (UE) and a Base Station (BS) to exchange scheduling requests (SRs) and grant messages before transmitting any data. This process alone already results in an average delay of up to 11.5 ms when considering a Transmission Time Interval (TTI) equal to 1 ms and an SR periodicity of 10 ms [3]. Reducing the slot duration can reduce this delay. However, additional scheduling changes have been necessary to sustain the URLLC requirements that characterize some vertical applications, such as those in Industry 4.0. In particular, Release 15 and 16 of the 3rd Generation Partnership Project (3GPP) standards have introduced the concept of grant-free scheduling (also referred to as Configured Grant for 5G New Radio [7]) to support URLLC.

With grant-free scheduling, the BS reserves resources for UL transmissions and informs the UEs of the reserved resources. When a UE wants to initiate a UL transmission, it directly utilizes the reserved resources, without sending an SR and waiting for the subsequent grant message from the BS. Recent studies have shown that grant-free scheduling in 5G NR considerably reduces the end-to-end latency [8]. The 3GPP standards introduce the possibility for grant-free scheduling to reserve resources to dedicated UEs, or to a group of UEs. In the first case, each resource is reserved for a specific UE, and only this UE can utilize the resource at any time. This approach is adequate for periodic traffic since the resource allocations can be planned, and resources can then be utilized efficiently. Such planning is not possible in the case of aperiodic, sporadic or uncertain traffic. Sharing dedicated resources by a group of UEs is hence an interesting option to optimize the usage of the radio resources in the presence of aperiodic traffic. In this case, UEs have to contend for their usage, and collisions are possible. 5G NR introduces the possibility to transmit K replicas of the same packet in consecutive slots to combat potential collisions. However, relying on the transmission of K replicas to achieve a target reliability level can result in additional delays. It is yet unknown whether 5G NR grant-free scheduling with K-repetitions and shared resources can satisfy critical applications and guarantee very high reliability levels with very low latency. In this context, this study presents an in-depth analysis of the reliability and latency levels that can be achieved with existing 5G NR grant-free scheduling solutions as a function of the number of UEs, the number of reserved radio resources, and the number of replicas K. To this aim, the study analytically quantifies the probability of successfully delivering a packet when using grant-free scheduling with K-repetitions and shared resources. In addition, the study analyzes the impact of self-collisions. Self-collisions occur when a UE has to transmit a new packet, and the transmission of the K replicas of the previous packet has not finished. If this happens, the new packet must be stored, and its transmission is delayed until all replicas of the previous packet have been transmitted. This study demonstrates for the first time that self-collisions have a non-negligible impact upon the capacity of 5G NR grant-free scheduling to support stringent URLLC reliability and latency levels.

2. Related Work

The 5G NR standard introduces the use of grant-free scheduling (also referred to as Configured Grant [7]). With grant-free scheduling, the network pre-configures the radio resources and assigns them to UEs without waiting for UEs to request resources. UEs can utilize the pre-assigned resources as soon as they have data to transmit. This is in contrast to grant-based scheduling, where UEs must request access to radio resources through the transmission of Scheduling Requests (SR). The BS assigns the radio resources to the UEs and notifies them using grant messages. UEs must wait to receive these grant messages before transmitting any data. Grant-free scheduling eliminates all delays introduced by the handshaking present in grant-based scheduling. Grant-free scheduling also improves the energy consumption of the UEs, reduces their complexity, and decreases the signaling overhead compared with grant-based scheduling ([8,9]). Grant-free scheduling can assign dedicated or shared resources to the UEs. The BS decides whether resources are dedicated to specific UEs, or are shared by a group of UEs [10]. Reserving resources to dedicated UEs is an interesting approach when we can plan ahead what is the demand for resources. This is for example the case of periodic traffic. However, reserving resources to dedicated users can be highly inefficient if the traffic demand is uncertain or aperiodic, and it is not possible to anticipate when these resources will be needed. In this case, it is possible to share radio resources by a group of UEs. This option ensures a more efficient utilization of resources, and the possibility to satisfy URLLC communication requirements. However, users must contend for the resources, and collisions can happen if two or more UEs simultaneously contend for the same resources. 5G NR introduces the possibility of transmitting K replicas of the same packet in consecutive slots to combat collisions and thus increase the probability of a correct reception [11,12].

The study in [13] analyzes the performance of the K replicas scheme. The authors propose transmitting the first copy of a packet using dedicated resources, and the following replicas using shared resources. The proposal also exploits shared diversity and advanced receiver processing techniques to reduce the impact of packet collisions. The proposal achieves adequate reliability levels and reduces the number of reserved (shared) radio resources, compared to a configuration that reserves resources to dedicated UEs. The study in [14] also transmits the first copy of a packet using dedicated resources. However, it does not consider the transmission of K replicas of a packet. Instead, the authors propose to retransmit the original packet in a shared resource only if the first transmission is not successful. This requires a handshaking between the UEs and the BS to exchange acknowledgement messages. This handshaking increases the latency, and can compromise the capability to adequately support URLLC applications with stringent latency requirements. In [15], the authors study the optimum number of replicas (K) necessary to achieve a target reliability level within a deterministic latency deadline. The study focuses upon aperiodic traffic and the case in which a group of UEs share resources. The authors show that randomly choosing the resource for each replica increases the probability of correctly delivering a packet. However, the study focuses on reliability levels up to $1-10^{-5}$ while some critical Industry 4.0 applications require higher reliability levels.

Previous studies have shown that transmitting K-repetitions of a packet increases the reception rate. However, this can be done at the expense of an inefficient use of the radio resources due to packet collisions or the unnecessary reservation of resources when the first replicas are correctly delivered. Latency requirements may also impose restrictions on the number of replicas that can be transmitted, and consequently on the reliability levels that may be achieved. In this context, several recent contributions have analyzed slight modifications to the K-repetitions scheme. For example, [16] proposed adaptively configuring the number of replicas transmitted based on the channel conditions. The objective is to utilize the radio resources efficiently by avoiding unnecessary retransmissions when the channel quality is good. A similar objective is sought in [17] where authors propose conditions to stop the transmission of replicas. Other interesting proposals in 3GPP standardization working groups include: the transmission of replicas within mini-slots (to reduce the latency) [18], the possibility for transmitting replicas across the slot border, or the concept of periodicity boundary [19]. These studies propose interesting variants of the K-repetitions scheme. However, it is yet unknown whether 5G NR grant-free scheduling with K-repetitions and shared resources can really support URLLC communications with strict reliability and latency requirements under the presence of aperiodic or sporadic traffic. This traffic is critical in many verticals, for example in Industry 4.0. In this context, this study conducts an in-depth evaluation of 5G NR grant-free scheduling with K-repetitions and shared resources in the presence of aperiodic or sporadic traffic. The study identifies the reliability and latency levels that can be achieved with 5G NR grant-free scheduling, and identifies its current limitations. The study analyzes the impact of the number of UEs in the network, the number of reserved radio resources, and the number of replicas K. The study also analyzes for the first time the impact of self-collisions. The conducted analysis helps to identify the reliability and latency levels that can be achieved based on network deployments and configuration options for 5G NR grant-free scheduling.

It should be noted that 3GPP standards define the possibility of utilizing grant-free scheduling and transmitting K replicas, but do not define a specific scheme to be implemented. This study is based on the implementation of 5G NR grant-free scheduling with K replicas and shared resources proposed in [15]. This implementation is chosen because it has been specifically designed to guarantee stringent URLLC latency and reliability requirements. To this aim, the implementation transmits original packets and all of the replicas using grant-free scheduling on shared radio resources. A different approach is proposed in [13] where dedicated resources are used to transmit the original packets, and shared resources are used for the following replicas. This approach can increase the delay compared to [15] if grant-based scheduling is utilized to allocate the dedicated resources. The efficient utilization of resources could also be compromised if dedicated resources were reserved for each UE when supporting applications with aperiodic traffic. The implementation of 5G NR grant-free scheduling with K–repetitions and shared resources proposed in [15] is therefore better suited to support URLLC applications with aperiodic or sporadic traffic.

3. Grant-Free Scheduling

This paper uses grant-free scheduling with K-repetitions and shared resources to evaluate the reliability and latency levels that can be achieved in the presence of aperiodic traffic. Following [20], reliability for URLLC services is defined as the percentage of data packets that are successfully delivered before the latency deadline L established by the service or application. Following 3GPP standards [11], UEs transmit the same data packet in K consecutive transmission slots with a duration T_{slot}. The UE randomly selects an RB (Resource Block) for each transmission from the U RBs available per T_{slot}. This is illustrated in Figure 1 that represents the time/frequency resource grid map in 5G NR, where the unit is an RB. In 5G NR, a wideband channel is divided into sub-frames, slots and RBs. An RB is the smallest unit of frequency resources that can be allocated to a UE. Without loss of generality, this study considers a numerology μ equal to 3 with a subcarrier spacing of 120 kHz [21]. An RB is then 1440 kHz (Δf) wide in frequency (12 sub-carriers of 120 kHz) and lasts for one time slot with the duration T_{slot} equal to 0.125 ms.

The reliability at the medium access level that can be achieved with grant-free scheduling with K-repetitions and shared resources depends upon two main factors. The first factor is the possibility that a packet is not correctly received due to the collision of all its K replicas with other transmissions; this is due to the random selection of the RB for the transmission of each replica. The study in [15] showed that the possibility to successfully deliver a packet increases with the number K of replicas. The second factor is the effect of self-collisions. A self-collision occurs when a UE has to transmit a new packet, and the transmission of the K replicas of the previous packet has not finished. If this happens, the new packet must be stored, and its transmission is delayed until all the replicas of the previous packet have been transmitted. This delay can result in the case that the new packet cannot be delivered within the latency limit, and hence self-collisions can impact the reliability of URLLC services. It is important then that the reliability (or probability that a packet is correctly received before the latency deadline) of grant-free scheduling with K-repetitions and shared radio resources is computed considering both the effect of collisions from other UEs, and the effect of self-collisions. In this case, the reliability or probability P_{rel} that a packet is correctly received by the BS must consider the probability P_{sc} that the transmission of the K replicas of a packet is not completed before the latency deadline L due to the effect of self-collisions. For the packets that are not affected by the effect of self-collisions, it must be considered the probability P_c that a packet is not correctly received due to the collision of all its K replicas with other transmissions. Hence, P_{rel} can be expressed as:

$$P_{rel} = 1 - (P_{sc} + (1 - P_{sc}) \cdot P_c) \quad (1)$$

In [15], its authors presented an expression to approximate the probability P_c of the collision of the K replicas of a packet with the transmission of other UEs. The expression was derived in scenarios where N UEs share the same pool of RBs. However, [15] did not analyze the impact of self-collisions, since the study only considered low values of K (equal to or lower than 4). For these low values, self-collisions might not have an impact upon the reliability, as will be later shown. In this paper, we analytically derive the exact probability of any collision of the K replicas of a packet with packets transmitted by other UEs (P_c). We also quantify the impact of self-collisions (P_{sc}), and analytically compute the reliability that can be achieved by grant-free scheduling with K-repetitions and shared resources (P_{rel}). These analytical expressions are a valuable contribution to the community since they can be easily utilized to evaluate 5G NR grant-free scheduling. The availability of these exact analytical expressions is particularly useful when considering applications with very demanding reliability and latency URLLC requirements. This is the case of certain Industry 4.0 applications. For example, motion control requires a maximum latency of 1 ms and a reliability of $1-10^{-6}$. Control-to-control applications require a maximum latency of 4 ms and a reliability of $1-10^{-8}$. Factory automation applications usually demand maximum latency values in the range 0.25–2.5 ms and reliability levels up to $1-10^{-9}$. In this case, simulations can be very computationally expensive if we want to compute the packet reception rate ($1 - P_c$) with reliability demands in the order of $1-10^{-6}$ to $1-10^{-9}$. In these scenarios, errors are very rare, and we need long and computationally expensive simulations to achieve accurate results. The analytical methodology utilized in this study is then an adequate and efficient tool for scenarios with demanding URLLC communication requirements.

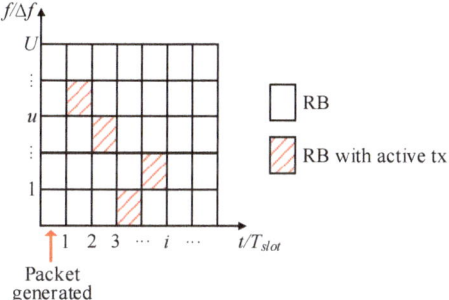

Figure 1. Illustration of the Fifth Generation New Radio (5G NR) resource grid map: Transmission of a data packet with four repetitions and a random selection of Resource Blocks (RBs) per slot.

3.1. Collisions with Other UEs

First, we focus on the probability P_c that a packet is not correctly received due to the collisions of its K replicas with the packets transmitted by other UEs. To this end, we consider UL transmissions and N UEs within a single cell with aperiodic traffic. Packets are generated by each UE following a Poisson distribution with exponential inter-arrival time. The average packet inter-arrival time is equal to $1/\lambda$, where λ is the average number of packets generated per second. We consider the transmission of small packets with a size of 32 bytes [22], and we assume without loss of generality that each packet requires only one RB.

The probability P_g that one or more packets are generated for a UE in a time period T_{slot} is equal to:

$$P_g = 1 - \exp(-T_{slot} \cdot \lambda) \quad (2)$$

We define R_i as the set of UEs for which a new packet could be generated in a slot s_i (the slot has a time duration equal to T_{slot}). Here, n_i is the number of UEs that do have a new packet to transmit in s_i. This n_i can then take any value between 0 and the cardinality of R_i. The probability $P_{tx}(n_i, R_i)$ that n_i UEs from the set R_i of UEs have new packets to be transmitted in s_i with duration T_{slot} is equal to:

$$P_{tx}(n_i, R_i) = \binom{|R_i|}{n_i} \cdot P_g^{n_i} \cdot (1 - P_g)^{|R_i| - n_i} \quad (3)$$

where $|R_i|$ represents the number of elements or the cardinality of the set R_i.

A packet will not be successfully delivered to the BS if all its K replicas collide with the trans-missions of other UEs. A UE has an active transmission in s_i if it generated a new data packet in the previous slots $s_{i-(K-1)}, \ldots, s_{i-1}$, and s_i. If this is the case, then the UE would be transmitting one of the K replicas in s_i. We denote as n_i^{act} the number of UEs with active transmissions in s_i. The probability $P_{nrc}(n_i^{act}, U)$ that n_i^{act} UEs do not collide with a given UE is equal to the probability that they do not select the same RB at a given slot for their next transmission as the UE under study. $P_{nrc}(n_i^{act}, U)$ is given by:

$$P_{nrc}(n_i^{act}, U) = \left(\frac{U - 1}{U}\right)^{n_i^{act}} \quad (4)$$

Equations (2)–(4) are necessary to compute the probability P_c that a packet is not correctly received at the BS due to the collision of all its K replicas with the transmissions of other UEs. To compute P_c, let us consider the case of a particular UE_1 that has to transmit the K replicas of a packet in slots s_i, $s_{i+1}, \ldots, s_{i+K-1}$. For the sake of clarity, we consider an example with $K = 4$, and s_i corresponding to s_3. P_c is then equal to the probability of collision of the 4 replicas transmitted in s_3, s_4, s_5, and s_6, which is represented by $P_{rc}(s_3, s_4, s_5, s_6)$:

$$P_c = P_{rc}(s_3, s_4, s_5, s_6) \quad (5)$$

To determine $P_{rc}(s_3, s_4, s_5, s_6)$, we first study the probability $P_{rc}(s_3)$ that the replica of the packet transmitted in s_3 collides with a transmission from any other UE. $P_{rc}(s_3)$ is given by the probability that one or more UEs (in addition to UE_1) have an active transmission in s_3 (i.e., $n_3^{act} \geq 1$), and that one or more of the n_3^{act} UEs select the same RB as UE_1 for their transmission. n_3^{act} is equal to $n_0 + n_1 + n_2 + n_3$, and the probability $P_{rc}(s_3)$ has to consider all possible combinations of n_0, n_1, n_2 and n_3 that result in $n_3^{act} \geq 1$. The probability $P(n_3^{act} \geq 1)$ can then be expressed as:

$$P(n_3^{act} \geq 1) = \sum_{n_0 = n_0^{min}}^{n_0^{max}} \left\{ P_{tx}(n_0, R_0) \cdot \sum_{n_1 = n_1^{min}}^{n_1^{max}} \left\langle P_{tx}(n_1, R_1) \cdot \sum_{n_2 = n_2^{min}}^{n_2^{max}} \left[P_{tx}(n_2, R_2) \cdot \sum_{n_3 = n_3^{min}}^{n_3^{max}} P_{tx}(n_3, R_3) \right] \right\rangle \right\} \quad (6)$$

where n_i^{max} and n_i^{min} represent the maximum and minimum possible values of n_i in each slot, and are equal to:

$$n_i^{max} = |R_i|, \forall i \leq 3 \quad (7)$$

$$n_i^{min} = \begin{cases} 1 & \text{if } i = 3 \ \& \ |R_i| = N - 1 \\ 0 & \text{otherwise} \end{cases}, i \leq 3 \quad (8)$$

where R_i is the set of UEs that could have a new packet to be transmitted in s_i. R_i is equal to the total number of UEs (N) minus UE_1 and all active UEs in the slot previous to s_i. The cardinality of R_i is then equal to:

$$|R_i| = N - 1 - \sum_{j=\max\{i-3, 0\}}^{i-1} n_j, \ i \leq 3 \quad (9)$$

It should be noted that n_i^{min} is equal to 0 or 1 in order to guarantee that n_3^{act} is equal to or higher than one. n_i^{act} can be expressed as:

$$n_i^{act} = \sum_{j=\max\{i-3,0\}}^{i} n_j, \quad i \leq 3 \tag{10}$$

To achieve finally the expression of $P_{rc}(s_3)$, we need to incorporate to the expression of $P(n_3^{act} \geq 1)$ in (6) the probability that one or more of the n_3^{act} UEs select the same RB as UE$_1$ for their transmissions. This probability is equal to $1 - P_{nrc}(n_3^{act}, U)$. $P_{rc}(s_3)$ is then calculated as:

$$P_{rc}(s_3) = \sum_{n_0=n_0^{min}}^{n_0^{max}} \left\{ P_{tx}(n_0, R_0) \cdot \sum_{n_1=n_1^{min}}^{n_1^{max}} \left\langle P_{tx}(n_1, R_1) \cdot \sum_{n_2=n_2^{min}}^{n_2^{max}} \left[P_{tx}(n_2, R_2) \cdot \sum_{n_3=n_3^{min}}^{n_3^{max}} \left\{ P_{tx}(n_3, R_3) \cdot (1 - P_{nrc}(n_3^{act}, U)) \right\} \right] \right\rangle \right\} \tag{11}$$

The probability of collision of the replica transmitted in s_4 depends upon the number n_4^{act} of UEs with active transmissions in s_4. This n_4^{act} depends on the number n_1, n_2, n_3 and n_4 of UEs that have new packets to transmit in s_1, s_2, s_3, and s_4, respectively. The probability that UEs have new packets to transmit in s_1, s_2, and s_3 is already included in (11) ($P_{tx}(n_1, R_1)$, $P_{tx}(n_2, R_2)$, and $P_{tx}(n_3, R_3)$ respectively). In this context, $P_{rc}(s_3)$ and $P_{rc}(s_4)$ are not independent, and they must be calculated jointly. We then compute the joint probability $P_{rc}(s_3, s_4)$ that the replicas transmitted in s_3 and s_4 collide with transmissions from other UEs. Computing $P_{rc}(s_3, s_4)$ only requires including in (11) the probability that there are UEs with new packets to be transmitted in s_4 (i.e., $P_{tx}(n_4, R_4)$), and the probability that one or more of the active n_4^{act} UEs in s_4 select the same RB for their transmission than UE$_1$. $P_{rc}(s_3, s_4)$ can then be expressed as:

$$P_{rc}(s_3, s_4) = \sum_{n_0=n_0^{min}}^{n_0^{max}} \left\{ P_{tx}(n_0, R_0) \cdot \sum_{n_1=n_1^{min}}^{n_1^{max}} \left\langle P_{tx}(n_1, R_1) \cdot \sum_{n_2=n_2^{min}}^{n_2^{max}} [P_{tx}(n_2, R_2) \cdot \right. \right.$$
$$\left. \left. \sum_{n_3=n_3^{min}}^{n_3^{max}} \left\langle \{P_{tx}(n_3, R_3) \cdot (1 - P_{nrc}(n_3^{act}, U))\} \cdot \sum_{n_4=n_4^{min}}^{n_4^{max}} \{P_{tx}(n_4, R_4) \cdot (1 - P_{nrc}(n_4^{act}, U))\} \right\rangle \right] \right\rangle \right\} \tag{12}$$

where n_4^{act}, $|R_4|$, n_4^{max} and n_4^{min} are defined as:

$$n_4^{act} = \sum_{j=1}^{4} n_j \tag{13}$$

$$|R_4| = N - 1 - \sum_{j=1}^{3} n_j \tag{14}$$

$$n_4^{max} = |R_4| \tag{15}$$

$$n_4^{min} = \begin{cases} 1 & \text{if } |R_4| = N - 1 \\ 0 & \text{otherwise} \end{cases} \tag{16}$$

The process followed to account for possible collisions of the replicas transmitted in s_5 and s_6 is similar to that considered for s_4. P_c can then be expressed as follows when $K = 4$:

$$P_c = \sum_{n_0=n_0^{min}}^{n_0^{max}} \left\{ P_{tx}(n_0, R_0) \cdot \sum_{n_1=n_1^{min}}^{n_1^{max}} \left\langle P_{tx}(n_1, R_1) \cdot \sum_{n_2=n_2^{min}}^{n_2^{max}} [P_{tx}(n_2, R_2) \cdot \right. \right.$$
$$\sum_{n_3=n_3^{min}}^{n_3^{max}} \left\langle \{P_{tx}(n_3, R_3) \cdot (1 - P_{nrc}(n_3^{act}, U))\} \cdot \sum_{n_4=n_4^{min}}^{n_4^{max}} [\{P_{tx}(n_4, R_4) \cdot (1 - P_{nrc}(n_4^{act}, U))\} \cdot \right. \tag{17}$$
$$\left. \left. \left. \sum_{n_5=n_5^{min}}^{n_5^{max}} \left\langle \{P_{tx}(n_5, R_5) \cdot (1 - P_{nrc}(n_5^{act}, U))\} \cdot \sum_{n_6=n_6^{min}}^{n_6^{max}} \{P_{tx}(n_6, R_6) \cdot (1 - P_{nrc}(n_6^{act}, U))\} \right\rangle \right] \right\rangle \right] \right\rangle \right\}$$

where n_i^{act}, $|R_i|$, n_i^{max} and n_i^{min} $\forall i \in [0, 2 \cdot K-1]$ are defined as:

$$n_i^{act} = \sum_{j=\max\{i-(K-1),0\}}^{i} n_j \tag{18}$$

$$|R_i| = N - 1 - \sum_{j=\max\{i-(K-1),0\}}^{i-1} n_j \tag{19}$$

$$n_i^{max} = |R_i| \tag{20}$$

$$n_i^{min} = \begin{cases} 1 & \text{if } i \geq K-1 \ \& \ |R_i| = N-1 \\ 0 & \text{otherwise} \end{cases} \tag{21}$$

The process illustrated for $K = 4$ can be followed to compute P_c for any value of K. As shown in (22), P_c can be computed using the auxiliary function $h_i(K, N, U)$ defined in (23) with i equal to cero. To simplify the notation, $h_i(K, N, U)$ is also represented as h_i in (22) and (23). As it can be observed in (23), h_0 depends on h_1, and in general, h_i depends on h_{i+1}, until h_{2K-1}.

$$P_c(K, N, U) = h_0(K, N, U) = h_0 \tag{22}$$

$$h_i = \begin{cases} \sum_{n_i=n_i^{min}}^{n_i^{max}} [P_{tx}(n_i, R_i) \cdot h_{i+1}] & \text{if } i \in [0, K) \\ \sum_{n_i=n_i^{min}}^{n_i^{max}} \left[P_{tx}(n_i, R_i) \cdot (1 - P_{nrc}(n_5^{act}, U)) \cdot h_{i+1}\right] & \text{if } i \in [K, 2 \cdot K-1) \\ \sum_{n_i=n_i^{min}}^{n_i^{max}} \left[P_{tx}(n_i, R_i) \cdot (1 - P_{nrc}(n_5^{act}, U))\right] & \text{if } i = 2 \cdot K-1 \end{cases} \tag{23}$$

The parameters n_i^{act}, $|R_i|$, n_i^{max} and n_i^{min} in (23) correspond to those expressed in (18)–(21).

3.2. Self-Collisions

The effect of self-collisions is illustrated in Figure 2. We may suppose that a UE starts transmitting a packet p_1 that was generated before t_0. Let us then suppose then that a second packet p_2 is generated before the K replicas of the previous packet p_1 have been transmitted. This is a self-collision. If a self-collision happens, p_2 can be stored, and its transmission will start after the UE has transmitted the Kth replica of p_1 (i.e., at t_1 in Figure 2). The transmission of the K replicas of p_2 will finish at t_2 that is equal to:

$$t_2 = 2 \cdot K \cdot T_{slot} + t_0 \tag{24}$$

The transmission of the K replicas of p_2 may finish after the latency deadline L, due to the time p_2 being stored as the K replicas of p_1 are being transmitted. We then analyze the probability P_{sc} that the transmission of K replicas of a packet is not completed before L due to the effect of self-collisions. This probability depends upon the number of replicas K and on the time instant at which p_2 was generated. Figure 2 illustrates how self-collisions affect the probability of completing the transmission of p_2 before L, with L equal to 1 ms. $L = 1$ ms implies that the maximum number of replicas K that can be transmitted per packet is 8. However, it is possible to transmit less than 8 replicas, and Figure 2 represents the case in which K is set equal to 4, 6 or 8. p_2 can be transmitted before the deadline L if it is generated at any time instant after $t_2 - L$, where t_2 is the time at which the transmission of the K replicas of p_2 is finished (the transmission of p_2 starts when the transmission of the K replicas of p_1 has finished at t_1). If p_2 is generated before $t_2 - L$, it is not possible to complete the transmission of the K replicas of p_2 before the latency deadline L. P_{sc} can then be computed as the probability that the time between the generation of two consecutive packets at a UE falls within the interval $[0, \Delta t]$, where

Δt represents the time difference between $t_2 - L$ and the time t_{p_1} at which p_1 is generated (see (26)). P_{sc} can then be expressed as:

$$P_{sc}(\Delta t) = \int_0^{\Delta t} \lambda \cdot e^{-t \cdot \lambda} \cdot dt \tag{25}$$

$$\Delta t = t_2 - L - t_{p_1} = 2 \cdot K \cdot T_{slot} - L - t_{p_1} \tag{26}$$

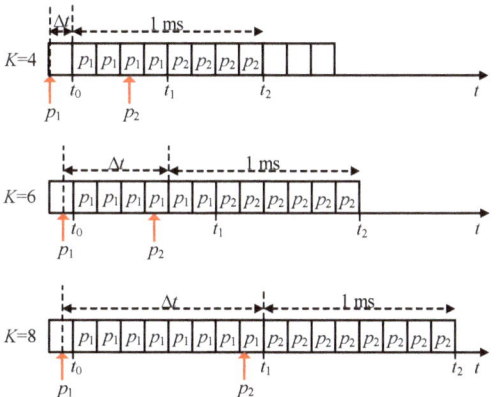

Figure 2. Scenarios with possible self-collisions ($L = 1$ ms and $K = 4$, 6 and 8).

As shown in (25) and (26), the negative effect of self-collisions increases with the value of K, since K influences the time a packet might be stored until the transmission of the previous packet is finished. However, increasing the number K of replicas transmitted for each packet is preferred, in order to combat possible collisions with other UEs sharing the same pool of radio resources. The next section will analyze both the effect of collisions from other UEs and the effect of self-collisions to analyze the reliability achievable with the grant-free scheduling with K-repetitions and shared radio resources.

4. Validation

This section validates the analytical expressions derived in Section 3.1 to calculate the probability P_c that a packet is not correctly received due to packet collisions with other UEs. To this aim, we compare the results achieved with the analytical expressions, with that obtained through simulations.

We have implemented a system level simulator in Matlab™ that accurately models the 5G NR grant-free scheduling process with K-repetitions and shared resources. The simulator emulates a single cell with N UEs that generate aperiodic traffic. Each UE models the packet traffic arrival, using a Poisson distribution with exponential inter-arrival time. The average packet inter-arrival time is equal to $1/\lambda$, where λ is the average number of packets generated per second. The simulator implements the time/frequency resource grid map of 5G NR. The time and frequency duration of RBs is configurable based on the considered 5G NR numerology μ. It is possible to also configure the number U of RBs available per time slot. The number K of replicas can also be configured in the simulation platform.

We have conducted a large number of simulations to ensure the accuracy of the simulation results, and compare them to those obtained with our analytical expressions and methodology. Simulations are here shown for K equal to 2, 4 and 8, λ equal to 0.1 packets, μ equal to 3, and U equal to 6 RBs per slot. UEs transmit small packets with a size of 32 bytes [22] that can be transmitted in a single RB. Figure 3 compares the value of P_c achieved analytically and through simulations for a varying number N of users in the cell. The figure shows that the results achieved analytically precisely match those obtained through the simulations. Similar trends have been observed for other values of the parameters. The results achieved clearly validate the proposed methodology and the analytical expressions presented in Section 3.1.

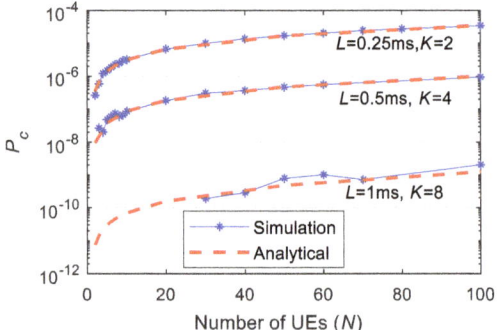

Figure 3. Comparison of analytical and simulation results for different latency requirements L and number of repetitions K ($U = 6$, $\lambda = 0.1$ packets).

It is important to highlight that this study focuses on URLLC applications that demand very high reliability levels. In simulations, we compute the number of packets for which the K replicas have collided with those packets transmitted by other UEs, and then compute the achieved reliability ($P_{rel} = 1 - P_c$). It is rare that all K replicas of a packet collide with transmissions from other UEs for low values of N. This is particularly the case when K increases. In this context, the computational cost of simulations significantly increases if we want to achieve accurate statistical results. This explains why simulation results are not shown for values of N below 30 when $K = 8$. It also highlights the value of our analytical expressions and methodology to estimate the performance of 5G NR grant-free scheduling for demanding URLLC applications and aperiodic traffic.

5. Performance Evaluation

This section evaluates the capacity of 5G NR grant-free scheduling with K-repetitions and shared resources to meet the reliability and latency requirements of URLLC services. To this aim, we use the analytical expressions that are derived in Section 3 and were validated in the previous section. Reliability for URLLC services is defined as the percentage P_{rel} of data packets that are successfully received by the BS before the latency deadline established by the service or application. In this study, we analyze first the reliability, considering only the effect of collisions from other UEs. This study analyzes then the impact of self-collisions on the capacity of 5G NR grant-free scheduling with K-repetitions and shared resources to achieve the reliability levels demanded by URLLC services. This is particularly relevant, as this study extends the state of the art by evaluating the capacity of 5G NR grant-free scheduling to sustain reliability levels even higher than $1–10^{-9}$. This study also evaluates the performance of 5G NR grant-free scheduling as a function of the number of UEs, the number of reserved radio resources, and the number K of replicas.

The performance of 5G NR grant-free scheduling is evaluated considering a single cell with N UEs. Packets are generated by each UE following a Poisson process with exponentially inter-arrival time. The average packet inter-arrival time is equal to $1/\lambda$, where λ is the average number of packets generated per second. UEs transmit small packets with a size of 32 bytes [22]. Radio resources are divided in 6×12 subcarriers (i.e., $U = 6$) with a subcarrier spacing of 120 kHz (i.e., $T_{slot} = 0.125$ ms). Figure 4 shows the probability P_c that a packet is not correctly received at the BS due to the collisions from other UEs experienced by all of the replicas of a packet (This would correspond to the reliability achieved with 5G NR grant-free scheduling if there were no self-collisions, i.e., $P_{sc} = 0$ and $P_{rel} = 1 - P_c$). The figure shows the value of P_c that can be achieved as a function of the number of UEs for latency requirements (L) of 0.25, 0.5, 0.75 and 1 ms. We focus on services with the most stringent latency requirements, given the challenge to satisfy high reliability levels when latency decreases [23]. For each value of L, the grant-free scheduling scheme is executed with the maximum possible number of replicas

K that can be transmitted within the required latency. For example, if the maximum latency L that can be tolerated is equal to 1 ms, the maximum number of replicas K that can be transmitted within 1 ms is equal to 8 ($L = 1$ ms corresponds to $8 \cdot T_{slot}$ when $T_{slot} = 0.125$ ms). Figure 4 also shows the performance achieved for two values of λ (0.1 and 1 packet(s)). The results depicted in Figure 4 clearly show that reducing the probability P_c of not receiving a packet to values as low as 10^{-9}, (and hence reaching reliability levels of 1–10^{-9} when the effect of self-collisions is not considered), can only be achieved with high values of K and values of L equal to 0.75 or 1 ms. Figure 4 also shows that the probability P_c increases with the number of UEs, since the risk of collision is higher. As a result, the capacity of 5G NR grant-free scheduling to support high reliability levels is significantly decreased as the number of UEs to be supported increases. Figure 4 also shows that the difficulty in supporting high reliability levels increases with λ, since the probability P_c increases as a result of a higher risk of collision between UEs.

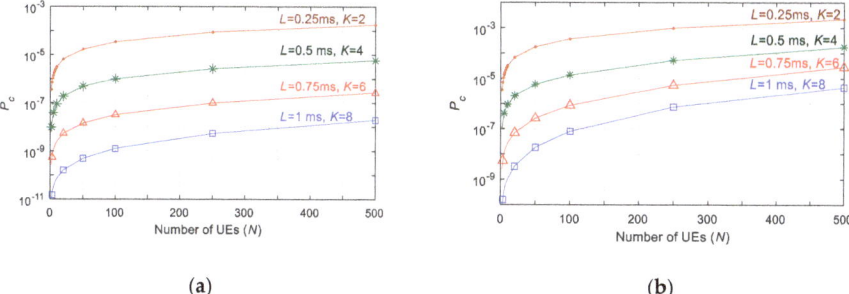

Figure 4. P_c as function of the number of User Equipments (UEs) and for different latency requirements L: (**a**) $\lambda = 0.1$ packets; (**b**) $\lambda = 1$ packet.

Figure 5 depicts the number of UEs that can be supported with a given latency requirement (L) and a reliability of $P_{rel} = 1 - P_c$ when $P_{sc} = 0$. It is important to remember that L establishes the maximum number of replicas K that can be transmitted. The results (the number of supported UEs) for each value of L in Figure 5 have been obtained for the maximum value of K permitted by L (K equal to 2, 4, 6 and 8 for L equal to 0.25, 0.5, 0.75 and 1 ms, respectively). The Release 15 of the 3GPP standards [22] establishes URLLC requirements with a latency of $L = 1$ ms and a reliability target of 1–10^{-5}. Figure 5 shows that grant-free scheduling with K-repetitions and shared resources can achieve a reliability equal to 1–10^{-5} with only $K = 2$ if we do not consider self-collisions. Grant-free scheduling with $K = 2$ can also guarantee a latency as low as 0.25 ms. For low values of the packet generation rate (i.e., $\lambda = 0.1$ packets), grant-free scheduling with 2 repetitions can support up to 34 UEs with a reliability of 1–10^{-5} and $L = 0.25$ ms if we do not consider self-collisions. The number of UEs that can be supported decreases with λ, since the risk of collision with other UEs increases when each UE transmits more packets per second. For example, only 4 UEs can be supported with $L = 0.25$ ms and a reliability of 1–10^{-5} when $\lambda = 1$ packet. If the latency requirement is relaxed to 0.5 ms or even higher, grant-free scheduling can support more than 500 UEs with only $K = 4$ when $\lambda = 0.1$ packets. If λ increases, grant-free scheduling can only guarantee the required reliability for 500 UEs if the latency requirement is 1 ms, and each UE can transmit 8 replicas of the same packet. These results show that the reliability and latency levels that can be achieved with grant-free scheduling depend upon configuration parameters (e.g., K), the traffic (e.g., λ) and the number of UEs supported. An adequate configuration and optimization of grant-free scheduling based on the network conditions could help support stringent reliability and latency levels. However, it is important to note that these results are achieved without considering self-collisions. The impact of self-collisions might be non-negligible when, for example, K and/or λ increase.

The Release 16 of 3GPP standards for 5G NR [2] defines use cases with higher reliability requirements (up to 1–10^{-6}). Some Industry 4.0 applications (e.g., factory automation) require even higher reliability levels (up to $1-10^{-9}$), as discussed in [3]. It is then important analyzing whether grant-free scheduling with K-repetitions and shared resources can guarantee reliability levels of the order of $1-10^{-9}$. Figures 4 and 5 show that grant-free scheduling can only guarantee very high reliability levels with high values of K, which limits the latency requirements (L) that can be satisfied. For example, a probability to correctly receive a packet equal to $1-10^{-7}$ cannot be guaranteed when $L < 0.5$ ms, even for the lower packet generation rates. If the reliability requirement increases to $P_{rel} = 1-10^{-9}$, grant-free scheduling can only support 5 UEs with $L = 0.75$ ms and $\lambda = 0.1$ packets. It can support 86 UEs if the latency requirement is relaxed to 1 ms. However, if λ increases to 1 packet then grant-free scheduling can only support 10 UEs with a reliability of $1-10^{-9}$ even if L is equal to 1 ms.

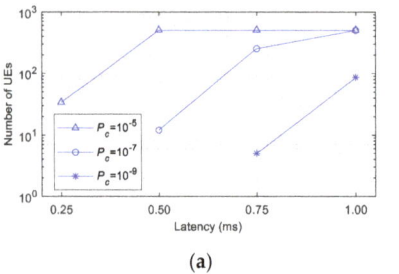
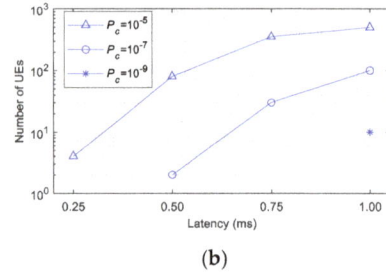

(a)   (b)

Figure 5. Number of UEs supported with different requirements (L and $P_{rel} = 1 - P_c$, when $P_{sc} = 0$): (a) $\lambda = 0.1$ packets; (b) $\lambda = 1$ packet.

Figure 6 shows the impact of U upon the performance of the grant-free scheduling scheme with K-repetitions and shared resources. U is the number of available RBs (Resource Blocks) per T_{slot}. In particular, Figure 6 depicts the number of UEs that can be supported with a given reliability and latency L when U decreases and λ is set equal to 0.1 packets (the reliability is equal to $P_{rel} = 1 - P_c$ when the effect of self-collisions is not taken into account, i.e., $P_{sc} = 0$). Figure 6 shows that the number of UEs that grant-free scheduling with K-repetitions can support for a given set of requirements strongly depends upon the number of RBs available. UEs randomly select an RB for each transmission from the U RBs available per slot. The probability that several UEs select the same RB for their transmissions increases when the number of RBs per slot decreases. Consequently, the probability P_c that a packet is not correctly received due to packet collisions, increases. In addition, the number of UEs that can achieve a target reliability level also decreases when the number of RBs per slot decreases. For example, 443 UEs can be supported with $L = 0.5$ ms (and hence $K = 4$) and $P_c = 10^{-5}$ when U is equal to 5 RBs. This number decreases to 69 UEs when U decreases to 3 RBs. This is a significant reduction of 84%. This reduction increases when the reliability demand increases. For example, 86 UEs can be supported with $P_c = 10^{-9}$ and $L = 1$ ms (and hence $K = 8$) when U is equal to 6. However, only 6 UEs can achieve these values of P_c and L if U decreases to 4 (i.e., a 93% reduction).

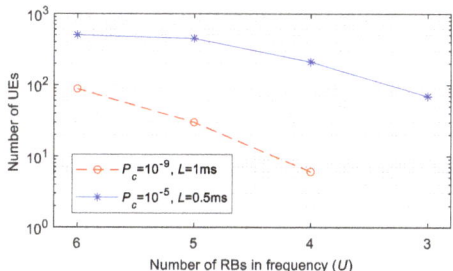

Figure 6. Number of UEs supported for a given L and $P_{rel} = 1 - P_c$ with $P_{sc} = 0$ as a function of the number U of available RBs per T_{slot} ($\lambda = 0.1$ packets).

All previous results have been derived without considering the effect of self-collisions. Self-collisions were illustrated in Figure 2, and the probability of self-collision was derived in Section 3.2. As previously described, if a packet p_2 is generated before the K replicas of the previous packet p_1 have been transmitted, p_2 will be stored and transmitted after completing the transmission of the K replicas of p_1. Due to the time that p_2 is stored, the transmission of its K replicas may finish after the latency deadline L. As presented in Section 3.2, it is not possible to complete the transmission of the K replicas of p_2 before the latency deadline L if p_2 is generated before $t_2 - L$ (t_2 is the time at which the transmission of the K replicas of p_2 is finished as shown in Figure 2). This results in that the probability P_{sc} (the probability that the transmission of K replicas of a packet is not completed before L due to the effect of self-collisions) is equal to the probability that the time between the generation of two consecutive packets at a UE falls within the interval $[0, \Delta t]$, where Δt represents the time difference between $t_2 - L$ and the time t_{p_1} at which p_1 is generated (see (25) and (26)).

We consider that packets are generated following a Poisson process with exponential inter-arrival time. As a result, Δt is homogeneously distributed between Δt_1 and Δt_2. For $K = 4$ in Figure 2, Δt_1 is equal to 0 and Δt_2 is equal to T_{slot}, since p_1 can be homogeneously generated between t_0 and $t_0 - T_{slot}$. When $K = 6$, Δt_1 is equal to $4 \cdot T_{slot}$, and Δt_2 is equal to $(4+1) \cdot T_{slot}$, since p_1 can be homogeneously generated between t_0 and $t_0 - T_{slot}$. Similarly, Δt_1 and Δt_2 are equal to $8 \cdot T_{slot}$ and $(8+1) \cdot T_{slot}$ for $K = 8$. Table 1 shows the value of P_{sc} given in (26) when Δt is equal to Δt_1 or Δt_2 considering $L = 1$ ms and $K = 4$, 6 and 8. $\Delta t = \Delta t_1$ corresponds to the scenario where self-collisions are less probable, while $\Delta t = \Delta t_2$ corresponds to the case in which they are more probable.

The results in Table 1 show that the probability of self-collision is non-negligible. For example, P_{sc} can reach values equal to 1.25×10^{-4} and 9.99×10^{-4} when K is equal to 4 and 8, respectively, and $\lambda = 1$ packet. It is also important to highlight that a comparison of results in Figure 4 and Table 1 shows that P_{sc} can be actually higher than P_c. This is for example the case when $K = 8$: P_c is lower than 10^{-7} and 10^{-5} for λ equal to 0.1 and 1 packet(s), respectively (Figure 4), while P_{sc} is approximately equal to 10^{-4} and 10^{-3} (Table 1). Grant-free scheduling can hence be limited by the effect of self-collisions, in particular when K increases. It is then important that the reliability (or probability that a packet is correctly received before the latency deadline) of grant-free scheduling with K-repetitions and shared radio resources is computed considering both the effect of collisions from other UEs and the effect of self-collisions following (1).

Table 1. P_{sc} for $L = 1$ ms.

K	$\Lambda = 0.1$ Packets		$\Lambda = 1$ Packet	
	$\Delta t = \Delta t_1$	$\Delta t = \Delta t_2$	$\Delta t = \Delta t_1$	$\Delta t = \Delta t_2$
4	0	1.25×10^{-5}	0	1.25×10^{-4}
6	5.00×10^{-5}	6.25×10^{-5}	5.00×10^{-4}	6.25×10^{-4}
8	9.99×10^{-5}	1.13×10^{-4}	9.99×10^{-4}	1.13×10^{-3}

Figure 7 plots $1 - P_{rel}$ for different values of K and L when considering both P_c and P_{sc}. The results are plotted considering $\Delta t = \Delta t_1$ for computing P_{sc}. $\Delta t = \Delta t_1$ corresponds to the case where self-collisions are less probable. Figure 4 shows that it is necessary to transmit a high number of replicas K within L to combat collisions from other UEs and correctly receive a packet at the BS. For example, Figure 4 shows that K must be equal to 8 in order to achieve $P_{rel} = 1-10^{-9}$ when $P_{sc} = 0$ and λ is equal to 1 packet. However, Table 1 showed that the effect of self-collisions increases with K even to the point that self-collisions limit the reliability that can be achieved. This is actually shown in Figure 7 when we consider $L = 1$ ms. In principle, it could be possible to satisfy a 1 ms latency requirement if we transmit 4, 6 or 8 replicas of a packet. Figure 7 shows that if $K = 4$ and $\Delta t = \Delta t_1$ (for computing P_{sc} in (26)), the impact of self-collisions is not relevant, and the reliability levels of $1-10^{-5}$ can be satisfied for more than 500 UEs and 80 UEs when λ is equal to 0.1 and 1 packet(s), respectively; these results are in line with those observed in Figure 4 for $K = 4$. However, when K is equal to 6 or 8, the effect of self-collisions becomes more relevant (Table 1), and Figure 7 shows that it can actually limit the maximum reliability that can be achieved independently of the number of UEs. In fact, the maximum reliability that can be achieved is approximately equal to $1 - P_{sc}$. In this case, for $K = 8$ and $\lambda = 1$ packet/s, the maximum reliability (when P_{sc} is computed considering $\Delta t = \Delta t_1$) that can be achieved is $1-10^{-3}$ when the latency requirement L is equal to 1 ms. It should be noted that reliability levels even higher than $1 - P_c = 1-10^{-9}$ were achieved when the effect of self-collisions was not considered (Figure 4). The results discussed so far correspond to the scenario where P_{sc} has been computed considering $\Delta t = \Delta t_1$. This corresponds to the scenario where self-collisions are less probable. Figure 7 also shows the reliability that can be achieved with $L = 1$ ms and $K = 4$ when $\Delta t = \Delta t_{avg}$. This Δt_{avg} is the average value of Δt. $\Delta t_{avg} = (\Delta t_1 + \Delta t_2)/2$, since Δt is homogeneously distributed between Δt_1 and Δt_2. Figure 7 shows that in this case it is not possible to achieve a reliability higher than $1-6.3 \times 10^{-5}$ and $1-6.3 \times 10^{-4}$ when λ is equal to 0.1 and 1 packet(s). Figure 7 also shows that the reliability becomes again nearly independent of the number of UEs that are being supported. The degradation of reliability experienced from $\Delta t = \Delta t_1$ to $\Delta t = \Delta t_{avg}$ is again due to a major relevance of the effect of self-collisions when we compute the reliability.

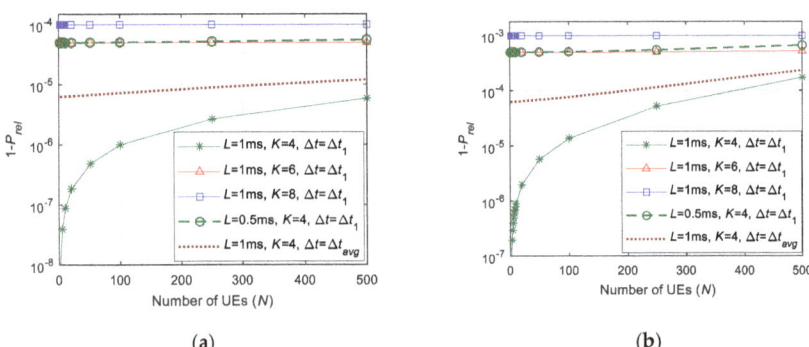

Figure 7. Reliability for different latency requirements L and number of repetitions K ($U = 6$): (a) $\lambda = 0.1$ packets; (b) $\lambda = 1$ packet.

Expressions in (25) and (26) show that P_{sc} also depends upon the latency requirement L. The effect of self-collisions is more relevant when the latency requirement is stricter. For example, Figure 7 shows that the effect of self-collisions already limits the maximum reliability that can be achieved when $K = 4$ if the latency requirement is equal to 0.5 ms. Latency requirements significantly influence the reliability levels that can be satisfied. This is the case because latency requirements limit the number K of replicas that can be sent for each packet. Figure 4 shows that the maximum reliability level that can be guaranteed depends on the latency requirements when only considering P_c. Figure 7 also shows that the effect of self-collisions becomes more relevant with stricter latency requirements. These results show that it is a challenge guaranteeing high reliability demands with very low latency levels.

The results in Figure 7 demonstrate that current 5G NR grant-free scheduling with K-repetitions and shared resources cannot guarantee some of the more demanding reliability and latency levels. However, it is important emphasizing that other proposals cannot meet such requirements either, and these actually perform worse than the implementation analyzed in this study. This is actually the case for the proposals that transmit the first copy of a packet in dedicated resources for the UEs. These resources can be reserved using grant-based scheduling (such as in [14]) or semi-persistent scheduling (such as in [13]). Grant-based scheduling requires the UE to send an SR to the BS, and wait for the BS to reply with a grant message. The exchange of these messages between the UE and the BS is illustrated in Figure 8 This handshaking generates a non-negligible T_{total} latency that is equal to:

$$T_{total} = 2\,T_{L1/L2} + T_{align} + 2\,T_{proc} + 3\,T_{tx} = 2.3 \text{ ms} \qquad (27)$$

where $T_{L1/L2}$ is the L1/L2 processing latency at the BS and the UE, T_{align} is the alignment latency (the alignment latency is the time elapsed from the moment the UE is ready to transmit to the actual time the transmission starts), T_{proc} is the processing latency (this latency represents the latency between the reception of the SR and the transmission of the grant message), and T_{tx} is the time required to transmit the SR and grant messages. Following [24], we consider $T_{L1/L2} = T_{align} = T_{tx} = 1$ TTI, and $T_{proc} = 2.33$ TTI. These values are a best-case scenario, since they represent reduced processing times that can be achieved with 3GPP Release 15 compared to Release 14. Equation (27) shows that the total latency (2.3 ms) introduced by the grant-based scheduling process to assign dedicated resources to UEs is higher than the latency achieved with the 5G NR grant-free scheduling implementation analyzed in this study. For example, Figure 7 shows that this implementation can guarantee latency levels below 1 ms (this latency is guaranteed with a reliability up to $1-10^{-5}$ when $K = 4$, $\lambda = 0.1$ packets, $U = 6$, and $\Delta t = \Delta t_{avg}$).

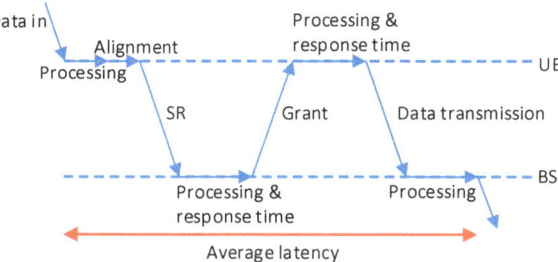

Figure 8. Latency introduced in grant-based scheduling.

The alternative to grant-based scheduling is Semi-Persistent Scheduling (SPS). In this case, UEs are assigned dedicated resources for a period of time. During this period, UEs can utilize the resources without requesting permission from the BS. This avoids the latency introduced by grant-based scheduling. However, semi-persistent scheduling inefficiently utilizes the radio resources when the traffic is aperiodic. This is the case, because it is not possible to predict when UEs will need resources. To illustrate this effect, let us consider a scenario with $N = 300$ users that generate aperiodic traffic ($\lambda = 0.1$ packets). We shall then suppose that users request a maximum latency of 1 ms and a reliability level equal to $1-10^{-5}$. Satisfying this demand requires reserving 300 RBs (one per UE) in a 1 ms time windowA lower number of resources would be necessary if traffic was periodic and we could estimate when each UE would require resources for their transmission. In this case, several UEs could share the same RB if they generate their packets at different time instants. This would reduce the total number of RBs necessary to serve all users. This is not possible in the case of aperiodic traffic, since we cannot predict when a UE would need radio resources. Figure 7 shows that our implementation of 5G NR grant-free scheduling with 4-repetitions and shared resources can support 300 UEs (with their latency and reliability demands) with only 48 RBs in a time window of 1 ms. This is 84% less radio resources than if we reserve dedicated resources per UE (with aperiodic traffic) for their first transmission using semi-persistent scheduling. These results clearly show that the implemented 5G NR grant-free scheduling with shared resources can better support URLLC applications with aperiodic traffic and stringent communication requirements than other existing proposals. However, the conducted analysis (e.g., Figure 7) has also shown that new solutions will be needed to guarantee very demanding reliability and latency levels such as those foreseen for some URLLC services in 3GPP Release 16.

6. Conclusions

This paper has analyzed the capacity of 5G NR grant-free scheduling to support URLLC services with strict reliability and latency levels such as those demanded by Industry 4.0. The study has focused on aperiodic or sporadic traffic and an implementation of 5G NR grant-free scheduling with K-repetitions and shared radio resources. This implementation has been chosen, since sharing radio resources is an attractive option for aperiodic traffic. In addition, the K-repetitions scheme can combat possible packet collisions between UEs that share radio resources. This study has analyzed the reliability and latency levels that can be achieved with existing 5G NR grant-free scheduling with shared radio resources as a function of the number of UEs, the number of reserved radio resources, and the number of replicas K. To this aim, this study has derived analytical expressions that quantify the exact probability of collision with packets transmitted by other UEs, and the impact of self-collisions. It is important to emphasize that this study is the first one that has evaluated the impact of self-collisions. Packet collisions and self-collisions have then been taken into account to derive analytically the reliability that can be achieved by existing 5G NR grant-free scheduling with shared resources. The derived analytical expressions have been validated against simulations. These expressions are a valuable contribution to the community, since they can be easily utilized to evaluate 5G NR grant-free scheduling.

This study has demonstrated that current 5G NR grant-free scheduling solutions cannot guarantee high reliability levels with strong latency requirements. This is partly due to the fact that strong latency requirements limit the number of replicas K that can be transmitted. In addition, self-collisions have a non-negligible impact that even limits the reliability that can be achieved when K increases. The impact of self-collisions also increases with the latency requirements. The obtained results demonstrate that new solutions are necessary for 5G NR grant-free scheduling to be able to support applications with stringent URLLC latency and reliability requirements under the presence of aperiodic traffic. In particular, the transmission of K replicas per packet might be inadequate to support aperiodic traffic with very low latency levels due to the impact of self-collisions. Consequently, other approaches should be designed to minimize collisions between UEs sharing radio resources. This study has shown that these new solutions cannot be based either on grant-based or semi-persistent scheduling. Grant-based scheduling introduces additional latency due to the exchange of messages between the UEs and the BS for assigning the radio resources. Semi-persistent scheduling with dedicated resources inefficiently utilizes the available resources when considering dedicated resources and aperiodic traffic. Innovative grant-free scheduling solutions are hence necessary to meet the URLLC requirements identified for 3GPP Release 16 and beyond. This could include, for example, the use of sensing mechanisms or full duplex techniques that can reduce packet collisions.

Author Contributions: Conceptualization, M.C.L.-E. and J.G.; methodology, M.C.L.-E. and J.G.; validation, M.C.L.-E. and J.G.; formal analysis, M.C.L.-E., J.G. and M.S.; investigation, M.C.L.-E.; writing—original draft preparation, M.C.L.-E.; writing—review and editing, J.G. and M.S.; funding acquisition, J.G. and M.S.

Funding: This work has been funded by the European Commission through the FoF-RIA Project AUTOWARE: Wireless Autonomous, Reliable and Resilient Production Operation Architecture for Cognitive Manufacturing

(No. 723909), and the Spanish Ministry of Economy, Industry, and Competitiveness, AEI, and FEDER funds (TEC2017-88612-R).

Conflicts of Interest: The authors declare no conflict of interest.

References

1. 3GPP. *Technical Specification Group Services and System Aspects*; Study on Communication for Automation in Vertical Domains; 3GPP: Sophia Antipolis, France, 2018; Release 16, 3GPP TR 22.804 V16.2.0.
2. 3GPP. *Technical Specification Group Radio Access Network*; Study on Physical Layer Enhancements for NR Ultra-Reliable and Low Latency Case (URLLC); 3GPP: Sophia Antipolis, France, 2018; Release 16, 3GPP TR 38.824 V1.0.0.
3. Klessig, H.; Ashraf, S.A.; Almeroth, B.; Riedel, I.; Puschmann, A.; Elste, T.; Simsek, M.; Schulz, P.; Matthe, M.; Fettweis, G.; et al. Latency Critical IoT Applications in 5G: Perspective on the Design of Radio Interface and Network Architecture. *IEEE Commun. Mag.* **2017**, *55*, 70–78.
4. 5G Alliance for Connected Industries and Automation (5G-ACIA). *5G for Connected Industries and Automation*, 2nd ed.; 5G Alliance for Connected Industries and Automation (5G-ACIA): Frankfurt, Germany, 2019.
5. Popovski, P.; Nielsen, J.J.; Stefanovic, C.; De Carvalho, E.; Ström, E.; Trillingsgaard, K.F.; Bana, A.-S.; Kim, D.M.; Kotaba, R.; Park, J.; et al. Wireless Access for Ultra-Reliable Low-Latency Communication: Principles and Building Blocks. *IEEE Netw.* **2018**, *32*, 16–23. [CrossRef]
6. 3GPP. *Technical Specification Group Radio Access Network*; Evolved Universal Terrestrial Radio Access (E-UTRA); Study on Latency Reduction Techniques for LTE; 3GPP: Sophia Antipolis, France, 2016; 3GPP TR 36.881 V14.0.0.
7. 3GPP. *Technical Specification Group Radio Access Network*; NR; Medium Access Control (MAC) Protocol Specification; 3GPP: Sophia Antipolis, France, 2018; Release 15, 3GPP TS 38.321 V15.4.0.
8. Berardinelli, G.; Mahmood, N.H.; Abreu, R.; Jacobsen, T.; Pedersen, K.; Kovacs, I.Z.; Mogensen, P. Reliability Analysis of Uplink Grant-Free Transmission Over Shared Resources. *IEEE Access* **2018**, *6*, 23602–23611. [CrossRef]
9. 3GPP. *Discussion on Configured Grant for NR-U*; 3GPP: Sophia Antipolis, France, 2018; 3GPP TSG-RAN, R1-1810329.
10. Li, Z.; Uusitalo, M.A.; Shariatmadari, H.; Singh, B. 5G URLLC: Design Challenges and System Concepts. In Proceedings of the 2018 15th International Symposium on Wireless Communication Systems (ISWCS), Lisbon, Portugal, 28–31 August 2018; pp. 1–6.
11. 3GPP. *Technical Specification Group Radio Access Network*; NR; Physical Layer Procedures for Data; 3GPP: Sophia Antipolis, France, 2018; Release 15, 3GPP TS 38.214 V15.4.0.
12. Wu, Y.; Zhang, L.; Wang, C.; Chen, Y. Performance Evaluation of Grant-Free Transmission for Uplink URLLC Services. In Proceedings of the 2017 IEEE 85th Vehicular Technology Conference (VTC Spring), Sydney, Australia, 4–7 June 2017; pp. 1–6.
13. Kotaba, R.; Manchon, C.N.; Balercia, T.; Popovski, P. Uplink Transmissions in URLLC Systems with Shared Diversity Resources. *IEEE Wirel. Commun. Lett.* **2018**, *7*, 590–593. [CrossRef]
14. Abreu, R.; Mogensen, P.; Pedersen, K.I. Pre-Scheduled Resources for Retransmissions in Ultra-Reliable and Low Latency Communications. In Proceedings of the 2017 IEEE Wireless Communications and Networking Conference (WCNC), San Francisco, CA, USA, 19–22 March 2017; pp. 1–5.
15. Singh, B.; Tirkkonen, O.; Li, Z.; Uusitalo, M.A. Contention-Based Access for Ultra-Reliable Low Latency Uplink Transmissions. *IEEE Wirel. Commun. Lett.* **2018**, *7*, 182–185. [CrossRef]
16. Jacobsen, T.; Abreu, R.; Berardinelli, G.; Pedersen, K.; Mogensen, P.; Kovacs, I.Z.; Madsen, T.K. System Level Analysis of Uplink Grant-Free Transmission for URLLC. In Proceedings of the 2017 IEEE Globecom Workshops (GC Wkshps), Singapore, 4–8 December 2017; pp. 1–6.
17. 3GPP. *Grant-Free Transmission for UL URLLC*; 3GPP: Sophia Antipolis, France, 2017; 3GPP TSG-RAN, R1-1706919.
18. 3GPP. *Enhancement of Uplink Grant-free transmission for NR URLLC*; 3GPP: Sophia Antipolis, France, 2018; 3GPP TSG-RAN, R1-1810176.
19. 3GPP. *Enhancement of Configured Grant for NR URLLC*; 3GPP: Sophia Antipolis, France, 2018; 3GPP TSG-RAN, R1-1812162.

20. 3GPP. *Technical Specification Group Radio Access Network*; NR; Service Requirements for the 5G System; 3GPP: Sophia Antipolis, France, 2018; Stage 1, Release 15, 3GPP TR 22.261 V15.7.0.
21. 3GPP. *Technical Specification Group Radio Access Network*; NR; Physical Channels and Modulation; 3GPP: Sophia Antipolis, France, 2018; Release 15, 3GPP TR 38.211 V15.4.0.
22. 3GPP. *Technical Specification Group Radio Access Network*; Study on Scenarios and Requirements for Next Generation Access Technologies; 3GPP: Sophia Antipolis, France, 2018; Release 15, 3GPP TR 38.913 V15.0.0.
23. Bennis, M.; Debbah, M.; Poor, H.V. Ultrareliable and Low-Latency Wireless Communication: Tail, Risk, and Scale. *Proc. IEEE* **2018**, *106*, 1834–1853. [CrossRef]
24. 3GPP. *Evaluation of Latency in LTE*; 3GPP: Sophia Antipolis, France, 2017; 3GPP TSG-RAN, R1-1720535.

© 2019 by the authors. Licensee MDPI, Basel, Switzerland. This article is an open access article distributed under the terms and conditions of the Creative Commons Attribution (CC BY) license (http://creativecommons.org/licenses/by/4.0/).

Article

Noninvasive Suspicious Liquid Detection Using Wireless Signals

Jiewen Deng [1], Wanrong Sun [1], Lei Guan [2], Nan Zhao [1], Muhammad Bilal Khan [1], Aifeng Ren [1], Jianxun Zhao [1], Xiaodong Yang [1,*] and Qammer H. Abbasi [3]

1. School of Electronic Engineering, Xidian University, Xi'an 710071, China; djw15529256085@163.com (J.D.); sunwanrong@xidian.edu.cn (W.S.); nan_zhao_@hotmail.com (N.Z.); engrmbkhan1986@gmail.com (M.B.K.); afren@mail.xidian.edu.cn (A.R.); jxzhao@xidian.edu.cn (J.Z.)
2. School of Life Sciences and Technology, Xidian University, Xi'an 710126, China; 15926395470@163.com
3. School of Engineering, University of Glasgow, Glasgow G12 8QQ, UK; Qammer.Abbasi@glasgow.ac.uk
* Correspondence: xdyang@xidian.edu.cn

Received: 20 July 2019; Accepted: 15 September 2019; Published: 21 September 2019

Abstract: Conventional liquid detection instruments are very expensive and not conducive to large-scale deployment. In this work, we propose a method for detecting and identifying suspicious liquids based on the dielectric constant by utilizing the radio signals at a 5G frequency band. There are three major experiments: first, we use wireless channel information (WCI) to distinguish between suspicious and nonsuspicious liquids; then we identify the type of suspicious liquids; and finally, we distinguish the different concentrations of alcohol. The K-Nearest Neighbor (KNN) algorithm is used to classify the amplitude information extracted from the WCI matrix to detect and identify liquids, which is suitable for multimodal problems and easy to implement without training. The experimental result analysis showed that our method could detect more than 98% of the suspicious liquids, identify more than 97% of the suspicious liquid types, and distinguish up to 94% of the different concentrations of alcohol.

Keywords: 5G; liquid detection; radio propagation; dielectric constant; WCI

---

1. Introduction

The illegal carrying and transportation of flammable and explosive liquids seriously affects public safety. Flammable and explosive liquids such as gasoline and alcohol are also commonly used in various terrorist activities. Therefore, the safety inspection of flammable and explosive liquids is of great significance for ensuring public safety. It has broad application prospects in the fields of public security, civil aviation, and customs. Liquids cannot be identified by the naked eye, and it is difficult to carry out dangerous liquid inspection in crowded places, which is a great challenge for security personnel. Moreover, there are still some places where the "taste liquid" method is used to determine whether the liquid is safe [1]. Forbid passengers to carry large amounts of liquid has become the main method to prevent terrorist attacks. For example, international civil aviation regulations prohibit carrying more than 100 mL of liquid, and trains and high-speed trains prohibit carrying flammable liquids and more than 120 mL of compressed spray. The traditional detection of suspicious items is either manual inspection (such as setting checkpoints at each entrance) or special equipment (such as surveillance cameras, X-ray machines, and ultra-wideband scanners), which is costly, expensive to deploy, and difficult to implement on a large scale [2]. It is necessary to introduce a new suspicious liquid detection scheme which is more economical and covers a wider range.

There are several mature liquid detection technologies. A traditional approach, the Raman spectrum analysis, uses the molecular structure to identify liquids according to their scattering spectroscopy. Raman spectroscopy has many unique advantages, such as wide detection range, sharp spectral peaks, and high resolution [3]. That is, when monochromatic light radiates on an object, the molecules of the substance will scatter, and the spectrum reflected by different substances will be different. Moreover, the method, based on X-ray image technology, is used to identify liquid substances, and can obtain the atomic number of the liquid through its X-ray [4]. However, this method has a certain error rate and the equipment is very expensive. Another method is to detect liquids based on the different absorption and attenuation characteristics of different substances via electromagnetic waves [5,6]. Microwave detection can achieve noncontact detection with a certain distance between the detected objects, and has high detection sensitivity. However, these approaches rely on expensive and specialized equipment, which does not facilitate wide deployment in practice.

Recently, Radio Frequency (RF) based sensing has drawn considerable attention. A couple of studies have explored the feasibility of using RF signals for remote monitoring and controlling during infusion. For instance, a wireless intelligent monitoring system, based on wireless communication and network technology, is put forward, which can monitor drip speed in real time and automatically alarm in abnormal conditions [7]. The ZigBee wireless sensor is used to detect the velocity of liquid droplets [8]. While these approaches mainly focus on exploiting the differences of wireless signal measurements to sense physical morphological changes of liquids, using fine-grained wireless channel information (WCI) to identify types of liquid remains an option.

5G communication technology is a data and information transmission technology developed by technicians based on 4G technology. Its advantages include sound transmission performance, fast transmission speed, high utilization rate of resources, and wide range of coverage. As such, it is favored in modern data and information transmission [9]. Based on ubiquitous wireless signals, the wireless sensing system will provide a variety of high precision, high reliability, high security, and convenient application services, among which the human behavior identification technology is at the core for public use [10]. As the information of the physical layer, the fine-grained WCI contains a lot of channel information which is invisible to the Medium Access Control (MAC) layer. The WCI can measure the frequency response of multiple subcarriers at the same time from a single packet, rather than the overall amplitude response superimposed by all subcarriers. Information about the WCI and frequency-selective channels is described [11].

In the field of wireless sensing, WCI has gradually become a popular area of research [12], such as indoor positioning [13–15], respiration detection [16–18], and behavior recognition [19,20]. Zhou et al. proposed using the WCI to detect the presence of people in the environment [21]. Although wireless signals have good applications in indoor positioning and fall detection, there are still relatively few studies on detecting the types and concentrations of liquids by using them. Liquid interferes with the path of radio signals. Different liquids have different degrees of interference to the radio signal propagation path due to their differences in dielectric constants, which leads to different changes in the WCI. These changes can be effectively observed at the signal receiver, so as to realize the detection and identification of various liquids. Table 1 shows the relative dielectric constants of common objects.

Table 1. The relative dielectric constants of common objects.

Object	Dielectric Constant
Water	80
Alcohol	24
Oil	2
Glycerol	37
Methanol	32
Sulfuric Acid	84

Following this introduction, the paper is organized into four sections. Section 2 describes the preparatory work. Section 3 details the band selection and method design. Results are analyzed in Section 4 and conclusions are drawn in Section 5.

2. Preparatory Work

2.1. Data Acquisition

The WCI represents the channel state of a communication link on Orthogonal Frequency Division Multiplexing (OFDM) technology. The WCI describes how links are transmitted from a transmitter to a receiver. It also combines the influence of scattering, fading, power attenuation, and other factors. The WCI reflects the performance of a link and the interference caused by other factors to a great extent [22]. The WCI is composed of an OFDM matrix containing 30 subcarriers. Data can be transmitted simultaneously on multiple subcarriers, greatly improving the efficiency and accuracy of the system [23]. The frequency domain of the wireless channel can be expressed as:

$$\vec{Y} = H \cdot \vec{X} + \vec{N}, \tag{1}$$

where $\vec{X}$ and $\vec{Y}$ are the transmitted and received vectors while $\vec{N}$ is Gaussian noise (AWGN) vector and H represents the frequency response of the channel.

The receivers receive packets for each subcarrier from a channel. The packet carrying the original amplitude and random phase information complex frequency domain can be expressed as:

$$H(n) = |H(n)| e^{\angle H(n)}. \tag{2}$$

In Equation (2), $H(n)$ is the data for subcarrier number n, where $n \in [1 \text{ to } 30]$. $|H(n)|$ is the raw amplitude information and $\angle H(n)$ denotes the random phase data.

2.2. Data Preprocessing

Our system uses a Hampel filter to eliminate the singular value in the data and construct a scale sequence with the median. Assuming that the median of the sequence is Z,

$$\{d(k)\} = \{|x_0(k) - Z|, \ldots, |x_{m-1}(k) - Z|\}. \tag{3}$$

The deviation scale of each data from the reference value is given. Suppose the median of $\{d(k)\}$ is D. The median has an absolute deviation of:

$$\text{MAD} = 1.4826 \times D. \tag{4}$$

MAD can replace the standard deviation σ. The Hampel filter uses m data in a mobile window to determine the validity of current data. If the data is valid, process it; otherwise, replace it with the median. The Hampel filter can protect the detailed information while filtering the singular value.

2.3. Classification Method

We used the K-Nearest Neighbor (KNN) algorithm to classify the amplitude information extracted from the WCI matrix to detect liquids. The KNN algorithm mainly relies on the surrounding adjacent samples to determine the category. If the k closest neighbors of a sample belong to a certain category in the feature space, the sample also belongs to the same category. In the KNN algorithm, the selected neighbors are all objects that have been correctly classified. Therefore, the KNN method is more suitable than other algorithms for the sample sets to be divided with a lot of crossover or overlap of the class domain.

The KNN algorithm takes the distance between objects as the nonsimilarity index to avoid the matching problem between objects. Using different distance calculation methods, there may be significant differences between the "neighbors" identified. We used the Euclidean distance method:

$$d(x, y) = \sqrt{\sum_{k=1}^{n}(x^k - y^k)^2}, \tag{5}$$

In the KNN algorithm, the choice of the K value will have a significant impact on the classification results. Generally, we can take a relatively small value of K, and cross validation is used to select the best value of K. Usually, K is an integer less than 20.

3. Method Design

3.1. Band Selection

The C-band is a frequency band from 4.0 to 8.0 GHz, which is used as the frequency band for downlink transmission of communication satellite signals. In the application of satellite television broadcasting and various small satellite ground stations, the band was first adopted and has been widely used. The Ministry of Industry and Information Technology (MITT) has announced the frequency band division of China's fifth generation of mobile communications in its latest official document [24]. The 4.8–5 GHz frequency band in China's 5G is located in the C-band, which is an important part of 5G communication in China.

The S-band refers to the electromagnetic wave band with a frequency range of 2–4 GHz, which is mainly used in relaying, satellite communication, radar, and so on. Now widely used in Bluetooth, ZigBee, wireless routing, and wireless mouse devices also use S-band electromagnetic waves.

Many researches have indicated that the longer the wavelength of the electromagnetic wave, the stronger its ability to diffract. For example, radio waves can be transmitted around tall buildings, and red light can travel far in fog to remind drivers, which is more effective than green light and yellow light. The shorter the wavelength of the electromagnetic wave, the greater the energy of the wave, and the stronger the penetrating capacity. As such, X-ray can penetrate through skin and bones, ultraviolet rays can kill bacteria, and strong ultraviolet rays can cause skin cancer.

Therefore, we selected electromagnetic waves of two frequency bands to study the influence of electromagnetic waves of different bands on liquid detection. We chose a 2.4 GHz signal located in the S-Band and a 4.8 GHz signal located in the C-band (5G frequency band) for comparison.

3.2. Method Design

The experimental scenario used to facilitate the detection and identification of suspicious liquids is shown in Figure 1. Our experiment was conducted in a conference room and used two sets of equipment placed on a desk to collect data. The transmitter and receiver were one meter apart. The liquid to be detected was placed statically between the transmitter and receiver at the same height. One set of equipment worked at 2.4 GHz, and the transmission and reception of signals were completed by the wireless network adapter of the computer and three omnidirectional antennas. The bandwidth was 20 MHz, and the output power of the transmitter was set at −5 dBm. The other worked at 4.8 GHz, which is consistent with the 5G standard in China, with the RF signal generator as the transmitter and spectrum analyzer as the receiver. The bandwidth was 100 MHz, and the output power of the transmitter was set at −5 dBm. The receiver collected the RF signal at the frequency of 4.8 GHz and 2.4 GHz, corresponding to the C-band (5G) and the S-band, respectively.

To facilitate the liquid detection and identification leveraging Wi-Fi signal, we exploited wireless channel information (WCI), the fine-grained description of the wireless channel, to capture the minute differences of the channel state change introduced by different liquids. Our method employed the amplitude information of WCI. First, we extracted the WSI from a pair of transmitters and receivers. Second, we preprocessed the data to remove the environmental noise and eliminate the singular value. Then the data were feature selected by a Principal Components Analysis (PCA) algorithm and classified by a K-Nearest Neighbor (KNN) algorithm. The flow chart of the method is shown in Figure 2.

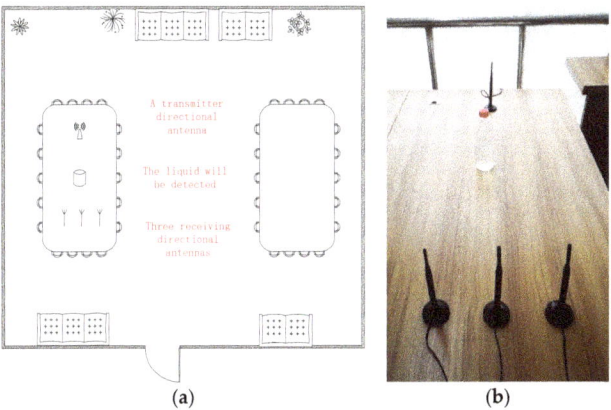

Figure 1. (a) Experimental scenario; (b) The actual scene.

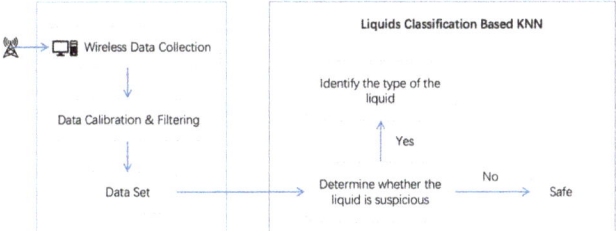

Figure 2. Method flow chart. KNN, K-Nearest Neighbor.

Using this method, we carried out three experiments. Experiment 1 was to distinguish between suspicious and nonsuspicious liquids, Experiment 2 was to identify the type of suspicious liquids, and Experiment 3 was to distinguish the three different concentrations of alcohol.

4. Evaluation and Analyses

In this work, containers made of three common materials were selected. Figure 3a shows our selection of containers, from left to right a paper cup, a plastic bottle, and a glass bottle. We used 50% alcohol, 75% alcohol, 95% alcohol, oil, and a compressed spray as the representative of suspicious liquid. We chose water as the representative of nonsuspicious liquid. Figure 3b shows the suspicious liquids selected for the experiment.

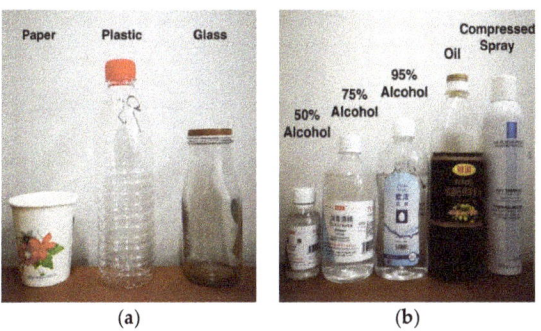

Figure 3. (a) The selection of containers; (b) Suspicious liquids selected from the experiment.

183

4.1. Detection of Suspicious and Nonsuspicious Liquids

In this section, we first analyze the detection of suspicious and nonsuspicious liquids.

Figure 4 shows the raw WCI amplitude information on 30 subcarriers when using C-band electromagnetic wave signals. We can see the difference of amplitude change between suspicious liquids and nonsuspicious liquid intuitively.

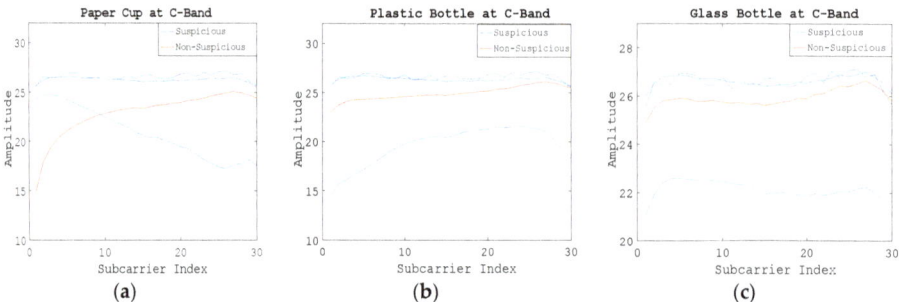

Figure 4. Amplitude information of 30 subcarriers of Step 1 at the C-band. (**a**) Using the paper cup; (**b**) Using the plastic bottle; (**c**) Using the glass bottle.

The conclusion of the S-band was the same, as shown in Figure 5.

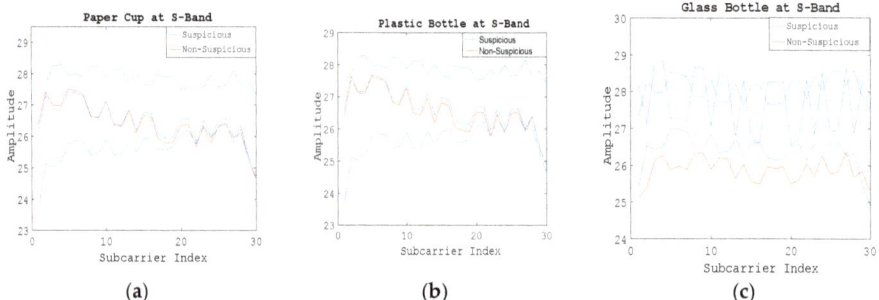

Figure 5. Amplitude information of 30 subcarriers of Step 1 at the S-band. (**a**) Using the paper cup; (**b**) Using the plastic bottle; (**c**) Using the glass bottle.

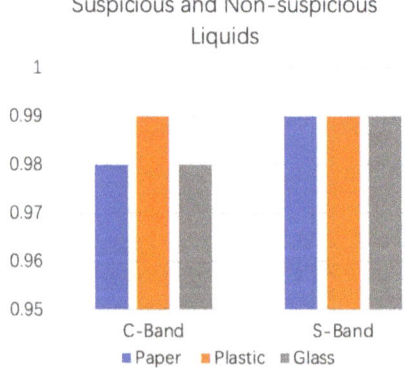

Figure 6. The KNN algorithm Classification results of Step 1 at the C-band and the S-band.

The KNN classification algorithm was used to further detect suspicious and nonsuspicious liquids. The classification results of the KNN algorithm for two bands of data are shown in Figure 6. Blue, orange, and grey represent the paper cup, the plastic bottle, and the glass bottle, respectively. In the C-band environment, the detection accuracy of Step 1 was over 98%, and that for S-band was 99%.

4.2. Identification of Suspicious Liquids

In this section, we will analyze the identification of suspected liquids.

Figure 7 shows the WCI amplitude information of 30 subcarriers. Blue, orange, and green represent alcohol, oil, and compressed spray, respectively. It can be seen that in the C-band environment, the WCI amplitude ranges of alcohol and oil were not very different, but the amplitude fluctuation trends were obviously different, and the amplitude ranges of these two categories were greatly different from those of oil.

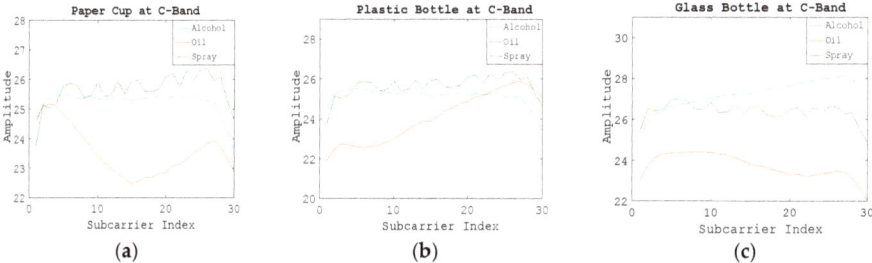

Figure 7. Amplitude information of 30 subcarriers of Step 2 at the C-band. (**a**) Using the paper cup; (**b**) Using the plastic bottle; (**c**) Using the glass bottle.

In the S-band environment, the WCI amplitude ranges of the three kinds of suspected liquids differed greatly, as shown in Figure 8.

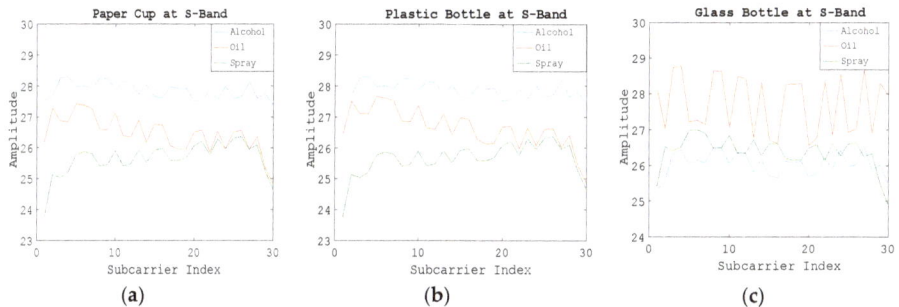

Figure 8. Amplitude information of 30 subcarriers of Step 2 at the S-band. (**a**) Using the paper cup; (**b**) Using the plastic bottle; (**c**) Using the glass bottle.

Figure 9 is the result of the KNN classification algorithm in identifying the types of suspicious liquids. In the C-band environment, the system can achieve more than 97% accuracy in identifying types of suspicious liquid (Step 2), and that for the S-band environment is 99%.

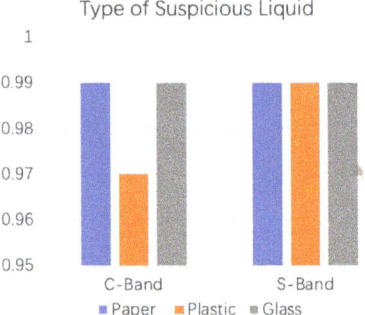

Figure 9. The KNN algorithm Classification results of Step 2 at the C-band and the S-band.

4.3. Detection of Different Concentrations of Alcohol

The dielectric constant of different liquids was quite different. Even for the same liquid, different concentrations had a certain effect on the dielectric constant. To verify this, we selected 50% alcohol, 75% alcohol, and 95% alcohol to carry out the experiment. Due to the different physical materials of the container, the WCI amplitude of each subcarrier was also affected by different containers in the same band environment. In addition, the amplitude of the WCI varied with the same kind of container at different wavelengths.

Figure 10 shows the WCI amplitudes of 30 subcarriers corresponding to different concentrations of alcohol in the C-band where the paper cup, plastic bottle, and glass bottle were detected. We can see that when the object is in the C-band environment, various containers performed differently when identifying different concentrations of alcohol, but they still accurately identified different concentrations of alcohol. This experiment verifies that the types of containers will not affect the system's identification of liquids, and further illustrates the reliability of our system in detecting and identifying different liquids. The conclusion is also applicable to the experimental measurement in the S-band environment.

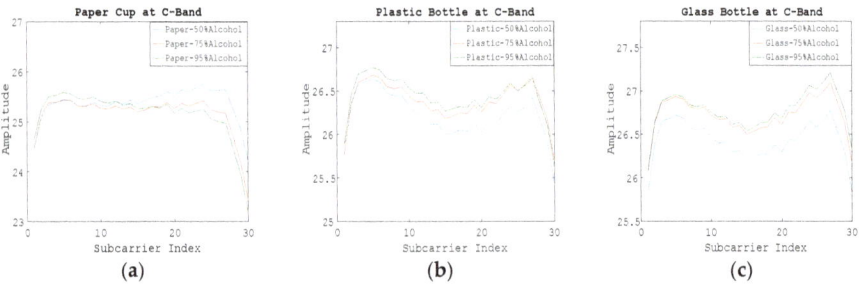

Figure 10. The wireless channel information (WCI) amplitudes of 30 subcarriers corresponding to different concentrations of alcohol at the C-band by using different containers. (**a**) Using the paper cup; (**b**) Using the plastic bottle; (**c**) Using the glass bottle.

Figure 11 shows the detection results of different concentrations of alcohol in C-band and S-band environments. From the figure, we can see that the detection accuracy of the C-band is higher than that of the S-band for the detection of alcohol with different concentrations. In the C-band frequency analysis experiment, the accuracy of the system for the detection of different concentrations of alcohol reached more than 91%, and that of S-band electromagnetic waves was up to 89%. Therefore, the C-band electromagnetic wave is superior to the S-band electromagnetic wave in the accurate detection of different concentrations of the same liquid.

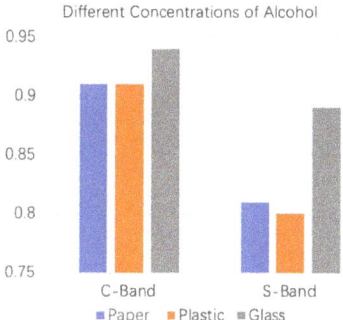

Figure 11. Detection results of different concentrations of alcohol at the C-band and the S-band.

As can be seen from Table 2, wireless sensing corresponding the 5G frequency band had excellent detection results for the detection and identification of suspicious liquids, no matter which container was selected. Moreover, our method had better performance and robustness in detecting different concentrations of alcohol, and had more subtle differences in dielectric constant than existing Wi-Fi technologies.

Table 2. The KNN algorithm classification results of the system.

Band Selection	C-Band			S-Band		
Container	Paper	Plastic	Glass	Paper	Plastic	Glass
Experiment 1	0.98	0.99	0.98	0.99	0.99	0.99
Experiment 2	0.99	0.97	0.99	0.99	0.99	0.99
Experiment 3	0.91	0.91	0.94	0.81	0.80	0.89

5. Conclusions

In this work, we explored the feasibility of using a wireless signal in a multi-band environment to detect suspicious liquids. Our work is novel because it demonstrates that it is possible to detect suspicious liquids accurately using radio signals without installing expensive liquid detection machines. Our system can not only detect whether the liquid is suspicious, but also further identify the types of suspicious liquids. In addition, we confirmed the feasibility of liquid concentration detection by using the WCI at a 5G frequency band.

The results analysis shows that our method can accurately detect suspicious and nonsuspicious liquids (Experiment 1) with more than 98% accuracy, regardless of the type and size of containers, and can identify the type of suspicious liquids (Experiment 2) with more than 97% accuracy. For the detection of alcohol with different concentrations, the accuracy can reach up to 94%. This provides better performance and robustness than existing Wi-Fi technology.

However, our method has limitations for liquids stored in metal containers. In that case, we recommend that security personnel intervene. Our method could be further improved by increasing the number of suspicious liquids prohibited in public and looking for ways to reduce the impact of metal containers on liquid detection.

Author Contributions: Conceptualization, J.D. and N.Z.; methodology, L.G.; software, J.D.; validation, L.G.; formal analysis, J.D.; writing—original draft preparation, J.D.; writing—review and editing, M.B.K.; supervision, W.S., A.R., J.Z., X.Y. and Q.H.A.; project administration, X.Y.; funding acquisition, X.Y.

Funding: The work was supported in part by the Fundamental Research Funds for the Central Universities (No. JB180205).

Conflicts of Interest: The authors declare no conflict of interest.

References

1. Hu, L. Research on Liquid Classification by CNN-SVM Model and WIFI Channel State Information. Master's Thesis, Hunan University, Changsha, China, 2018.
2. Wang, C.; Liu, J.; Chen, Y.; Liu, H.; Wang, Y. Towards in-baggage Suspicious Object Detection Using Commodity WiFi. In Proceedings of the 2018 IEEE Conference on Communications and Network Security (CNS), Beijing, China, 30 May–1 June 2018.
3. Jiang, X.; Qian, Y.T. Research on The Improvement of Raman Spectrum Qualitative Technique and Its Application in Airport Inspection. *Shandong Inoustrial Technol.* **2017**, *351*, 220–222.
4. Wang, Y.S.; Li, B.L. Recognition of Liquid Substances in Containers Based on X-Ray Images. In Proceedings of the 15th National Conference on Image and Graphics, Guangzhou, China, 10 December 2010; p. 5.
5. Xu, X. Study on Microwave Detection Method for Quality of Liquid Food. Master's Thesis, University of Electronic Science and Technology of China, Sichuang, China, 2015.
6. Ma, Y. Microwave Detection of Mixed Liquid Components. Master's Thesis, Taiyuan University of Technology, Taiyuan, China, 2013.
7. Xiao, W.; Tu, Y.; Wang, J.; Mao, Y.W. Design and Implementation of Wireless Intelligent Monitoring System for Liquid Droplets. *Electron. Meas. Technol.* **2008**, *1*, 133–136.
8. Yu, H.; Ding, B.; Sun, X.; Ding, Z. Design and Implementation of Infusion Monitoring System Based on ZigBee Technology. *Piezoelectrics Acoustooptics* **2013**, *35*, 756–762.
9. Chengming, Z. Practical Thinking of 5G Communication Technology Based on Internet of Things. *Telecom World* **2019**, *26*, 95–96.
10. Wang, C.; Chen, S.; Yang, Y.; Hu, F.; Liu, F.; Wu, J. Literature Review on Wireless Sensing Wi-Fi Signal-Based Recognition of Human Activities. *Tsinghua Sci. Technol.* **2018**, *23*, 203–222. [CrossRef]
11. Yang, Z.; Liu, Y. Wi-Fi Radar: From RSSI to CSI. *Commun. CCF* **2014**, *10*, 11.
12. Li, X.; Zhang, D.; Xiong, J.; Zhang, Y.; Li, S.; Wang, Y.; Mei, H. Training-Free Human Vitality Monitoring Using Commodity Wi-Fi. *ACM Interact. Mob. Wearable Ubiquitous Technol.* **2018**, *2*, 121. [CrossRef]
13. Xiang, L.; Li, S.; Zhang, D.; Jie, X.; Hong, M. Dynamic-MUSIC: Accurate device-free indoor localization. In Proceedings of the ACM International Joint Conference on Pervasive & Ubiquitous Computing, Heidelberg, Germany, 12–16 September 2016.
14. Wang, J.; Jiang, H.; Xiong, J.; Jamieson, K.; Xie, B. LiFS: Low Human-Effort, Device-Free Localization with Fine-Grained Subcarrier Information. In Proceedings of the International Conference on Mobile Computing & Networking, New York, NY, USA, 3–7 October 2016.
15. Xiong, J.; Sundaresan, K.; Jamieson, K. ToneTrack: Leveraging Frequency-Agile Radios for Time-Based Indoor Wireless Localization. In Proceedings of the 21st Annual International Conference on Mobile Computing and Networking, Paris, France, 7–11 September 2015.
16. Liu, X.; Cao, J.; Tang, S.; Wen, J.; Guo, P. Contactless Respiration Monitoring Via Off-the-Shelf WiFi Devices. *IEEE Trans. Mob. Comput.* **2016**, *15*, 2466–2479. [CrossRef]
17. Wang, H.; Zhang, D.; Ma, J.; Wang, Y.; Wang, Y.; Wu, D.; Gu, T.; Xie, B. Human respiration detection with commodity WiFi devices: Do user location and body orientation matter? In Proceedings of the ACM International Joint Conference on Pervasive & Ubiquitous Computing, Heidelberg, Germany, 12–16 September 2016.
18. Zhang, F.; Zhang, D.; Xiong, J.; Wang, H.; Niu, K.; Jin, B.; Wang, Y. From Fresnel Diffraction Model to Fine-Grained Human Respiration Sensing with Commodity Wi-Fi Devices. *ACM Interact. Mob. Wearable Ubiquitous Technol.* **2018**, *21*, 53. [CrossRef]
19. Wei, W.; Liu, A.X.; Shahzad, M.; Kang, L.; Lu, S. Understanding and Modeling of WiFi Signal Based Human Activity Recognition. In Proceedings of the International Conference on Mobile Computing & Networking, Paris, France, 7–11 September 2015.
20. Yan, W.; Jian, L.; Chen, Y.; Gruteser, M.; Liu, H. E-eyes: Device-free location-oriented activity identification using fine-grained WiFi signatures. In Proceedings of the 20th Annual International Conference on Mobile Computing and Networking, Maui, HI, USA, 7–11 September 2014.
21. Zhou, Z.; Zheng, Y.; Wu, C.; Shang, G.L.; Liu, Y. Towards Omnidirectional Passive Human Detection. In Proceedings of the 2013 Proceedings IEEE INFOCOM, Turin, Italy, 14–19 April 2013.

22. Wu, K. Wi-metal: Detecting metal by using wireless networks. In Proceedings of the IEEE International Conference on Communications, Kuala Lumpur, Malaysia, 22–27 May 2016.
23. Zhong, S.; Huang, Y.; Ruby, R.; Lu, W.; Qiu, Y.X.; Wu, K. Wi-fire: Device-free fire detection using WiFi networks. In Proceedings of the IEEE International Conference on Communications, Paris, France, 21–25 May 2017.
24. Technology, M. The MIIT has Issued a Notice Concerning the Use of the 3300–3600 MHz and 4800–5000 MHz Frequency Bands in The Fifth-generation Mobile Communication System. 2017. Available online: http://www.miit.gov.cn/n1146295/n1652858/n1652930/n3757020/c5907905/content.html (accessed on 15 July 2018).

© 2019 by the authors. Licensee MDPI, Basel, Switzerland. This article is an open access article distributed under the terms and conditions of the Creative Commons Attribution (CC BY) license (http://creativecommons.org/licenses/by/4.0/).

Article

A Distributed Testbed for 5G Scenarios: An Experimental Study

Mohammad Kazem Chamran [1], Kok-Lim Alvin Yau [1,*], Rafidah M. D. Noor [2] and Richard Wong [1]

1. School of Science and Technology, Sunway University, Subang Jaya 47500, Malaysia; mohamma.c@imail.sunway.edu.my (M.K.C.); richardwtk@sunway.edu.my (R.W.)
2. Department of Computer System and Technology, University of Malaya, Kuala Lumpur 50603, Malaysia; fidah@um.edu.my
* Correspondence: koklimy@sunway.edu.my

Received: 16 September 2019; Accepted: 29 October 2019; Published: 19 December 2019

Abstract: This paper demonstrates the use of Universal Software Radio Peripheral (USRP), together with Raspberry Pi3 B+ (RP3) as the brain (or the decision making engine), to develop a distributed wireless network in which nodes can communicate with other nodes independently and make decision autonomously. In other words, each USRP node (i.e., sensor) is embedded with separate processing units (i.e., RP3), which has not been investigated in the literature, so that each node can make independent decisions in a distributed manner. The proposed testbed in this paper is compared with the traditional distributed testbed, which has been widely used in the literature. In the traditional distributed testbed, there is a single processing unit (i.e., a personal computer) that makes decisions in a centralized manner, and each node (i.e., USRP) is connected to the processing unit via a switch. The single processing unit exchanges control messages with nodes via the switch, while the nodes exchange data packets among themselves using a wireless medium in a distributed manner. The main disadvantage of the traditional testbed is that, despite the network being distributed in nature, decisions are made in a centralized manner. Hence, the response delay of the control message exchange is always neglected. The use of such testbed is mainly due to the limited hardware and monetary cost to acquire a separate processing unit for each node. The experiment in our testbed has shown the increase of end-to-end delay and decrease of packet delivery ratio due to software and hardware delays. The observed multihop transmission is performed using device-to-device (D2D) communication, which has been enabled in 5G. Therefore, nodes can either communicate with other nodes via: (a) a direct communication with the base station at the macrocell, which helps to improve network performance; or (b) D2D that improve spectrum efficiency, whereby traffic is offloaded from macrocell to small cells. Our testbed is the first of its kind in this scale, and it uses RP3 as the distributed decision-making engine incorporated into the USRP/GNU radio platform. This work provides an insight to the development of a 5G network.

Keywords: D2D communication; 5G; sensor network; sensor; end-to-end delay; USRP; distributed mechanism; Raspberry Pi

1. Introduction

Fifth generation (5G) is a promising next-generation cellular network armed with new features, particularly device-to-device (D2D) communication that enables direct communication between devices without going through base stations (BSs). This helps to offload traffic from macrocell (MC) BSs to small cell (SC) (i.e., femtocell) BSs, as well as user equipment and devices, including sensors, while increasing network cell coverage via multihop transmission [1–3]. In 5G, a node can operate either as a licensed user (or a primary user, PU) to utilize its licensed channels (or cellular channels), or as an unlicensed user (or a secondary user, SU) to explore and utilize white spaces, which are the underutilized licensed channels (or cognitive channels) [4]. D2D enables nodes to access both cellular and cognitive channels to improve spectrum efficiency in order to improve data transmission rate and quality of service (QoS) [5–7].

1.1. Our Contributions

At present, the majority of the research related to 5G presents theoretical analysis [8–12] and simulation studies [11–16]. In general, various theoretical state of the art and open issues are presented in [8], the effects of ultra-densification are investigated in [9], various network architectures, medium access mechanisms, and open issues are presented in [10], as well as routing algorithms to achieve lower interference and a balanced traffic load amoung routes in 5G environment are investigated in [11,12], respectively. In addition, traffic offloading from backbone routes and the central controller to distributed nodes is investigated in [13,14], the transmission delay is predicted based on channel states in [15], and the feasiblity of D2D in 5G environment is investigated in [16]. Some researchers conduct proof of concept experiments; however, the focus is primarily on the physical layer, particularly spectrum management in [2], interference mitigation in [17], channel sensing in [18], as well as on the data link layer, particularly channel hopping (or switches) in [19].

This study focuses on the networking aspect over a 5G-based platform using universal software radio peripheral with GNU radio (USRP/GNU radio) units and Raspberry Pi3 B+ (RP3) processors [20]. GNU radio is an open source software that serves as development toolkit [21] in the platform (for more details see Section 2.3). There are two types of testbeds under investigation in this paper: (a) the traditional testbed comprised of BSs or nodes connected to a single traditional processing unit using a wired medium via a switch [2]; and (b) our distributed testbed in which each BS and node is embedded with a separate processing unit, namely Raspberry Pi3 B+ (RP3). We consider a 5G scenario with: (a) D2D communication; and (b) heterogeneous MC and small cell BSs with different sensing and transmission capabilities, as well as processing capabilities (i.e., using operating systems with different capabilities). Comparison is made on the performance measures of routes via D2D and MC BS. The MC BS selects a route, and informs FC BSs and nodes about the route; subsequently, the FC BSs and nodes setup the route accordingly. Therefore, FC BSs and nodes can be relaxed from performing route selection and channel sensing. Our testbeds are sufficient for the investigation of our contributions, although further extension is suggested in Section 7.

Our contributions are twofold:

- Performance comparison achieved in our distributed testbed based on proof of concept experiments involving multihop transmission, which is necessary in next-generation wireless sensor networks. The BSs and nodes are heterogeneous from MC and SCs with different sensing and transmission capabilities, as well as processing capabilities (i.e., using operating systems with different capabilities).
- Performance analysis of the software and hardware processing delays for communication via D2D and going through BSs over the testbed, which is required for route selection.

1.2. Significance of This Paper

There are two main investigations in this paper. Firstly, comparison is made of the performance measures achieved by the traditional testbed and our distributed testbed. Our proposed testbed is distributed in nature and it has a closer resemblance to a real deployed network. The software and hardware processing delays, which are generally ignored in theoretical analysis and simulation, are investigated. Secondly, using the testbeds, comparison is made on the performance measures achieved by: (a) the traditional direct communication with MC BS; and (b) the D2D communication. This comparison is useful for the MC BS to make decision on route selection. This is because, while D2D communication can offload traffic from MC BSs to SC BSs, the end-to-end delay over a multihop transmission increases, and so the direct communication with MC BS can be favorable. The end-to-end delay of a route changes with its operating environment (e.g., the processing capability) and its operation (e.g., the lower read and write rates of RP3 contribute to a higher end-to-end delay and lower packet delivery ratio in our distributed testbed compared to the traditional testbed). Lower end-to-end delay is favorable to support real-time applications integrated with sensors, such as driverless vehicles. To the best of our knowledge, this is the first USRP/GNU radio platform incorporated with RP3 implementation with this scale and functionality.

1.3. Organization of This Paper

The rest of this paper is organized as follows. Section 2 presents research background. Section 3 presents related work. Section 4 presents system model and delay measurement. Section 5 presents experimental setup. Section 6 presents experimental results. Section 7 presents our conclusion and future work.

2. Background

This section presents an overview of 5G, USRP, GNU radio, and RP3.

2.1. 5G

The 5G network is a heterogeneous network that consists of different kinds of network cells, including MC and femtocell (FC). The transmission of the BSs and nodes are characterized by different frequency bands and transmission power levels. In Figure 1, a 5G network consists of two main planes: (a) *control plane* consists of MC BSs, which use higher transmission power levels at lower frequency bands, contributing to larger transmission ranges; and (b) *data plane* consists of FC BSs and nodes, which use lower transmission power levels at higher frequency bands, contributing to smaller transmission ranges [22]. The control plane communicates with the cloud, which consists of a central controller (CC) that manages and coordinates global functions, such as route selection. MC BSs can coordinate and communicate among themselves via the cloud [12,23], and this helps them to determine the nodes that each of them must cover. The FC BSs can coordinate and communicate among themselves via D2D if they are within each other's transmission range, and this helps them to: (a) use a route established from a FC source node FC_s to a FC destination node FC_d by the CC; and (b) offload traffic from MC BSs. Both MC and FC overlap, and FC BSs can communicate with each other directly. Hence, the MC BS in the control plane can select a route, and inform FC BSs and nodes in the data plane about the route; subsequently, the FC BSs and nodes setup the D2D route accordingly. Therefore, FC BSs and nodes can be relaxed from performing complex tasks, such as route selection.

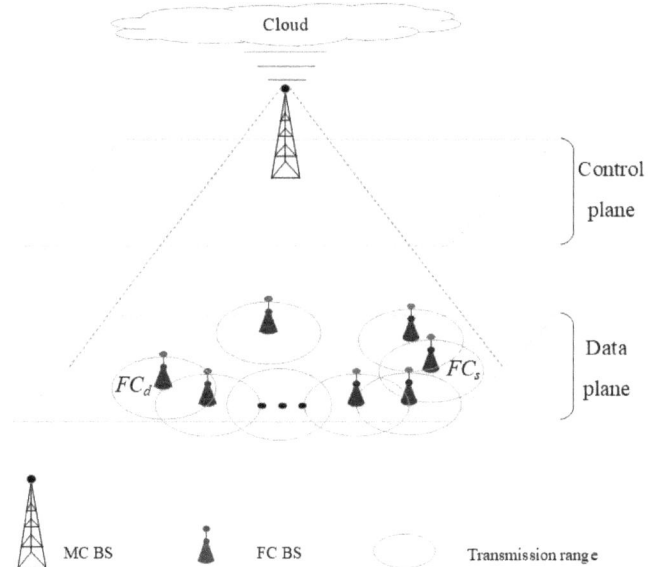

Figure 1. A 5G network that consists of a single MC BS and a number of FC BSs.

2.2. Universal Software Radio Peripheral

USRP is an off-the-shelf wireless device that can be configured with a wide range of operating parameters, such as the types of modulation schemes and the channel frequency bands. Our testbed uses USRP N200 series that provides high processing capability. Figure 2 shows a USRP unit that has a set of two omni-directional VERT900 antennas—one for transmission and one for reception—for simultaneous transmissions in two different operating channels within channel frequency bands 850–890 MHz and 2.3–2.4 GHz. The selected channel frequency bands include the television and global system for mobile communication (GSM) bands. The antennas are connected to a daughterboard. There are two types of paths: (a) *receive path* in which analogue signals are received and moved from the radio frequency (RF) front end towards RP3 for reception; and (b) *transmit path* in which digital signals move from RP3 towards the RF front end for transmission.

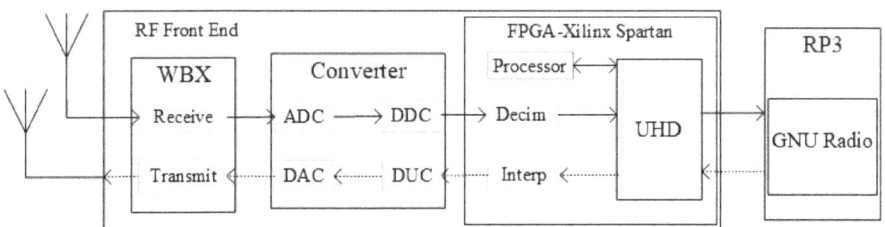

Figure 2. Transmit and receive paths between an antenna and a RP3 via a USRP/GNU radio unit. Solid arrow line is part of a receive path, and dotted arrow line is part of a transmit path.

The USRP consists of three main sections as follows:

- *Wide bandwidth transceive (WBX)* is the RF front end that provides access to different operating channels within a range of 50 MHz of RF bandwidth with 8 bit samples, or 25 MHz of RF bandwidth with 16 bit samples. The maximum transmission power is 100 mW (or 20 dBm) with a noise figure of 5 dB.

- *Converter* consists of: (a) an analogue-to-digital converter (ADC) and a digital down converter (DDC) in the receive path; and (b) a digital-to-analogue converter (DAC) and a digital up converter (DUC) in the transmit path. DDC selects desired signals from an array of signals captured by ADC, while DUC increases the bandwidth of baseband signals so that they are compatible with DAC.
- *Field-programmable gate array (FPGA)*, specifically the Xilinx Spartan 3A-DSP 1800 board [24] used in this platform, consists of: (a) a decimation filter for achieving the required interface bandwidth in the receive path, and an interpolation filter for achieving the opposite in the transmit path; (b) a USRP hardware driver (UHD) block with a software interface that enables various components to communicate among themselves; and (c) a processor block that performs encoding/decoding, modulation/demodulation, timing synchronization, and other signal processes required for software defined radio (SDR) operations. The FPGA communicates with RP3 via power over Ethernet (PoE). It provides connection between: (a) gigabit Ethernet CAT 5E-350 MHz cables, which provide a maximum data rate of 1000 megabits per second (Mbps) connected to a Gigabit switch; and (b) USB3, which provides a maximum data rate of 1600 Mbps. During system initialization, the kernel, which is the fundamental part of an operating system, of GNU radio controls and monitors programs and systems, as well as performs default functions, such as checking and assigning memory space to FPGA [24].

2.3. GNU Radio

GNU radio, together with its extended version called GNU radio companion (GRC), is an open source SDR that enables users to design: (a) configurable blocks to perform communication tasks using the C++ language; and (b) flow graphs to connect the blocks using the Python language. As shown in Figure 3, the blocks and flow graphs define the roles of the source, intermediate, and destination nodes as follows:

- *Source node*, which is a RP3 unit with an Internet protocol (IP) address (e.g., 192.168.10.2) and a port number (e.g., 1234), generates and sends a data or video stream in the form of frames encapsulated in user datagram protocol (UDP). In GRC, the frames pass through three main components: (a) an *encoder* that converts the frames into packets with a predefined payload length (e.g., 1472 bytes); (b) a *Gaussian minimum shift keying (GMSK) modulator* that converts the packets into modulated signals at baseband (e.g., the minimum non-zero frequencies); and (c) a *USRP sink block* that sets the center frequency (e.g., 850 MHz), channel gain (e.g., 1dB), and sample rate (e.g., 1 MHz). Finally, the signals are broadcasted.
- *Intermediate node* receives signals from a transmitter, which can be a source node or an upstream intermediate node, and transmits them to the next-hop node, which can be a destination or a downstream intermediate node. There are two processes that help to improve the quality of packets before forwarding them in order to reduce interference and address poor channel quality [25]: (a) to demodulate signals to packets, and then to decode packets to frames; and (b) to encode frames to packets, and then to modulate packets to signals. The demodulation and decoding processes are performed at the receiver unit, and then modulation and encoding processes at the transmitter unit.
- *Destination node*, which is a RP3, receives and demodulates signals to packets, and then decodes packets to frames. Then, a UDP sink block sends the frames to an application (e.g., a VLC media player) through a port (e.g., port number 1236 or udp://@:1236).

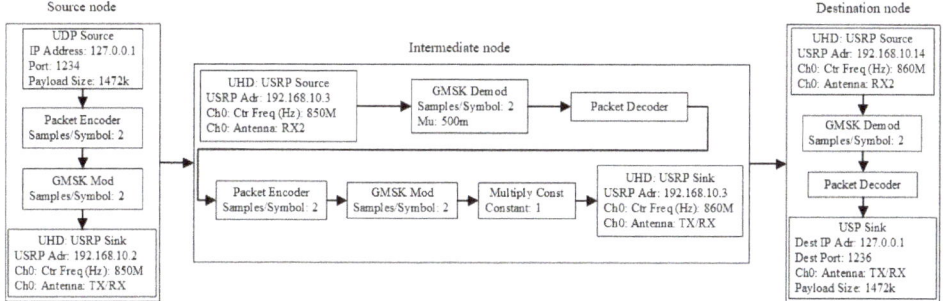

Figure 3. GNU radio flow graph that consists of source, intermediate and destination nodes.

2.4. Raspberry Pi3 B+

Conventionally, a testbed consists of BSs or nodes connected to a single traditional processing unit (e.g., a personal computer or a laptop) using a wired medium via a switch [2] (see Section 5). The BSs and nodes exchange control messages and data packets over the wired and wireless media, respectively. In this paper, each BS and node is embedded with a separate processing unit, namely the RP3 unit. Both control messages and data packets are exchanged over the wireless medium.

There are *three* main advantages. Firstly, *ease of implementation* because nodes can be placed further apart from each other rather than being constrained by physical cables and connections to a single switch. Secondly, *higher cost efficiency (or lower overhead)* because nodes do not communicate with a traditional processing unit. Thirdly, *lower energy consumption* with the use of RP3 compared to personal computers, laptops, and a switch.

However, there are three main disadvantages. Firstly, *lower processing capability*. The RP3 processor (e.g., 1.4 GHz 64-bit quad-core processor with 1 GB non-expandable on-board RAM) is suffice to perform basic tasks and support simple applications (e.g., running GRC in the Linux environment). Secondly, *lower data rate*. The network interface of RP3 has approximately 324 Mbps data rate only, which is low compared to 761 Mbps offered by the gigabit Ethernet of a CORE i7 personal computer. This increases the end-to-end delay of the communication between a RP3 and a USRP. In addition, RP3 is embedded with an SD card, such as a high capacity HC-I class 10 SD card that offers a data rate of 10 megabytes per second (MBps), which is low compared to 550 MBps offered by a solid-state hard drive. This increases the internal delay of the communication between a RP3 and an SD card. This is significant because such communication is commonplace as an operating system (e.g., Ubuntu) is stored as an image in the SD card. It is worth mentioning that the speed (or rate) of read and write on SD cards reduces with the increase of occupied space, and the read rate is generally higher than the write rate as shown in Section 5.3.2. Thirdly, *lower storage space*. The SD card provides low storage space (e.g., 32 GB) for operating systems and software applications.

3. Related Work

This section presents related works on testbeds, particularly USRP/GNU radio platforms, for investigating the networking aspect of 5G. It covers two main topics. *Firstly*, the communication delay between nodes along a route. The routes are assumed to be readily available, and they are selected and provided by the central controller. *Secondly*, the testbeds, particularly USRP/GNU radio and RP3 platforms.

3.1. Communication Delay between Nodes

In [26], the hardware and software processing delays are investigated on a testbed comprised of ten USRP/GNU radio nodes connected to a single traditional processing unit (i.e., a personal computer) via a switch. The end-to-end delay has shown to reduce since the personal computer can pre-process route selection prior to data transmission.

In [27], the response delay (or the round-trip time) is investigated on a testbed comprised of two USRP/GNU radio nodes connected to a personal computer. The response delay is the duration between the moment the first byte of a packet passes the digital signal processing block of a sender and the moment the first byte of an acknowledgement packet arrives at the sender. The delay is incurred in: (a) the initiation process that includes modulation, sampling, encoding, as well as packet transfer between GNU radio and kernel (or the operating system); (b) the buffering process that collects and stores packets in a buffer (e.g., the buffer of a VLC media player); and (c) the transmission process that receives and sends packets to the FPGA unit of USRP so that they are interpolated before being transmitted via antenna. Measurement shows that the initiation process has the highest time delay, and the transmission process has the lowest time delay.

In [28], the hardware and software processing delays, as well as the response delay, are investigated on a testbed comprised of two USRP/SDR nodes, which serve as the source and destination nodes, connected to a single traditional processing unit (i.e., a personal computer). The source node transmits a data packet to the destination node; and subsequently, the destination node returns a response packet to the source node. The delays are incurred in: (a) processes run in a USRP/SDR node (e.g., operating system and the modulation process); and (b) communication between the two nodes. Measurement shows that the software processing delay incurred in SDR is significantly higher than the hardware processing delay incurred in USRP and the communication delay incurred between the two nodes.

In [29], the response delay, which includes the waiting time of a packet in a queue, is investigated on a testbed comprised of four USRP/GNU radio nodes connected to a single traditional processing unit (i.e., another USRP/GNU radio unit). There are a pair of PU transmitter and receiver, and another pair of SU transmitter and receiver. The SU transmitter must sense the operating channels before transmission so as not to interfere with the PUs. Up to 30% of the delay incurred in the SU transmitter is attributed to channel sensing, which can be reduced to increase throughput at the expense of higher interference level to PUs. Hence, there is a tradeoff between the delay and throughput performances.

In [30], the hardware and software processing delays of different processes are investigated on a testbed comprised of two USRP/GNU radio nodes embedded with separate processing units (i.e., personal computers). The nodes are connected to each other via Ethernet. Examples of the USRP processes are the operating system processes in the kernel, and the decimation filtering in FPGA; and an example of the GNU radio process is the modulation process. During measurement, a 1 μs guard time is included between the processes. A node transmits a ping packet to another node. The packet moves through the transmit and receive paths, and the timestamps for different processes in the USRP/GNU radio node are recorded. Measurement shows that the hardware and software processing delays are highest for processes running in the Kernel. This indicates that the USRP/GNU radio platform has low efficiency providing low network performance, particularly high end-to-end delay.

In this paper, the testbed is comprised of five USRP/GNU radio nodes embedded with separate processing units. Investigation is conducted on multihop transmissions in the network layer.

3.2. Testbed of USRP/GNU Radio and Raspberry Pi

This section presents related works on USRP/GNU radio with and without Raspberry Pi.

3.2.1. USRP/GNU Radio without Raspberry Pi

In [2,31–36], a testbed consists of BSs or nodes connected to a single traditional processing unit (e.g., a personal computer or a laptop) via a switch (i.e., Gigabit D-link) [2] (see Figure 4a). This allows the central controller to exchange control messages with BSs and nodes via a switch in a centralized manner, while the BSs and nodes can exchange data packets using the wireless medium [37] in a distributed manner. Examples of control messages include those that carry information about channel sensing and selection, route discovery and selection, and handshaking (e.g., request, acknowledgement, and response messages). Therefore, the response delay of a D2D type of communication in a real testbed is a cause of concern because of the sensitivity of wireless communication and the delay incurred due to the distance between a node pair. The response delay is important in D2D communication because if it may not fulfill the delay requirement (or higher than a threshold), MC BS must be used. This paper focuses on the response delay, which is end-to-end in nature, between a source node from first transmitted packet up to last received one.

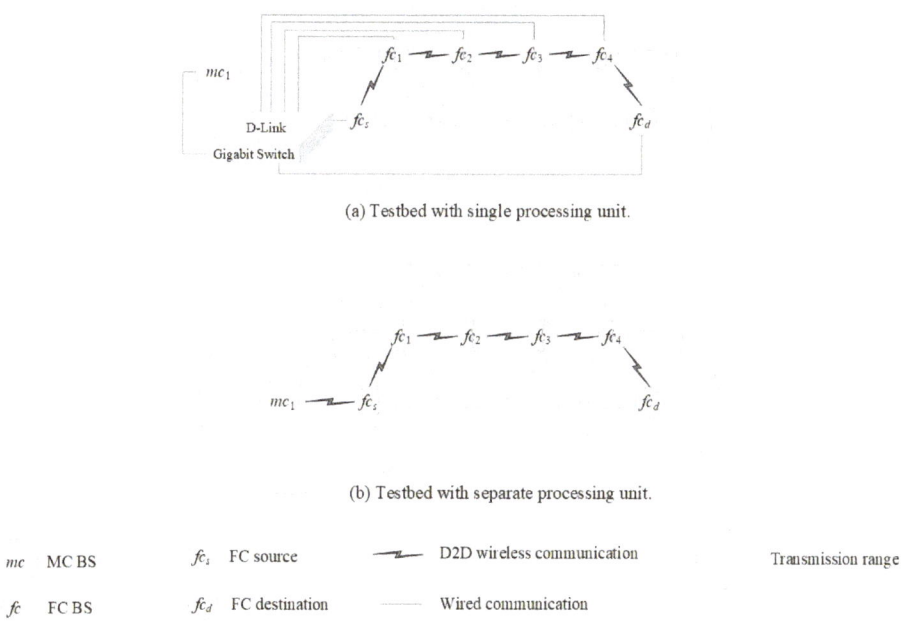

Figure 4. Two scenarios in experimental setup.

In this paper, each BS or node is embedded with a separate processing unit, particularly RP3 as the core processing unit, to provide a more realistic wireless testbed, and so a single traditional processing unit is not needed.

3.2.2. USRP/GNU Radio with Raspberry Pi

In [38], a single USRP/GNU radio embedded with Raspberry Pi3 is used to generate signals in the range of FM radio frequency bands. In addition, a radio station also generates signals in the frequency bands. Subsequently, the signals generated by the Raspberry Pi3 and radio station are measured using a spectrum analyzer, and a comparison is made. The quality of signals generated by the USRP/GNU radio, despite using a lower transmission power at lower frequency bands, has shown to be close to that from a radio station. In general, the received signals from USRP/GNU radio has a lower throughput and energy consumption.

In [39], a single USRP/GNU radio node embedded with Raspberry Pi3 is used to perform energy-based channel sensing in order to detect PUs activities. There are two main sources of energy consumption in Raspberry Pi3: (a) software initiation; and (b) the calculation of the available memory of the kernel running GNU radio libraries. Channel sensing has shown to incur the highest delay. Similar investigation is performed in [20]. Energy consumption in Raspberry Pi3 has shown to be considerably lower compared to that in personal computer.

In [40], a single register transfer level (RTL) dongle embedded with a Raspberry Pi transmits and receives in the frequency bands 24–1850 MHz. The general-purpose input/output (GPIO) pins of the Raspberry Pi is used to generate and transmit pulse width modulation (PWM). The energy consumption of the RTL dongle embedded with Raspberry Pi has shown to be less than 3 watts, and so the dongle and Raspberry Pi can be powered by portable batteries. In addition, the use of Raspberry Pi has shown to enable the detection of a wide range of frequency bands while incurring low energy consumption.

In [41], a single USRP/GNU radio node connected to a personal computer, which serves as the transmitter, broadcasts signals to a RTL dongle embedded with a Raspberry Pi, which serves as the receiver. The testbed consists of a low-cost radio community that transmit at two frequency bands, namely 915 MHz (or the ISM band) and 40.68 MHz (or the FM radio frequency band). The testbed has demonstrated the capability of Raspberry Pi for transmitting and receiving signals in these frequency bands, and the quality of reception depends on the transmission power and the height of the antenna of the transmitter.

In [42], a testbed consists of two USRP/GNU radio nodes: (a) a static node, which is connected to a personal computer, serves as the ground BS; and (b) a dynamic node, which is embedded with Raspberry Pi3, is installed on an unmanned aerial vehicle (or a drone). The ground BS receives location information from the drones so that it can monitor the location of the drone. The ground BS and drone exchange messages in the frequency bands 400–4400 MHz. The testbed has demonstrated the capability of Raspberry Pi3 for setting up communication and processing information with lower energy consumption.

In [43], a testbed consists of three main USRP/GNU radio nodes: (a) a SU source node, which is connected to a personal computer; (b) a SU intermediate node, which is embedded with Raspberry Pi3, that performs energy-based channel sensing; and (c) a SU destination node, which is embedded with Raspberry Pi3. The rest of the nodes are PUs. The SU source node transmits data packets to the SU destination node in multiple hops without interfering with the random PUs' activities. The channel sensing delay incurred by the SU intermediate node embedded with Raspberry Pi3 has shown to be twice of that incurred by the SU source node connected to a personal computer.

In this paper, our testbed consists of five USRP/GNU radio nodes embedded with Raspberry Pi3 B+ (RP3) that constitutes a source node and four intermediate nodes. In addition, a personal computer serves as the destination node. The USRP/GNU radio performs communication, and the RP3 performs processes. While existing works in the literature [20,38–43], focus on the capability and compatibility of USRP/GNU radio and RP3, this paper focuses on end-to-end hardware and software processing delays between a source node and a destination node, and the use of the delay measurement for route selection (i.e., either via D2D or MC BS).

4. System Model and Delay Measurement

This study measures network performance, particularly end-to-end delay, packet delivery ratio, and throughput, under different scenarios characterized by the characteristics of 5G, and compares results obtained from two types of testbeds, namely: (a) a testbed with a single traditional processing unit via a switch (see Figure 4a), (see Section 5.3.1); and (b) a testbed with separate processing unit (see Figure 4b), particularly RP3, embedded in each node and BS without using a switch (see Section 5.3.2).

The rest of this section presents system model in Section 4.1, D2D link delay in Section 4.2.1, and D2D end-to-end delay in Section 4.2.2.

4.1. System Model

The system topology consists of femtocell nodes $fc_f \in \{fc_1, fc_2, ..., fc_{|F|}\}$ located within the transmission range of a MC BS. Nodes can transmit UDP packets in one of the routes in a route set, specifically $k_k \in K = \{k_1, k_2, ..., k_{|K|}\}$. Each link $l_n \in L = \{l_1, l_2, ..., l_{|L|}\}$ uses one of the channels $c_c \in$

$C = \{c_1, c_2, ..., c_{|C|}\}$. Each route $k_k = \cup l_n \in L$ consists of a set of links from a femtocell source node to a femtocell destination node. a femtocell source node fc_s sends packets to a femtocell destination node fc_d along a primary route $k_k = (fc_s, fc_1, fc_2, fc_3, fc_4, fc_d)$, which is D2D and multihop in nature. In a testbed with a single traditional processing unit, the femtocell nodes $fc_s, fc_1, fc_2, fc_3, fc_4, fc_d$ are connected to a MC BS, which serves as the centralized processor, via a switch as shown in Figure 4a (see Section 5.3.1 for more descriptions). On the other hand, in a testbed with separate processing units, each femtocell node is embedded with a separate processing unit, namely RP3, as shown in Figure 4b (see Section 5.3.2 for more descriptions). In this paper, a primary route has up to five hops. The primary route uses cognitive channels (or white spaces in licensed channels), and the secondary route uses cellular channels (or the licensed channels). The use of primary routes helps to reduce the congestion level of MC BS [4]. However, when the primary route becomes unavailable or broken, then a secondary route $k_k = (fc_s, mc_1, fc_d)$, which passes through the macrocell BS mc_1. The route selection between primary and secondary routes is shown in the form of a flowchart in Figure 5 and an algorithm in Algorithm 1.

Algorithm 1 Route selection between the primary route (via D2D) and the secondary route (via MC BS)

1: **procedure** ROUTE SELECTION
2:    **for** $k_1 = (fc_s, fc_1, fc_2, fc_3, fc_4, fc_d)$ and $k_2 = (fc_s, mc_1, fc_d)$ **do**
3:       **if** route k_1 is available **then**
4:          fc_s send packet to fc_1
5:          **if** fc_1 is not available **then**
6:             packet goes through mc_1 (secondary route) to fc_d
7:          **end if**
8:          **if** fc_2 is available **then**
9:             check the second condition:
10:             **if** $t^{fc_1 to fc_2}$ is $\leq \alpha$ **then**
11:                fc_1 send packet to fc_2
12:             **end if**
13:          **end if**
14:          **if** fc_2 is not available **then**
15:             packet goes through mc_1 to fc_d
16:          **end if**
17:          **if** fc_3 is available **then**
18:             check the second condition:
19:             **if** $t^{fc_2 to fc_3}$ is $\leq \alpha$ **then**
20:                fc_2 send packet to fc_3
21:             **end if**
22:          **end if**
23:          **if** fc_3 is not available **then**
24:             packet goes through mc_1 to fc_d
25:          **end if**
26:          **if** fc_4 is available **then**
27:             check the second condition:
28:             **if** $t^{fc_3 to fc_4}$ is $\leq \alpha$ **then**
29:                fc_3 send the packet to fc_4
30:                fc_4 send the packet to destination fc_d
31:             **else** packet goes through mc_1 to fc_d
32:             **end if**
33:          **end if**
34:       **end if**
35:    **end for**
36: **end procedure**

The end-to-end delay t^{k_k} of a primary route k_k increases with the number of hops [44], and it must be less than a threshold $t^{k_k} < \alpha$, where $\alpha = 10$ ms is imposed by the IEEE 802.15.4 standard [14,45]. The secondary route is selected if the threshold is not fulfilled. The threshold α is imposed due to the need to reduce end-to-end delay in order to support and deploy real-time applications, including applications integrated with sensors such as driverless vehicles, in 5G. Long software and hardware processing delays can increase the queue size of base stations and nodes, and so they affect network performance, such as reducing packet delivery ratio [46,47].

In this paper, route selection is made by a central controller, and so the underlying routes, as well as the channels of the links in the routes, are readily available. There is a single MC BS that selects a route, and informs FC BSs and nodes about the route; subsequently, the FC BSs and nodes setup the route accordingly. Further extension to the testbed, such as increasing the number of MC BSs, is suggested in Section 7. The investigation takes into account the effects of the characteristics of 5G, including heterogeneity that involves nodes with different features and characteristics (i.e., different transmission power, frequency range, and strength of operating system).

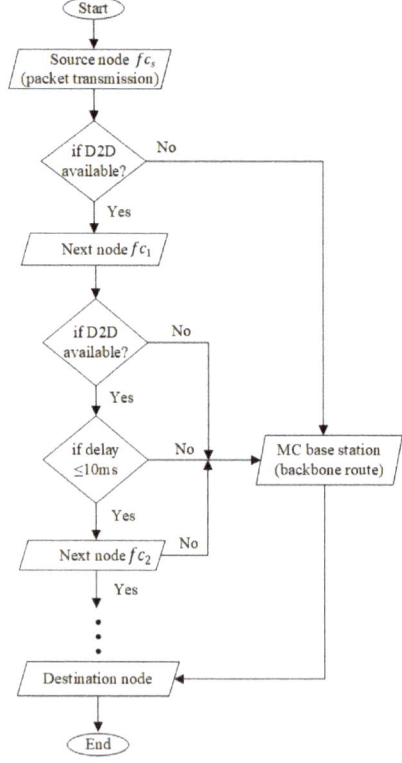

Figure 5. Conditions for selecting a primary route (via D2D).

4.2. Delay Measurement

This section presents D2D link and end-to-end delay, respectively.

4.2.1. D2D Link Delay

The D2D link delay (or per-hop delay) consists of three kinds of delays as shown in Figure 6. Firstly, the *software processing delay* $t^s_{l_n}$ is incurred in GNU radio running on processing unit, such as personal computer and RP3, to process packets, such as encoding and modulating packets (see Section 2.3), over a link l_n. Secondly, *hardware processing delay* $t^h_{l_n}$ is incurred in USRP to convert electrical packets to RF signals for transmission in the transmit path, and to convert RF signals to electrical packets upon reception in the receive path, over a link l_n. Thirdly, the propagation delay $t^p_{l_n}$ is incurred for the RF signal to travel from one USRP/GNU radio node to another over a link l_n; however, it is negligible compared to software and hardware processing delays [27].

The D2D link or per-hop delay t_{l_n} for link $l_n \in L$ is as follows:

$$t_{l_n} = t_{l_n}^s + t_{l_n}^h \tag{1}$$

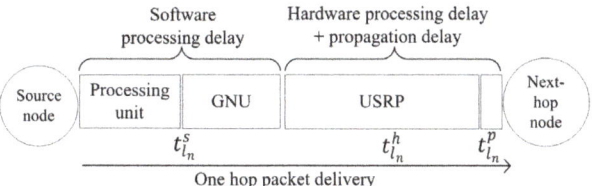

Figure 6. D2D link delay for a single-hop transmission over a link l_n. The processing unit can be either a personal computer or a RP3.

4.2.2. D2D End-to-End Delay

The D2D end-to-end delay t^{k_k} of a route $k_k \in K = \{k_1, k_2, ..., k_{|K|}\}$ is the accumulation of the D2D link delay at each link, which consists of software and hardware processing delays, as follows:

$$t^{k_k} = \sum_{l_n \in k_k} t_{l_n} \tag{2}$$

5. Experimental Setup

This experiment investigates the link (or per hop) and end-to-end delays of a route via D2D communication among heterogeneous BSs and nodes (i.e., MC and FCs) (see Section 4 for more details).

5.1. Experiment Parameters

The experiment uses: (a) licensed channels, including the TV frequency bands (i.e., 850–890 MHz) and long term evolution (LTE) frequency bands (i.e., 2.3–2.4 GHz) [48]; and (b) unlicensed channels, particularly the industrial, scientific and medical (ISM) frequency bands (i.e., 2.4 GHz). The experimental parameters are summarized in Table 1. In addition, the USRP parameters are presented in Section 2.2, and the GNU radio parameters and flow graph are presented in Section 2.3 and Figure 3, respectively. In GNU radio, the bandwidth can be represented by sample rate (or the number of samples per second). The transmission power used for the range of frequency bands within 850 and 890 MHz with a 1 dB set gain is 10 mW for 10 dBm.

Table 1. Experimental parameters.

Category	Parameter	PCU	RPU
Experiment	Duration	300 s	300 s
USRP	Number of channels	6	6
	Transport layer	UDP	UDP
	Bandwidth	1.6 Mbps	1.6 Mbps
	Transmission power	10 dBm	10 dBm
Antenna	Carrier frequency	850 MHz	850 MHz
Computer	Operating system	Ubuntu	-
Switch	Number of units	1	-
	Number of inputs	6	-
RP3	Operating system	-	Ubuntu-Mate
PoE	Number of units	-	5
	Number of inputs	-	1

5.2. Experiment Measurement

The software and hardware processing delays are measured using Wireshark [49], which is an open source packet analyzer (or a packet sniffer) software. For each packet transmission, Wireshark is used to measure the delays incurred by the processing unit (i.e., the time period for a media stream to be transformed into packets), the USRP (i.e., the time period for a packet to traverse from the USRP sink block to the antenna for transmission), and the propagation from one USRP unit to another. The IP address of the source node of the packet can be identified using Wireshark. The delay incurred by GNU radio (i.e., time period for a packet to traverse from the USRP source block to the USRP sink block) can be measured using Python, whereby the delay is given by the time difference between the USRP source block and the USRP sink block.

Meanwhile, there are *two* types of delays that are negligible: (a) the delay incurred for the initialization of different software prior to converting frames into packets; and (b) the delay incurred for signal propagation because the same transmit and receive components are used in both PCU and RPU.

5.3. Experiment Testbeds

The testbed has a route with five hops as shown in Figure 7. A video stream in the form of UDP packets is generated at a source node, forwarded along a route with four intermediate nodes, and received at a destination node. Video stream is chosen in this experiment due to its stringent QoS compared to data packets. Two types of testbeds are considered. Figure 4a shows a testbed with a single processing unit (PCU), in which a number of heterogeneous MC and FC BSs connect to a single traditional processing unit (i.e., a personal computer) using a wired medium via a switch. Figure 4b shows a testbed with separate processing units (RPU), in which the MC and FC BSs are embedded with separate processing units, namely RP3, and the FC BSs are located within the MC BS proximity. The rest of this section presents the two types of testbeds and their differences.

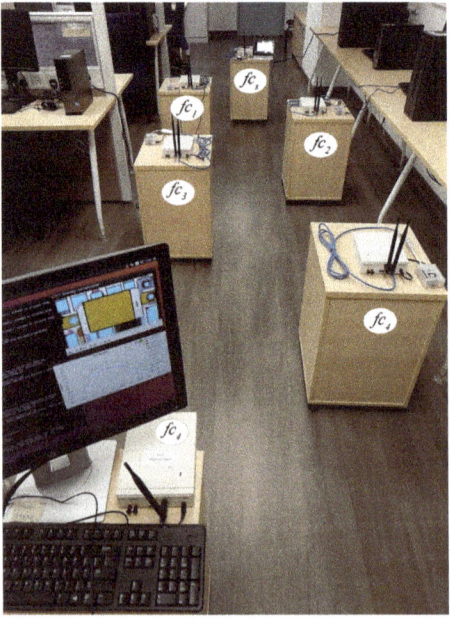

Figure 7. An experimental setup for a D2D route with five hops using RP3 in the RPU testbed, which is equivalent to Figure 4b.

5.3.1. Testbed with a Single Processing Unit (PCU)

Figure 8a shows a testbed in which BSs or nodes are connected to a single traditional processing unit (PCU) (i.e., a personal computer) via a switch [2]. The personal computer has a CORE i7 processor, a 8 GB RAM, and a 1 terabyte of storage space in hard disk, and it provides a centralized approach to handle D2D communication. In other words, a single GNU radio flow graph (see Figure 3 in Section 2.3) is installed in the personal computer. The BSs and nodes exchange control messages with the personal computer via Gigabit Ethernet CAT 5E-350 MHz cables connected to a Gigabit switch, and data messages among themselves via wireless transmission. The control message has a packet size of 826 megabytes and a payload size of 1472 bytes, and it contains the source node IP address, destination node IP address, available channels, acknowledgement, packet types, and timestamp. The source node incurs *three* types of delays: (a) the software processing delay incurred in the personal computer (i.e., to generate and segment video stream into UDP packets); (b) the software processing delay incurred in GNU radio (see Section 2.3 for the processes); and (c) the hardware processing delay incurred in USRP (see Section 2.2 for the processes) and the propagation delay in the transmission incurred from the current node to the next-hop node. Subsequently, the intermediate nodes do not incur the software processing delay in the personal computer. Based on Equation (2), the end-to-end delay of a route $k_k \in K$ is as follows:

$$t^{k_k} = (t^{PC}_{l_1} + t^{GNU}_{l_1} + t^{h}_{l_1}) + (t^{GNU}_{l_2} + t^{h}_{l_2}) + ... + (t^{GNU}_{l_n} + t^{h}_{l_n}) \qquad (3)$$

where $t^{PC}_{l_*}$ is the single processing unit delay, $t^{GNU}_{l_*}$ is the GNU radio processing delay, and $t^{h}_{l_*}$ is the hardware processing delay (see Figure 6). The software processing delay is $t^{s}_{l_n = l_1} = t^{PC}_{l_1} + t^{GNU}_{l_1}$ for the first hop and $t^{s}_{l_n \neq l_1} = t^{GNU}_{l_*}$ for the subsequent hops. The $t^{GNU}_{l_*}$ includes the response time incurred for a node of a route to request for next-hop node information (e.g., the next-hop node IP address and operating channel) from the personal computer.

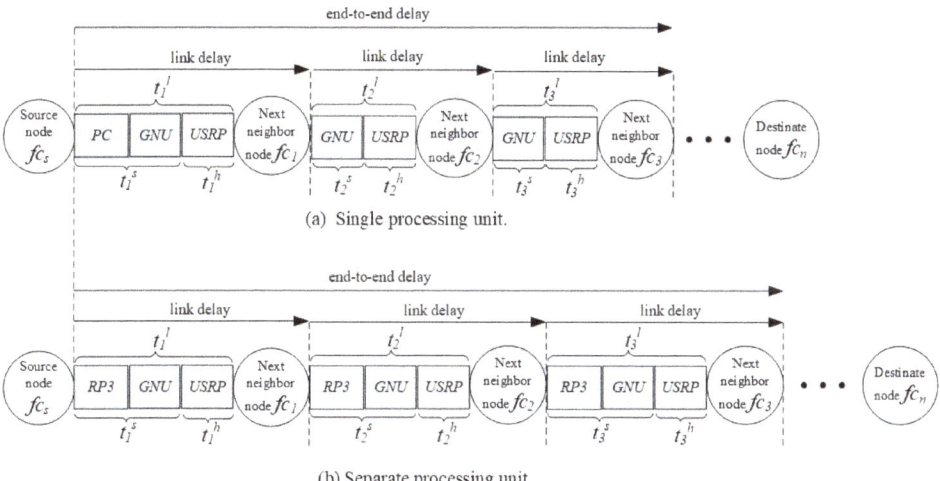

Figure 8. The end-to-end delay of a route in the two testbeds. The delays are shown in same sized blocks although the time period of each block may be different.

5.3.2. Testbed with Separate Processing Units (RPU)

Figure 8b shows a testbed in which each BS or node is connected to a separate processing unit (RPU) (i.e., a RP3), and so a switch is not required. RP3 provides a distributed approach to handle D2D communication. In the route, fc_s transmits to next-hop fc_1 in 850 MHz. Then, fc_1 receives in

850 MHz and transmits in 860 MHz; fc_2 receives in 860 MHz and transmits in 870 MHz; fc_3 receives in 870 MHz and transmits in 880 MHz; fc_4 receives in 880 MHz and transmits in 890 MHz to the destination node fc_d.

The BSs and nodes exchange control and data messages among themselves via wireless transmission. Each node incurs three types of delays: (a) the software processing delay incurred in RP3; (b) the software processing delay incurred in GNU radio; and (c) the hardware processing delay incurred in USRP and the propagation delay. In contrast to the testbed with a PCU (see Section 5.3.1), the intermediate nodes incur the software processing delay in the RP3. Unlike PCU, in RPU, every single node (either source, intermediate, or destination node) receives decision on the next-hop node and the transmission channel, which incurs $t_{l_n}^{RP3}$ (see Figure 8), from the personal computer. The node can also receive such information from the personal computer. Since RP3 has limited processing capability, the software processing delay for the software processes is non-negligible at each node in RPU, causing longer end-to-end delay compared to that in PCU. Based on Equation (2), the end-to-end delay of a route $k_k \in K$ is as follows:

$$t^{k_k} = \sum (t_{l_n}^{RP3} + t_{l_n}^{GNU} + t_{l_n}^{h}) \qquad (4)$$

where the software processing delay is $t_{l_*}^{s} = t_{l_*}^{RP3} + t_{l_*}^{GNU}$ for all hops.

The software processing delay for personal computer $t_{l_*}^{PC}$ and RP3 $t_{l_*}^{RP3}$ are different. There are two types of access rates: (a) the *read rate* is the speed in which a node reads information and reading is needed during transmission; and (b) the *write rate* is the speed in which a node writes information and writing is needed during reception. For both read and write, the access time defines their rates. Access time is the average duration for the kernel to access a partition and perform a task, including read and write. In this evaluation, the kernel performs read and write operations from the boot partition of the Linux Ubuntu MATE operating system installed in RP3. The read rate is considered in a source node, both read and write rates are considered in an intermediate node, and the write rate is considered in a destination node. In general, the average read rate is higher than the average write rate. In personal computer, the read and write rates are much higher; specifically the hard disk drive (HDD) of a personal computer has a drives spin of 7200 revolutions per minute (RPM) in our testbed, and the read and write rates are approximately 80 MBps and 50 MBps, respectively. In RP3, the read and write rates are approximately 22.6 MBps and 16.6 MBps, respectively, for the memory card (i.e., the SD card) on RP3. Due to the lower write rate, the packet queue size at intermediate and destination nodes increase with the number of packets. Hence, RP3 has a limited performance. Figure 9 shows the average read and write rates for reading and writing 1000 samples from/to the boot partition of RP3 with an average access time of 0.48 ms. The average read rate hovers between 22 and 23 MBps, and the average write rate varies between 9 and 20 MBps. The average access time refers to the duration of reading a packet from or writing a packet to the boot partition, and it has an average value of 0.56 ms. The significant lower read and write rates of RP3 can contribute to a higher end-to-end delay and lower packet delivery ratio in RPU compared to PCU.

For an intermediate node, it receives and processes (i.e., decodes and demodulates) packets before it writes them in its memory, then it reads them from the memory and processes (i.e., encodes and modulates) the packets for transmission. The intermediate node must write an entire packet before it can read the packet again for transmission, which is a phenomenon called buffering that occurs during initiation. This allows packets to be fully converted (or digitized) before transmission; however, it causes a higher end-to-end delay as the number of hops increases.

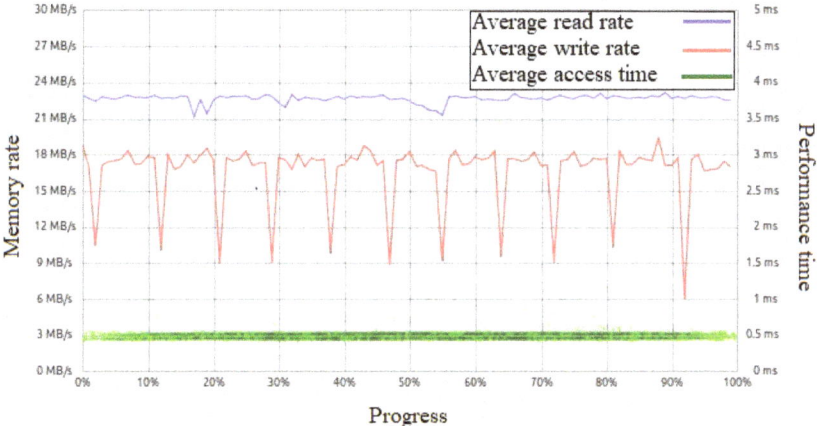

Figure 9. Comparison between read and write performance of SD card on RP3.

6. Experimental Results

This section compares the PCU and RPU performance measures in Section 6.1, and the packet delivery via primary route (i.e., via D2D) and secondary route (i.e., via MC BS) in Section 6.2.

6.1. Performance Comparison between PCU and RPU

Figure 10a presents a logarithmic graph that shows the total delay, as well as the software and hardware processing delays, for a route from the source node to the destination node. For each number of hops, the results are presented using a pair of bars: the left represents PCU, and the right represents RPU. The total delay is higher in RPU, and it increases as the number of hops increases, specifically the total delay increases from 0.024102s for a single hop to 0.10566s for five hops in PCU, and from 0.02992 s for a single hop to 0.18924s for five hops in RPU. Compared to the hardware processing delay, the software processing delay is significantly lower due to the high processing capability of the personal computer in PCU; specifically, the software processing delay is 0.00018s (or 0.7468%) for a single hop, 0.00023s (or 0.6189%) for two hops, 0.00038s (or 0.6316%) for three hops, 0.00057s (or 0.6958%) for four hops, and 0.00079s (or 0.7476%) for five hops. However, the software processing delay is significantly higher than that in PCU (i.e., approximately five times higher); specifically, the software processing delay is 0.00141s (or 4.7125%) for a single hop, 0.00433 s (or 6.252%) for two hops, 0.00628s (or 6.8484%) for three hops, 0.00771s (or 6.5844%) for four hops, and 0.01074s (or 5.675%) for five hops. This is because, in RPU, RP3 has a lower processing capability, causing a higher total delay in each hop.

Figure 10b shows that the packet delivery ratio reduces as the number of hops in a route from the source node to the destination node increases because more intermediate nodes are affected by the ambient noise in the operating environment. Specifically, the packet delivery ratio reduces from 99.7% for a single hop to 94.8% for five hops in PCU, and it reduces from 97.49% for a single hop to 88.73% for five hops in RPU.

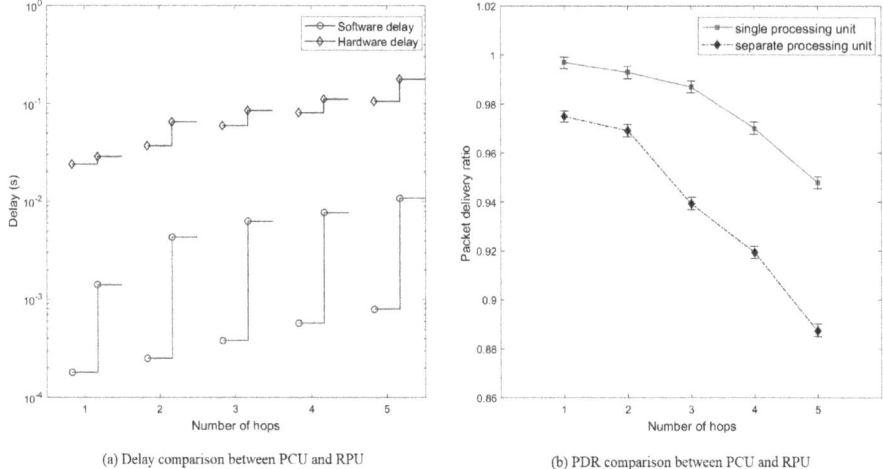

(a) Delay comparison between PCU and RPU

(b) PDR comparison between PCU and RPU

Figure 10. End-to-end delay and packet delivery ratio for PCU and RPU. (**a**) Comparison of software and hardware processing delays between PCU and RPU. (**b**) Comparison of packet delivery ratio between PCU and RPU.

6.2. Comparison of Packet Delivery via Primary and Secondary Routes

This section compares the performance of a communication that uses: (a) a multihop primary route (via D2D); and (b) a secondary route (via MC BS). This is because a source node can communicate with a destination node via either a multihop D2D route or going through a BS. Nevertheless, the delay in a multihop D2D route must be less than 10 ms (see Section 4). Figure 11 shows that a packet delivered via a D2D route has a higher end-to-end delay than its counterpart route using MC BS.

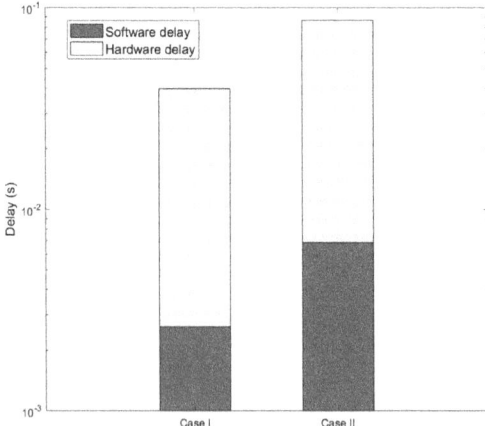

Figure 11. Comparison of software and hardware processing delays between Case I and Case II. In Case I, a two-hop communication is performed via MC BS in PCU. In Case II, a two-hop communication is performed via D2D in RPU.

This section investigates the end-to-end delay of a primary route when nodes are embedded with RP3, and compares it with that of a secondary route, in which only source and destination nodes are embedded with RP3, and the BS is implemented using a personal computer. In Figure 12, a RP3 source node fc_s can communicate with a RP3 destination node fc_d via either: (a) Case I which is a direct communication with MC BS MC_1 (i.e., a personal computer), specifically $fc_s - MC_1 - fc_d$ in PCU; or (b) Case II which is a two-hop route using D2D communication, specifically $fc_s - fc_1 - fc_d$ in RPU, whereby fc_1 is an intermediate node embedded with RP3. The total delay of Cases I and II are 0.039736 s and 0.08666 s, respectively, and so the total delay of Case II is more than twice higher than that in Case I. In RPU, the total software and hardware processing delay of a D2D communication from the source node up to the intermediate node is 0.03384s; while in PCU, the total software and hardware processing delay from the source node up to the MC BS is 0.01655 s. In RPU, the total software and hardware processing delay from the intermediate node fc_1 to the destination node is 0.05282 s; while in PCU, the total software and hardware processing delay from the MC BS to the destination is 0.02318 s. Hence, Case I has a lower total delay as compared to Case II due to its greater processing capability. Figure 11 shows the total delay, which includes software and hardware processing delays, of a direct communication with MC BS and a multihop D2D route. Figure 13 shows the packet delivery ratio via D2D and MC BS. A source node transmits a packet towards a destination node, the packet goes through intermediate node as the destination node is beyond the transmission range of the transmitter. For this packet transmission, the intermediate node is first selected (i.e., Case I), and then the MC BS is selected (i.e., Case II). From the source node to the intermediate node in Figure 13, the MC BS has a 99.65% packet delivery ratio, while the node with RP3 has 97.32%. From the intermediate node to the destination node, MC BS delivers 99.59% of the packets, while the node with RP3 delivers 96.63% of the packets. Therefore, in Figure 13 the total packet delivery ratio from a source node to a destination node via MC BS is 99.24%. However, the packet delivery ratio from the source node to the destination node via RP3 is 93.95%, which is about 7% lesser.

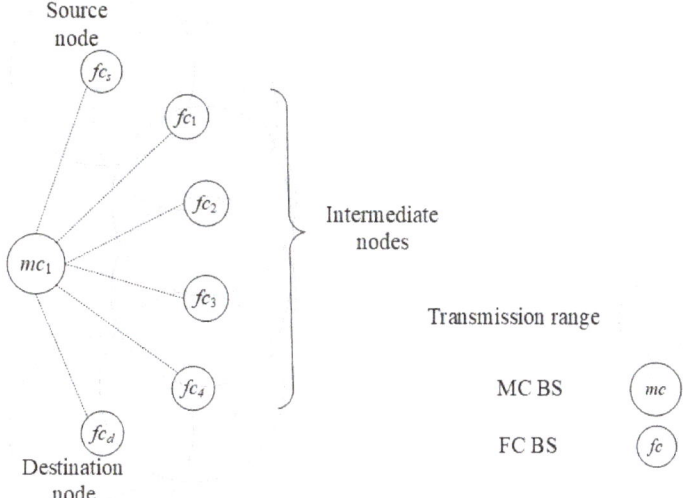

Figure 12. A D2D route from a FC source node fc_s to a FC destination node fc_d.

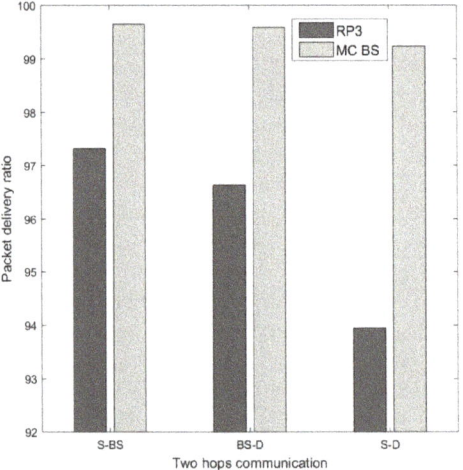

Figure 13. Comparison of the two-hop packet delivery via D2D and MC BS.

In Figure 14, a throughput comparison is made between PCU and RPU. Higher throughput refers indicates a higher successful packets transmission rate [26]. Throughput reduces as the number of hops increases for both PCU and RPU; however, PCU achieves a higher throughput compared to RPU because of higher packet delivery ratio (see Figure 10b).

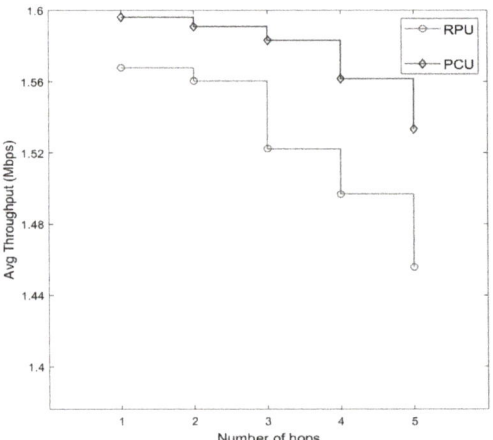

Figure 14. Throughput comparison between RPU and PCU.

7. Conclusions and Future Work

This paper presented an experimental study to compare performance measures in a testbed with a single processing unit (PCU) and a testbed with separate processing units (RPU). The testbed consists of nodes implemented using universal software radio peripheral with GNU radio. In PCU, base stations and user equipment, including sensors, connect to a single centralized traditional processing unit (e.g., a personal computer or a laptop) via physical cables and a switch. On the other hand, in RPU, each BS or node is embedded with a separate processing unit, particularly Raspberry Pi3 B+.

While PCU is a widely used testbed in the literature, nodes are constrained to be located at close proximity to the centralized processing unit. Meanwhile, RPU has a closer resemblance to a real deployed network, and it has not been investigated in the literature, and so it is the focus of this paper. Our experimental results showed that: (a) the end-to-end delay is lower in PCU as control messages are exchanged via a switch using gigabit Ethernet; and (b) the per-hop and end-to-end delays increase with the number of hops in RPU. However, in RPU, device-to-device communication between nodes from a source node to a destination node can offload traffic from BS, which is one of the promising features of 5G. Therefore, this paper presents a case study in which the intermediate node of a two-hop route can be: (a) a node (via D2D); or (b) a macrocell BS. While the preceding case can reduce the traffic amount at a macrocell BS, it can increase end-to-end delay and reduce packet delivery ratio compared to the latter case due to its lower processing capability.

As for future work, we aim to relax the assumptions made in this article to enable a macrocell base station (MC BS) to receive updates from femtocell base stations and nodes. Examples of such updates are the packet delivery ratio and per-hop delay, which allows MC BS to make decision on route selection based on the updates under unpredictable and dynamic operating environment.

Author Contributions: M.K.C. designed and implemented testbed. K.-L.A.Y. contributed to system model and literature review. R.M.D.N. and R.W. contributed to paper review and formatting. All authors have read and agreed to the published version of the manuscript.

Funding: This work was part of the project entitled "A Novel Clustering Algorithm based on Reinforcement Learning for the Optimization of Global and Local Network Performances in Mobile Networks" funded by the Malaysian Ministry of Education under Fundamental Research Grant Scheme FRGS/1/2019/ICT03/SYUC/01/1, as well as the Partnership Grant CR-UM-SST-DCIS-2018-01 and RK004-2017 between Sunway University and University of Malaya.

Conflicts of Interest: The author declares no conflict of interest.

Abbreviations

The following abbreviations are used in this manuscript:

USRP	Universal software radio partnership
RP3	Raspberry Pi3 B+
5G	Fifth generation
D2D	Device to device
MC	Macrocell
FC	Femtocell
PC	Picocell
CC	Central controller
BC	Base station
WBX	Wide bandwidth transceiver
FPGA	Field-programmable gate array
RF	Radio frequency
ADC	Analogue-to-digital converter
DAC	Digital-to-analogue converter
DUD	Digital up converter
DDC	Digital down converter
UHD	USRP hardware driver
SDR	Software defined radio
PoE	Power over Ethernet
Mbps	Megabits per second
USP	Universal serial bus
CAT	Category
GRC	GNU radio companion
IP	Internet protocol
UDP	User datagram protocol

GMSK   Gausian minimum shift keying
SD   Secure digital
RAM   Random access memory
RPU   Separate processing unit
PCU   Single processing unit
HDD   Hard disk drive

References

1. Ashraf, M.I.; Tamoor-Ul-Hassan, S.; Mumtaz, S.; Tsang, K.F.; Rodriquez, J. Device-to-device assisted mobile cloud framework for 5G networks. In Proceedings of the IEEE International Conference on Industrial Informatics (INDIN), Emden, Germany, 24–26 July 2017; pp. 1020–1023. [CrossRef]
2. Raschellà, A.; Umbert, A. Implementation of Cognitive Radio Networks to evaluate spectrum management strategies in real-time. *Comput. Commun.* **2016**, *79*, 37–52. [CrossRef]
3. Okasaka, S.; Weiler, R.J.; Keusgen, W.; Pudeyev, A.; Maltsev, A.; Karls, I.; Sakaguchi, K. Proof-of-concept of a millimeter-wave integrated heterogeneous network for 5G cellular. *Sensors* **2016**, *16*, 1362. [CrossRef] [PubMed]
4. Asadi, A.; Member, S.; Wang, Q.; Member, S.; Mancuso, V. A Survey on Device-to-Device Communication in Cellular Networks. *IEEE Commun. Surv. Tutor.* **2014**, *16*, 1801–1819. [CrossRef]
5. Aly, A.A.; ELAttar, H.M.; ElBadawy, H.; Abbas, W. Aggregated Throughput Prediction for Collated Massive Machine-Type Communications in 5G Wireless Networks. *Sensors* **2019**, *19*, 3651. [CrossRef]
6. Chang, B.; Zhao, G.; Imran, M.A.; Chen, Z.; Li, L. Dynamic Wireless QoS Analysis for Real-Time Control in URLLC. In Proceedings of the 2018 IEEE Globecom Workshops, GC Wkshps 2018—Proceedings, Abu Dhabi, UAE, 9–13 December 2018; pp. 1–5. [CrossRef]
7. Zhang, H.; Liao, Y.; Song, L. D2D-U: Device-to-Device Communications in Unlicensed Bands for 5G System. *IEEE Trans. Wirel. Commun.* **2017**, *16*, 3507–3519. Available online: http://xxx.lanl.gov/abs/1610.04982 (accessed on 1 August 2019). [CrossRef]
8. Yau, K.l.A.; Qadir, J.; Wu, C.; Imran, M.A.L.I.; Ling, M.E.E.H. Cognition-inspired 5G cellular networks: A review and the road ahead. *IEEE Access* **2018**, *6*, 35072–35090. [CrossRef]
9. Ge, X.; Tu, S.; Mao, G.; Wang, C.X.; Han, T. 5G Ultra-Dense Cellular Networks. *IEEE Wirel. Commun.* **2016**, *23*, 72–79. Available online: http://xxx.lanl.gov/abs/1512.03143 (accessed on 22 December 2018). [CrossRef]
10. Agiwal, M.; Roy, A.; Saxena, N. Next generation 5G wireless networks: A comprehensive survey. *IEEE Commun. Surv. Tutor.* **2016**, *18*, 1617–1655. [CrossRef]
11. Kim, J.; Molisch, A.F. Quality-aware millimeter-wave device-to-device multi-hop routing for 5G cellular networks. In Proceedings of the 2014 IEEE International Conference on Communications, ICC 2014, Sydney, Australia, 10–14 June 2014; pp. 5251–5256. [CrossRef]
12. He, J.; Song, W. Evolving to 5G: A Fast and Near-optimal Request Routing Protocol for Mobile Core Networks. *IEEE Wirel. Netw.* **2014**, 4586–4591. [CrossRef]
13. Tran, T.X.; Hajisami, A.; Pompili, D. ULTRA-DENSE HETEROGENEOUS SMALL CELL DEPLOYMENT IN 5G AND BEYOND Cooperative Hierarchical Caching in 5G Cloud Radio Access Networks. *IEEE Netw.* **2017**, *31*, 35–41. [CrossRef]
14. Chávez-Santiago, R.; Szydełko, M.; Kliks, A.; Foukalas, F.; Haddad, Y.; Nolan, K.E.; Kelly, M.Y.; Masonta, M.T.; Balasingham, I. 5G: The Convergence of Wireless Communications. *Wirel. Pers. Commun.* **2015**, *83*, 1617–1642. [CrossRef] [PubMed]
15. Mahmood, A.M.; Al-yasiri, A.; Alani, O.Y. Cognitive Neural Network Delay Predictor for High Speed Mobility in 5G C-RAN Cellular Networks. In Proceedings of the 2018 IEEE 5G World Forum (5GWF), Silicon Valley, CA, USA, 9–11 July 2018; pp. 93–98. [CrossRef]
16. Abd, S.A.; Manjunath, S.; Abdulhayan, S. Direct Device-to-Device Communication in 5G Networks. In Proceedings of the 2016 International Conference on Computation System and Information Technology for Sustainable Solutions (CSITSS), Bangalore, India, 6–8 October 2016; pp. 216–219. [CrossRef]
17. Wassie, D.A.; Berardinelli, G.; Tavares, F.M.L. Experimental Verification of Interference Mitigation techniques for 5G Small Cells. In Proceedings of the 2015 IEEE 81st Vehicular Technology Conference (VTC Spring), Glasgow, UK, 11–14 May 2015; pp. 1–5. [CrossRef]

18. Quaresma, J.; Ribeiro, C.; Gameiro, A.; Zelenak, M.; Duplicy, J. Distributed RF sensing framework with radio environment emulation. In Proceedings of the 2013 IEEE 14th International Symposium on a World of Wireless, Mobile and Multimedia Networks, WoWMoM 2013, Madrid, Spain, 4–7 June 2013; pp. 1–6. [CrossRef]
19. Jain, A.; Sharma, V.; Amrutur, B. Soft real time implementation of a Cognitive Radio testbed for frequency hopping primary satisfying QoS requirements. In Proceedings of the 2014 20th National Conference on Communications, NCC 2014, Kanpur, Uttar Pradesh, India, 28 February–2 March 2014; pp. 1–6. [CrossRef]
20. Barrak, S.E.; Lyhyaoui, A.; Puliafito, A.; Serrano, S. Implementation of a low cost SDR-based Spectrum Sensing Prototype using USRP and Raspberry Pi board. *Int. Conf. Autom. Control Eng. Comput. Sci. (ACECS)* **2017**, *20*, 54–58.
21. Briand, A.; Albert, B.B.; Gurjao, E.C. Complete software defined RFID system using GNU radio. In Proceedings of the 2012 IEEE International Conference on RFID-Technologies and Applications, RFID-TA 2012, Nice, France, 5–7 November 2012; pp. 287–291. [CrossRef]
22. Öhlén, P.; Skubic, B.; Rostami, A.; Ghebretensaé, Z.; Mårtensson, J.; Fiorani, M.; Monti, P.; Wosinska, L. Data Plane and Control Architectures for 5G Transport Networks. *J. Light. Technol.* **2016**, *34*, 1501–1508. [CrossRef]
23. Maksymyuk, T.; Gazda, J.; Yaremko, O.; Nevinskiy, D. Deep Learning Based Massive MIMO Beamforming for 5G Mobile Network. In Proceedings of the 2018 IEEE 4th International Symposium on Wireless Systems within the International Conferences on Intelligent Data Acquisition and Advanced Computing Systems (IDAACS-SWS), Lviv, Ukraine, 20–21 September 2018; pp. 241–244. [CrossRef]
24. Ettus, M.; Braun, M. The universal software radio peripheral (usrp) family of low-cost sdrs. In *Opportunistic Spectrum Sharing and White Space Access: The Practical Reality*; Wiley Telecom: Hoboken, NJ, USA, 2015; pp. 3–23.
25. Ali, S.A.; Umrani, F.A.; Umrani, N.A. Performance Evaluation Of Decode and Forward Cooperative Communication Protocol. *Int. J. Inf. Technol. Electr. Eng.* **2012**, *1*, 321–325.
26. Syed, A.; Yau, K.L.; Qadir, J.; Mohamad, H.; Ramli, N.; Keoh, S. Route selection for multi-hop cognitive radio networks using reinforcement learning: An experimental study. *IEEE Access* **2016**, *4*, 6304–6324. [CrossRef]
27. Truong, N.B.; Suh, Y.J.; Yu, C. Latency analysis in GNU radio/USRP-based software radio platforms. In Proceedings of the IEEE Military Communications Conference MILCOM, San Diego, CA, USA, 18–20 November 2013; pp. 305–310. [CrossRef]
28. Van den Bergh, B.; Vermeulen, T.; Verhelst, M.; Pollin, S. CLAWS: Cross-Layer Adaptable Wireless System enabling full cross-layer experimentation on real-time software-defined 802.15. 4. *EURASIP J. Wirel. Commun. Netw.* **2014**, 1–13. [CrossRef]
29. Khandakar, A.; Mohamed, A.M.S. Understanding probabilistic cognitive relaying communication with experimental implementation and performance analysis. *Sensors* **2019**, *19*, 179. [CrossRef]
30. Nychis, G.; Hottelier, T.; Yang, Z.; Seshan, S.; Steenkiste, P. Enabling MAC Protocol Implementations on Software-Defined Radios. *Netw. Syst. Des. Implement.* **2009**, *9*, 91–105.
31. Chen, W.T.; Chang, K.T.; Ko, C.P. Spectrum monitoring for wireless TV and FM broadcast using software-defined radio. *Multimed. Tools Appl.* **2016**, *75*, 9819–9845. [CrossRef]
32. Byun, S.S. TCP over scarce transmission opportunity in cognitive radio networks. *Comput. Netw.* **2016**, *103*, 101–114. [CrossRef]
33. Gameiro, A.; Ribeiro, C.; Quaresma, J. Selective reporting—A half signalling load algorithm for distributed sensing. *Eurasip J. Wirel. Commun. Netw.* **2013**, *2013*, 1–14. [CrossRef]
34. Marko, H.; Korpi, J.; Hiivala, M. Predictive Channel Selection for over-the-Air Video Transmission Using Software-Defined Radio Platforms. In *International Conference on Cognitive Radio Oriented Wireless Networks*; Springer: Cham, Switzerland, 2016; Volume 172, pp. 569–579. [CrossRef]
35. Berardinelli, G.; Buthler, J.L.; Tavares, F.M.L.; Tonelli, O.; Wassie, D.A.; Hakhamaneshi, F.; Sørensen, T.B.; Mogensen, P. Distributed Synchronization of a testbed network with USRP N200 radio boards. In Proceedings of the IEEE International Conference on Communications, Pacific Grove, CA, USA, 2–5 November 2014; pp. 563–567. [CrossRef]
36. Nagaraju, P.B.; Ding, L.; Melodia, T.; Batalama, S.N.; Pados, D.A.; Matyjas, J.D. Implementation of a Distributed Joint Routing and Dynamic Spectrum Allocation Algorithm on USRP2 Radios. In Proceedings of the IEEE International Conference on Communications, Boston, MA, USA, 21–25 June 2010; pp. 26–27. [CrossRef]

37. Chang, B.; Zhao, G.; Chen, Z.; Li, L.; Imran, M.A. Packet-Drop Design in URLLC for Real-Time Wireless Control Systems. *IEEE Access* **2019**. [CrossRef]
38. Martínez-quintero, J.C.; Estupiñán-cuesta, E.P.; Rodríguez-ortega, V.D. Raspberry PI 3 RF signal generation system. *Visión Electrónica* **2019**, *13*, 2.
39. Zhang, N.; Sun, W.; Lou, W.; Hou, Y.T.; Trappe, W. ROSTER: Radio context attestation in cognitive radio network. In Proceedings of the 2018 IEEE Conference on Communications and Network Security, CNS 2018, Beijing, China, 30 May–1 June 2018; pp. 1–9. [CrossRef]
40. Tomar, V.S.; Bhatia, V. Low Cost and Power Software Defined Radio Using Raspberry Pi for Disaster Effected Regions. *Procedia Comput. Sci.* **2015**, *58*, 401–407. [CrossRef]
41. Al-safi, A.; Narasimhan, L.; Bazuin, B. Software Defined Community Radio Using Low Cost Hardware and Free Software. In Proceedings of the Universal Technology Management Conference (UTMC), Bemidji, MN, USA, 26–28 May 2016; pp. 51–55.
42. Shi, Y.; Wensowitch, J.; Ward, A.; Badi, M.; Camp, J. Building UAV-Based Testbeds for Autonomous Mobility and Beamforming Experimentation. In Proceedings of the 2018 IEEE International Conference on Sensing, Communication and Networking, SECON Workshops 2018, Hong Kong, China, 11–13 June 2018; pp. 1–5. [CrossRef]
43. Park, H.J.; Lee, G.M.; Shin, S.H.; Roh, B.H.; Oh, J.M. Implementation of Multi-Hop Cognitive Radio Testbed using Raspberry Pi and USRP. *Int. J. Interdiscip. Telecommun. Netw.* **2017**, *9*, 37–48. [CrossRef]
44. Habib, M.A.; Moh, S. Robust Evolutionary-Game-Based Routing for Wireless Multimedia Sensor Networks. *Sensors* **2019**, *19*, 3544. [CrossRef] [PubMed]
45. Shariatmadari, H.; Iraji, S.; Anjum, O.; Riku, J.; Li, Z.; Wijting, C. Delay Analysis of Network Architectures for Machine-to-Machine Communications in LTE System. In Proceedings of the 21st International Conference on Telecommunications (ICT)—Workshop on M2M Solutions and Services, Lisbon, Portugal, 4–7 May 2014; pp. 502–506. [CrossRef]
46. Homaei, M.H.; Salwana, E.; Shamshirband, S. An Enhanced Distributed Data Aggregation Method in the Internet of Things. *Sensors* **2019**, *19*, 3173. [CrossRef]
47. Jin, Z.; Ma, Y.; Su, Y.; Li, S.; Fu, X. A Q-learning-based delay-aware routing algorithm to extend the lifetime of underwater sensor networks. *Sensors* **2017**, *17*, 1660. [CrossRef]
48. Zhang, W.; Wang, C.X.; Ge, X.; Chen, Y. Enhanced 5G Cognitive Radio Networks Based on Spectrum Sharing and Spectrum Aggregation. *IEEE Trans. Commun.* **2018**, *PP*, 492–496. [CrossRef]
49. Orebaugh, A.; Ramirez, G.; Beale, J. *Wireshark & Ethereal Network Protocol Analyzer Toolkit*; Elsevier: Amsterdam, The Netherlands, 2006.

© 2019 by the authors. Licensee MDPI, Basel, Switzerland. This article is an open access article distributed under the terms and conditions of the Creative Commons Attribution (CC BY) license (http://creativecommons.org/licenses/by/4.0/).

MDPI
St. Alban-Anlage 66
4052 Basel
Switzerland
Tel. +41 61 683 77 34
Fax +41 61 302 89 18
www.mdpi.com

Sensors Editorial Office
E-mail: sensors@mdpi.com
www.mdpi.com/journal/sensors

www.ingramcontent.com/pod-product-compliance
Lightning Source LLC
LaVergne TN
LVHW071944080526
838202LV00064B/6674